Elements of Marine Ecology

Elements of Marine Ecology

Fourth Edition

R.V. Tait
F. A. Dipper

Butterworth-Heinemann
Linacre House, Jordan Hill, Oxford OX2 8DP
225 Wildwood Avenue, Woburn, MA 01801-2041
A division of Reed Educational and Professional Publishing Ltd

℞ A member of the Reed Elsevier plc group

OXFORD   BOSTON   JOHANNESBURG
MELBOURNE   NEW DELHI   SINGAPORE

First published 1968
Reprinted 1970
Second edition 1972
Reprinted 1975, 1977, 1978
Third edition 1981
Reprinted 1983, 1988, 1992
Fourth edition 1998

British Library Cataloguing in Publication Data
Tait, R. V. (Ronald Victor)
  Elements of marine ecology. – 4th ed.
  1 Marine ecology
  I Title   II Dipper, Frances
  577.7

ISBN 0 7506 2088 9

Typeset by Keyword Publishing
Printed and bound in Great Britain by Clays Ltd, St Ives plc

Contents

Preface

It is now sixteen years since the third edition of this student textbook was published, and during this time advances in the field of marine biology have been many and varied. Continuing demand for the book has therefore led to this fourth, completely revised and updated edition. Developments in technology have been particularly rapid, especially in the fields of diving, precision instruments, satellite communications and computers, and these advances have greatly improved and expanded our ability to explore the oceans and their marine life. At the same time, the impact of man on the marine environment is ever increasing and cannot be ignored in any study of marine ecology. Therefore the chapters on 'Measuring and sampling' and 'Sea fisheries' have been extensively revised, and a new chapter on the 'Human impact on the marine environment' has been added. The aim of the latter is to introduce the student to the various ways in which our activities are changing the ecology of the oceans, with potentially far-reaching effects. Additional material on coral reefs, hydrothermal vents, phenomena such as 'El Niño' and ocean processes, including the important 'microbial loop' of the food web, have been incorporated into existing chapters.

Although the book has been extensively revised, this edition retains its original aim of presenting marine ecology as a coherent science and its scope derives from the original, broad definition of ecology as the study of organisms in relation to their surroundings. The text has been compiled as introductory reading for students undertaking courses in marine biology. It provides information and ideas over the general field of marine ecology with reading lists and references from which additional material can be sought. With the busy student in mind, the text has been re-arranged into clearly labelled and numbered sections. The field course book list in the appendices has been extensively revised and updated as have the references and student texts listed at the end of each chapter.

This new edition has been prepared by myself Dr Frances Dipper, with the guidance, help and unbounded enthusiasm of Ronald Tait. It was therefore with great sadness that I received the news of his untimely death last year. He will be sorely missed by his family and friends and by all those past students whom he has introduced to the fascination of the marine environment. It has been a privilege and a pleasure to work on this new edition of his book.

Our thanks are due to our many colleagues who have provided us with helpful comments, material and encouragement throughout the preparation of this edition.

Frances Dipper

1 The oceans

1.1 Introduction

Ecology is the study of relationships between organisms and their surroundings. This study is fundamental to an understanding of biology because organisms cannot live as isolated units. The activities which comprise their lives are dependent upon, and closely controlled by, their external circumstances, by the physical and chemical conditions in which they live and the populations of other organisms with which they interact. In addition, the activities of organisms have effects on their surroundings, altering them in various ways. Organisms therefore exist only as parts of a complex entity made up of interacting inorganic and biotic elements, to which we apply the term *ecosystem*.

All life on earth constitutes a single ecosystem divisible into innumerable parts. This book is concerned with the greatest of these divisions, *the marine ecosystem*, occupying a larger volume of the biosphere than any other. The marine ecosystem can be further subdivided into many component ecosystems in different parts of the sea.

Living processes involve energy exchanges. Energy for life is drawn primarily from solar radiation, transformed into the chemical energy of organic compounds by the photosynthetic processes of plants; thence transferred through the ecosystem by movements of materials within and between organisms, mainly through the agencies of feeding, growth, reproduction and decomposition. An ecosystem is therefore essentially a working, changing and evolving sequence of operations, powered by solar energy. In the long term, the intake of energy to the system is balanced by energy loss as heat. Most scientists believe that life began within the solar-powered environment of the ocean ecosystem. With the discovery in the 1970s of complex animal communities powered by geothermal energy at deep-sea vents (see Section 6.4.4), some scientists now argue that these vents may have allowed the first primitive organisms, such as bacteria, to evolve.

The aim of marine ecological studies is to understand marine ecosystems as working processes. At present our knowledge is incomplete and only speculative analyses can be made. This book provides general information about how marine organisms are influenced by, and have effects on, their environment, and describes some of the methods of investigation which may eventually provide the necessary information for a better understanding of marine ecosystems. A good starting-

1

point is to describe briefly some of the major physical features of the oceans. There are several excellent introductory oceanography texts to which the student should refer for more detailed information. These are given in the references at the end of Chapter 1.

1.2   Extent and depth of the oceans

Seawater covers approximately 71 per cent of the earth's surface, an area of about 361 million square kilometres (139 million square miles) comprising the major ocean areas shown in Figure 1.1. In the deepest parts the bottom lies more than 10 000 m from the surface, and the average depth is about 3700 m. Although marine organisms are unevenly distributed, they occur throughout this vast extent of water and have been brought up from the deepest places.

1.2.1   The continental shelf

Close to land the sea is mostly shallow, the bottom shelving gradually from the shore to a depth of about 200 m. This coastal ledge of shallow sea bottom is the *continental shelf* (see Figure 1.2). About 8 per cent of the total sea area lies above it. Its seaward margin is termed the *continental edge*, beyond which the water becomes much deeper. The steeper gradient beyond the continental edge is termed the *continental slope*.

The width of the continental shelf varies very much in different parts of the world from less than 100 metres to more than 100 kilometres, with an average of about 65 km. It is extensive around the British Isles, where the continental edge runs to the west of Ireland and the north of Scotland. The English Channel, Irish Sea and almost the entire North Sea lie above the shelf. The shelf is also broad beneath the China Sea, along the Arctic coast of Siberia, under Hudson Bay, and along the Atlantic coast of Patagonia where the shelf extends out to the Falkland Islands.

Many of the shelf areas are of special economic importance because geographically the major fisheries are concentrated here. Northern hemisphere temperate and sub-polar continental shelves are particularly important in this respect. Shelf areas are also widely exploited as sources of oil and gas.

Several processes contribute to the formation of the continental shelf. It is formed partly by wave erosion cutting back the coastline. It may be extended seawards by accumulations of material eroded from the coast, or by river-borne silt deposited on the continental slope. Parts of the shelf appear to consist largely of material held against the continents by underwater barriers formed by reef-building organisms or by tectonic folding. In other places the shelf has been formed chiefly by sinking and inundation of the land; for example, under the North Sea. It is possible that in some regions the shelf has been broadened by increments of materials thrust up the continental slope by pressures between the continental blocks and the deep ocean floor.

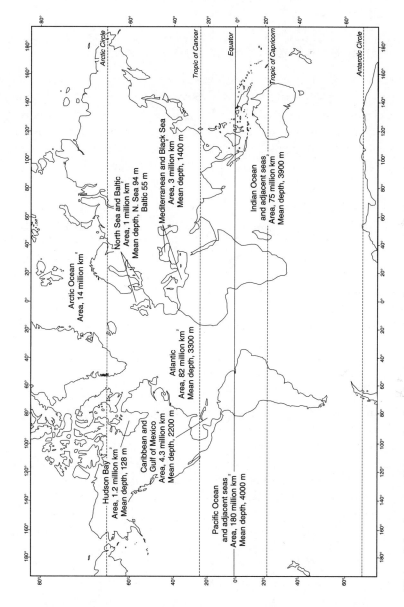

Figure 1.1   Areas and mean depths of major oceans and seas.

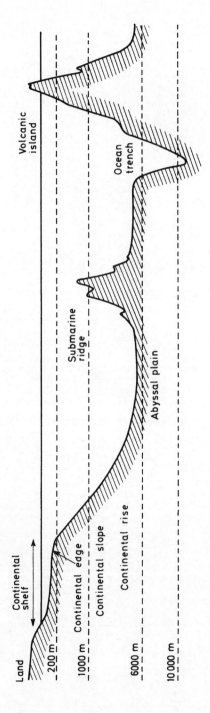

Figure 1.2   Terms applied to parts of the sea bottom.

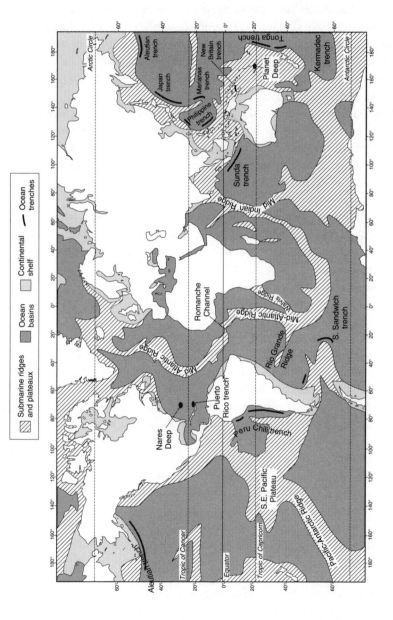

Figure 1.3   Main areas of continental shelf, submarine ridges and plateaux, and ocean trenches.

1.2.2   The ocean basins

Beyond the continental edge the gradient of the sea bottom becomes steeper (the *continental slope*) and descends to the floor of the ocean basins, often reaching a depth of 3000–6000 m and even deeper in some places. The angle and extent of the continental slope vary with locality, averaging a gradient of about 7 per cent (a drop of 70 m in 1 km horizontal distance), but may be as steep as 50 per cent. The slope is seldom an even descent, and is much fissured by irregular gullies and steep-sided submarine canyons.

At the bottom of the continental slope the gradient becomes less steep due to the accumulation of sediment. This zone is termed the *continental rise*. It merges gradually with the deep ocean floor, which in some areas may be virtually flat over great areas, forming *abyssal plains*, extending for hundreds of miles with only slight changes of level. But in places the ocean floor rises to form ranges of submarine mountains with many summits ascending to within 2000–4000 m of the surface and the highest peaks breaking the surface as oceanic islands. These submarine oceanic ridges and plateaux are a major feature of the earth's crust, covering an area approximately equal to that of the continents. There are other parts where the ocean floor is furrowed by deep troughs, the *ocean trenches*, in which the bottom descends to depths of 7000–11 000 m.

1.2.3   Oceanic ridges

At one time the sea-bed was thought to be flat and featureless and it is only in relatively recent times that its mountainous nature has been revealed. The system of oceanic mountain ridges is vast and its extent was not fully charted until the 1950s and 1960s. Full details of all the known ridges, together with ocean trenches, depth contours and other physical features, are shown in a series of maps in *The Times Atlas and Encyclopedia of the Sea* (Couper, 1989).

One part of the submarine ridge system forms a barrier separating the deep levels of the Arctic basin from those of the Atlantic. Much of the crest of this ridge is within 500 m of the surface, extending from the north of Scotland and the Orkneys and Shetlands to Rockall and the Faroes (the Wyville–Thompson ridge), and then to Iceland (the Iceland–Faroes rise), and across to Greenland and Labrador (the Greenland–Iceland rise).

The bottom of the Atlantic is divided into two basins by the mid-Atlantic ridge which extends from 70°N to 55°S. It runs from the Arctic through Iceland, and then follows a roughly S-shaped course from north to south, touching the surface at the islands of the Azores, St Paul, Ascension, Tristan da Cunha and Bouvet. A branch of the mid-Atlantic ridge, the Walvis ridge, extends from Tristan da Cunha to Walvis Bay on the west coast of Africa. South of South Africa the mid- Atlantic ridge trends eastwards and links with a north–south submarine ridge, the mid-Indian ridge, bisecting the Indian Ocean between Antarctica and the Indian and Arabian Peninsulas, and extending into the Arabian Sea (the Carlsberg ridge).

In the Pacific a very broad submarine plateau extends in a north-easterly direction from Antarctica to the west coast of North and Central America, its peaks forming some of the East Pacific Islands. The more numerous islands of the Central and West Pacific appear mainly to be of separate volcanic origin. A peculiar feature of the Pacific Ocean is the large number of flat-topped, underwater hills known as guyots. Although the summits of some of them now lie beneath as much as 800 m of water, they have the appearance of having been worn flat by wave erosion. It seems likely that at some earlier time these volcanic mounds reached above the surface, but have subsequently subsided.

1.2.4 Ocean trenches

These are the deepest parts of the ocean floor where depths exceed 7000 m. They occur mainly beneath the western Pacific Ocean close to oceanic islands; for example, east of the Philippines and the Mariana Islands is the Mariana Trench, where the deepest known soundings have been made at 11 034 m. This is part of a great line of trenches extending north from the Philippines, along the east of Japan and on to the Aleutians. The bottom is also very deep in the New Britain Trench near the Solomon Islands, and in the Tonga Trench and Kermadec Deep to the north-east of New Zealand. In the eastern Pacific the Peru–Chile Trench lies close to the west coast of South America.

In the Indian Ocean the deepest water has been found in the Sunda Trench south of Java and also in an area south-east of the Cocos Islands. In the Atlantic, water of comparable depth occurs in a pit north-east of Puerto Rico, and a trench near the South Sandwich Islands.

1.2.5 Plate tectonics

The theory of plate tectonics was formulated in the late 1960s and brings together the theories of sea floor spreading and continental drift. The continents and ocean basins are believed to have evolved over the past 200 million years or so and plate tectonics provides an explanation for the way in which this may have happened.

According to current theories of plate tectonics and sea floor spreading, the outer crust of the earth (the *lithosphere*) is made up of about 20 separate *lithospheric plates* which cover the molten *mantle* rather like a cracked shell. Most plates carry continental masses plus adjacent ocean floor whilst some carry only ocean floor. The rigid, cool plates are not fixed in position, but ride on top of the hot partly molten asthenosphere, moving continents and ocean basins with them. These movements over the mantle are not yet fully understood but the main driving force is thermal energy. A complex convective flow deep within the mantle is probably involved (Whitmarsh *et al.*, 1996).

Submarine ridges, including the mid-oceanic ridges, are believed to mark the

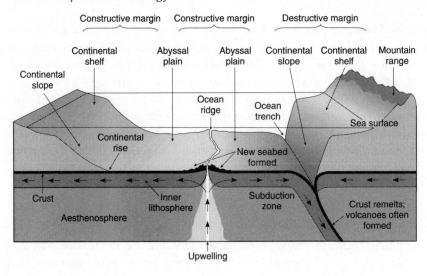

Figure 1.4   Diagram to illustrate the theory of sea floor spreading. Submarine ridges mark lines of tectonic plate separation, where new crust is created as mantle material moves to the surface. Ocean trenches are formed and crust is lost where continental and oceanic plates collide.

lines where lithospheric plates are moving apart. To fill the gaps between separating plates, molten basalt wells up from the interior to the surface, forming submarine ridges which gradually subside laterally to become new ocean floor. As the new crust is formed, the earlier crust spreads away. The spreading rate has been calculated as ranging from less than 1 cm per year from the mid-Atlantic ridge to 16 cm per year from the East Pacific Rise between the Pacific and Nazca plates. Along the centre line of each ridge there is a depression which marks the actual line of division from which lateral spreading is taking place (Figure 1.4).

Where the edges of moving plates collide one plate is forced below the other to form a deep oceanic trench with adjacent volcanic islands. This is known as *subduction*, and the edge of the plate is forced down into the mantle and resorbed. Where an oceanic plate collides with a continental plate, mountain ranges may be thrust up consisting of volcanic material and folded sediments. Where two oceanic plates collide, a volcanic island arc develops. Subduction is the basis of the so-called 'ring-of-fire' in the Pacific Ocean which is an almost continuous chain of volcanoes surrounding it.

Submarine ridges and trenches are therefore associated with areas of volcanic activity resulting from these movements of the earth's crust. The ridges are essentially different from mountain ranges on land because they are formed entirely of extrusions of igneous rock into the sea floor, whereas mountains on land consist mainly of folded upthrusts of sedimentary rock.

On this theory, the Atlantic and part of the Indian Oceans are thought to be younger than the Pacific, and to have originated in Triassic times when a splitting of the lithosphere was followed by a break-up and separation of continental blocks. The Atlantic and Indian Oceans may still be enlarging at the expense of the Pacific by a westward drift of North and South America and by a combination of northward and eastward drifts of the crustal plates bearing Africa, Eurasia and Australia (Dietz and Holden, 1970; Anderson, 1992).

The margins and relative movements of the major plates are shown in Figure 1.5. Comparison with Figure 1.3 will reveal the coincidence of ridges and trenches with lines of separation and subduction between lithosphere plates.

1.3 Ocean currents

The major currents of the oceans are caused by the combined effects of wind action and barometric pressures on the surface, and density differences between different parts of the sea. The density differences exist mainly because of inequalities of heat exchange between atmosphere and water at various parts of the sea surface, and also because of differences of evaporation and dilution. The course taken by currents is influenced by the rotation of the earth and by the shape of the continents and ocean floor. To-and-fro and rotatory oscillations generated by tidal forces are superimposed on the movements resulting from interactions of atmosphere and ocean (see Section 8.1). The flow of ocean currents is consequently meandering rather than steady, complicated by innumerable ever-changing eddies, comparable in some respects to the movements of the atmosphere but generally proceeding much more slowly (Perry and Walker, 1979; McWilliams, 1977).

Mostly, the ocean currents move slowly and irregularly. In the Equatorial currents the surface water usually flows at some 8–14 km/day. The North Atlantic Drift transports water from the region of Nova Scotia towards the British Isles at an average speed of approximately 19 km/day. Parts of the Gulf Stream move exceptionally rapidly, speeds of up to 180 km/day having been recorded. Less is known about flow rates below the surface. Some measurements at deep levels indicate speeds of 2–10 km/day, sometimes much faster.

1.3.1 Coriolis effect

Wind action on the surface does not simply blow the water in the same direction as the wind, except in very shallow depths. The earth's rotation causes a deflecting effect so that surface water is moved at an angle to the wind. This deflection, generally known as the Coriolis effect after the French engineer and mathematician who first derived an equation for it, influences any object moving on the earth's surface, and is due to the rotational movement of the earth beneath the moving body. In most cases the effect is so small compared with other forces

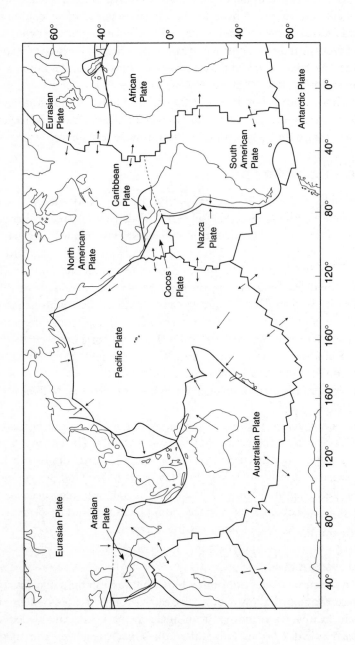

Figure 1.5 Major tectonic plate boundaries. Arrows indicate relative directions of movement. Submarine ridges develop at separating boundaries; ocean trenches or new mountains at colliding boundaries. Dotted lines indicate uncertain plate boundaries.

involved that it can be ignored, but in movements of the atmosphere and oceans the Coriolis effect has a magnitude comparable with the forces producing the motion, and must certainly be taken into account in understanding the course of ocean currents (Ingmanson and Wallace, 1995).

The Coriolis effect is equivalent to a force acting at 90° to the direction of movement, tending to produce a right-handed or clockwise deflection in the northern hemisphere and a left-handed or counterclockwise deflection in the southern hemisphere. It is proportional to the speed of movement and to the sine of latitude, being zero at the Equator. The Coriolis deflecting force, *F*, acting on a body of mass *m* moving with velocity *V* in latitude ϕ can be expressed

$$F = 2\omega V \sin \phi m$$

ω being the angular rate of rotation of the earth.

Thus the Coriolis deflection acts to the right of wind direction north of the Equator and to the left south of the Equator. Theoretically, in deep water of uniform density it results in a deflection of 45° to wind direction at the surface. This deflection increases with depth. The speed of the wind-generated current decreases logarithmically with depth and becomes almost zero at the depth at which its direction is opposite to the surface movement. The deflecting effect is less in shallow or turbulent water. A sharp temperature gradient near the surface has an effect similar to a shallow bottom. The warmer surface layer tends to slide over the colder water below, following wind direction more closely than it would if the temperature were uniform throughout the water column.

1.3.2 Surface currents

The chief surface currents and their relation to prevailing winds are shown in Figure 1.6. In the Equatorial belt between the Tropics of Cancer and Capricorn, the North-East and South-East Trade Winds blow fairly consistently throughout the year, setting in motion the surface water to form the great North and South Equatorial Currents which flow from east to west in the Atlantic, Indian and Pacific Oceans. Across the path of these currents lie continents which deflect the water north or south.

In the Atlantic, the Equatorial Currents are obstructed by the coast of Brazil, and the greater part of the water flows northwards into the Caribbean and Gulf of Mexico. The main surface outflow from the Gulf of Mexico flows strongly northwards past the coast of Florida (the Florida Current), and then out into the North Atlantic as the Gulf Stream. In the South Atlantic, water from the Equatorial Current is deflected southwards as the Brazil Current.

As the water moves away from equatorial regions, its course is influenced by the rotation of the earth. In the Gulf Stream, which at first flows in a north-easterly direction, the Coriolis effect gradually turns the water towards the right until,

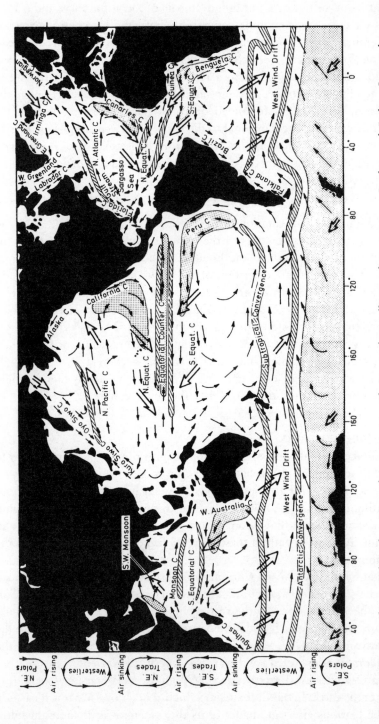

Figure 1.6 Prevailing winds and major surface currents, with zones of upwelling and convergence. In the north Indian Ocean the summer circulation is shown. ■ zones of upwelling; ▓ zones of convergence; ⇒ prevailing winds; → surface currents.

between latitudes 40–45°N, it is flowing eastwards across the Atlantic and becomes known as the North Atlantic Current. By the time it reaches the eastern part of the Atlantic it has been further deflected until it is flowing towards the south as the Madeira and Canaries Currents, eventually merging once again with the North Atlantic Equatorial Current. This vast circulation of surface water in a clockwise gyre surrounds an area of relatively little surface movement, the Sargasso Sea.

Where the North Atlantic Current moves eastwards across the Atlantic it is influenced by winds, the Westerlies, blowing from the south-west. These winds deflect some of the surface water towards the north-east to form the North Atlantic Drift, which flows into the Bay of Biscay and along the west and north of the British Isles, some eventually entering the northern part of the North Sea. North Atlantic Drift water also flows far up into the Arctic along the west and north coasts of Norway (the Norwegian Coastal Current), while some turns westwards south of Iceland (the Irminga Current). Part of this water flows clockwise around Iceland, and part flows on to the west and eventually reaches the west coast of Greenland (the West Greenland Current).

The inflow of water into the Arctic must be balanced by an equivalent outflow. Some of the surface water cools and sinks, and leaves the Arctic as a deep current (see page 16). There is also an outflow of cold surface water from the Arctic which enters the Atlantic as the East Greenland and Labrador Currents, and flows along the coast of Labrador and down the eastern seaboard of the United States, eventually sinking below the warm waters of the Gulf Stream flowing in the opposite direction.

In the South Atlantic, water from the Brazil Current under the influence of the Coriolis effect makes a counter-clockwise rotation, flowing eastwards across the Atlantic between latitudes 30–40°S, and turning in a northerly direction along the west coast of Africa. Here it is known as the Benguela Current which eventually merges with the Equatorial Current. In the southern hemisphere the Westerlies blow from the north-west, and deflect surface water into the Southern Ocean where there are no intervening land masses to interrupt the flow. Here the sea is driven continually in an easterly direction by the prevailing winds and becomes a great mass of moving water, the Antarctic Circumpolar Current, which encircles Antarctica. The surface current is termed the West Wind Drift.

The surface movements of the Pacific Ocean have a broadly similar pattern to those of the Atlantic. The Kuro Siwo Current, flowing in a north-easterly direction past the south island of Japan, is the counterpart of the Gulf Stream in the Atlantic. This water moves eastwards across the North Pacific towards the coast of British Colombia (the North Pacific Current), and then mostly turns south as the California Current. A cold current, the Oyo Siwo, flows down the western side of the Pacific towards the north Japanese island.

The surface circulation of the northern part of the Indian Ocean is complicated by seasonal changes in the direction of the monsoons. During the winter the ocean is warmer than the Asian land mass. Air overlying the sea rises and is replaced by cool air flowing off the land. This prevailing wind blows from the north-east between November and April and is termed the North-East Monsoon. It corresponds with the North-East Trade Winds and sets up a North Equatorial Current flowing from east to west, turning south along the African coast. At this time surface water between Arabia and India moves mainly in a west or south-west direction, and a clockwise gyre develops between India and Burma.

In summer, when the land becomes hot, air rises above the land and is replaced by the inflow of the South-West Monsoon, starting in April and usually blowing strongest in August to September. This carries water-saturated air over the land and causes the Monsoon rains. The reversal in direction of wind changes the direction of flow of surface water, driving water between Arabia and India eastwards and setting up a Monsoon Current from west to east in place of the North Equatorial Current. On reaching Indonesia this current turns south to join the South Equatorial Current. In the South Indian and South Pacific Oceans there is a counter-clockwise surface gyre and a deflection of water into the Southern Ocean, similar to the South Atlantic.

In the Equatorial belt between the latitudes of the North-East and South-East Trade Winds there is a calm zone, the Doldrums, where the effects of wind are minimal. In the Pacific and Indian Oceans, a certain amount of backflow of surface water towards the east occurs in this region, forming the Equatorial Countercurrents. There is relatively little backflow of surface water in the Atlantic, but a short distance below the surface there is an appreciable movement of water from west to east. In the Pacific, there is an even more extensive subsurface current, the Cromwell Current, which transports a large volume of water in an easterly direction between latitudes 2°N and 2°S at depths of 20–200 m.

The distribution of warm or cold surface water has a great influence on climate, and accounts for the climatic differences at equal latitudes on the east and west sides of the oceans. For example, the British Isles lie to the north of Newfoundland but here we have a temperate climate whereas Newfoundland is subarctic. This is because warm water from low latitudes moves across to high latitudes towards the eastern side of the ocean, and cold water moves towards low latitudes along the western side. Consequently, mild conditions extend further north in the east. The Bristol Channel in England lies slightly to the north of the Gulf of St Lawrence in Canada but certainly does not experience the freezing winter conditions of the latter.

1.3.3   Water movements below the surface

Oceanic circulation should be visualized in three dimensions. We have already mentioned that wind action on the surface sets different layers of water in movement in different directions. Where the wind causes a surface current, the moving water must be replaced by a corresponding inflow from elsewhere. This may be surface water from other regions or deep water rising to the surface, often both. Also, when surface water flows from low to high latitudes, cooling leads eventually to sinking and this causes movement of water at deep levels.

The replacement of the water of the North and South Equatorial Currents is derived partly from surface water from higher latitudes and partly from upwelling deep water. The temperature of the Canaries and Benguela Currents is low compared with other surface water at these latitudes because of mixing with enormous volumes of cold water from below. Similarly, upwelling into the California, Peru and West Australia Currents cools the surface water.

Water movements also arise from density changes due to differences of temperature or salinity. In low latitudes the surface is warm, and has a low density. We have already seen how this water is carried by surface currents into high latitudes, and there it loses heat and increases in density until it eventually becomes heavier than the underlying water. It then sinks and returns towards the Equator at deep levels. However, these density changes occurring as water moves from place to place are often modified by the effects of alterations in salinity. In low latitudes, although warming reduces the density, this is offset to some extent by evaporation raising salinity and thus increasing density. Heavy rainfall in some tropical areas reduces surface density by dilution. In high latitudes water density is increased by cooling and also by the greater salinity which occurs when ice crystals separate in the formation of sea ice, whereas density is lowered by dilution of the water by snow, rainfall, land drainage and melting of ice. The effects on water density of interactions between ocean and atmosphere are therefore extremely complex. It is a generalization to say that the density of surface water at high latitudes increases to the point at which water sinks and subsequently flows to lower latitudes below the surface.

Several factors influence the course of subsurface currents. They are subject to the Coriolis effect and to tide-generating forces. They are deflected or obstructed by submarine ridges. Their direction may be modified by the presence and movements of other water masses. Atmospheric interactions at the surface may have remote effects on the deep levels.

The relationships of these influences are intricate and not fully understood. Evidently water movements below the surface are subject to much variation from time to time associated with deep turbulence and eddies. In many areas it is possible to distinguish three main systems of subsurface water movements, the Bottom Current, the Deep Current and the Intermediate Current. These must not be regarded as steady progressions but rather as representing an overall transport of water within which turbulence, eddies and gyres move different parts in different and changing directions.

In the Atlantic, Indian and Pacific Oceans the Bottom Currents result mainly

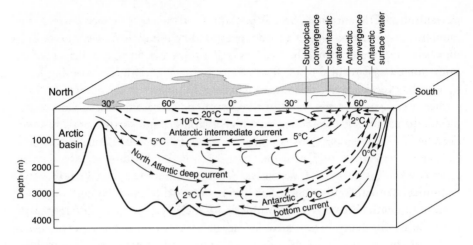

Figure 1.7   Some of the main movements of water below the surface of the Atlantic.

from the sinking of cold water around the Antarctic continent. The spread of cold bottom water from the Arctic Ocean is obstructed by the series of submarine ridges between Scotland and Labrador (see Section 1.2.3) and by the shallow Bering Straits. Cold water sinking in the Arctic is therefore trapped in the Arctic Basin. Beneath the Southern Ocean the cold water can escape, and creeps slowly northwards along the bottom, initially at a temperature of about 0°C but gradually becoming warmer as heat is gained by admixture with other warmer water, and perhaps a little by conduction through the sea-bed.

The bottom current in the Atlantic, deriving from the Antarctic and usually termed the Antarctic Bottom Current, flows mainly up the western basin to the west of the mid-Atlantic ridge, being held back from the eastern basin by the Walvis ridge. A corresponding ridge between Tristan da Cunha and the Brazil coast, the Rio Grande ridge, is incomplete and permits the passage of the bottom water. Just south of the Equator the mid-Atlantic ridge is cut by the Romanche Channel (Figure 1.3), through which some bottom water from the Antarctic eventually enters the eastern basin.

The Antarctic Bottom Current flows northwards across the Equator and has been traced to about latitude 40°N. Here it gradually loses its identity as it merges with water flowing in the opposite direction, the North Atlantic Deep Current (Figure 1.7). This water comes mainly from the cooling and sinking of surface water carried into the Arctic by the North Atlantic Drift. We have seen that the coldest water is held back within the Arctic basin by submarine ridges, but a large volume of water spills over the crest and this becomes the North Atlantic Deep Current. It has at first a temperature of 7–8° and is characterized by relatively high salinity and oxygen content. It flows down the southern face of the ridge and along the bottom until it meets the colder Bottom Current moving northwards from Antarctica. It continues its progress southwards getting colder as it goes, flowing above the Antarctic Bottom Current at levels between about 1500 and 3000 m, and is joined by part of the deep water outflow from the Mediterranean.

Near latitude 60°S, the Deep Current rises to the surface, upwelling to replace surface water, some of which is spreading northwards under wind influence as a surface drift while some is sinking due to cooling to become the Antarctic Bottom Current. The cold surface water spreading to the north has a temperature of 0–4°C and the salinity is reduced to about 34‰ (see Section 4.3.1) by melting ice. At approximately latitude 50°S it reaches an area of warmer and lighter surface water and sinks below it, continuing to flow northwards as the Antarctic Intermediate Current at depths between about 800 and 1200 m. This water can be traced to about 20°N.

Regions where surface currents meet, and surface water consequently sinks, are termed *convergences*. The Antarctic Intermediate water sinks at the Antarctic Convergence, and this occurs all round the Southern Ocean, mainly between latitudes 50 and 60°S. Further north at about 40°S is the Subtropical Convergence, another zone where surface water sinks and mixes with the Intermediate water. In the North Atlantic, a southward-flowing Intermediate Current is formed where the Labrador Current dips beneath the Gulf Stream (Figure 1.6).

In the Indian and Pacific Oceans the Intermediate and Bottom Currents both flow in a northerly direction, as in the Atlantic, but are not traceable so far north. The Deep Currents seem to be derived largely from the backflow of the Intermediate and Bottom Currents, and flow southwards until they rise to the surface in the Southern Ocean. All around Antarctica the deep water upwells and the surface water sinks, and at the same time most of the water of the Southern Ocean at all depths is also flowing eastwards as the Antarctic Circumpolar Current. This brings about a continual transport of water from the Atlantic to the Indian Ocean, from the Indian to the Pacific and the Pacific to the Atlantic, whereby the waters of the major ocean basins are intermixed.

1.3.4 Ocean eddies

Within the large-scale circulation described in Section 1.3.3, are more localized water movements known generally as *eddies*. These may be likened to our atmospheric weather systems of cyclones, anticyclones and fronts. Some of the best-known eddies form when the boundary of a strong current such as the Gulf Stream starts to meander rather like a river. As the amplitude of the meanders increases, the current flow may pinch off a meander (like an ox-bow lake along a river) forming isolated rotating eddies of water either side of the current boundary. In the case of the western boundary of the Gulf Stream, warm water eddies pinch off into the cold water outside the Gulf Stream and cold water eddies pinch off into the warm Gulf Stream. Eddies range in size from tens to hundreds of kilometres in diameter and may maintain their identity for some considerable time. A relatively new discovery is the existence of Meddies which are eddies of Mediterranean outflow water which track across the Atlantic but do not come to the surface. Only a few kilometres across, they may last two years or more and move considerable distances (Richards and Gould, 1996; Robinson, 1983).

1.3.5 World Ocean Circulation Experiment

In 1989, a comprehensive 10-year international programme called the World Ocean Circulation Experiment (WOCE) was initiated, with the primary aim of studying the ways in which the oceans may affect climate (Fofonoff and Holliday, 1994). It will provide a baseline description of the present state of the oceans such that future changes may be recorded. The study is being co-ordinated by Britain's Institute of Oceanographic Sciences (IOS).

The core of the project is the Hydrographic Programme. This is providing measurements of temperature, salinity and various dissolved constituents that conform to strict observational standards. Zonal and meridinal sections are being worked across the world oceans. The field programme collects data through a variety of instruments. The programme was timed to make use of satellites launched in the early 1990s. Fixed current meter mooring arrays have been deployed in all the major ocean basins to measure subsurface currents. A drifter programme was begun in 1991 using Lagrangian surface drifters and subsurface floats (see Section 3.1.1) to provide track charts for the upper ocean.

1.3.6 El Niño

Occasional major disruptions occur to the ocean current systems. These cause widespread changes to climate and sea level conditions especially in the tropics. The best known of these are the El Niño southern oscillation events. Under normal conditions, the cold, nutrient-rich Peru current flows northward along the western coast of South America (see Figure 1.6). This is accompanied by a coastal upwelling of nutrients caused by southerly winds. As a result, these waters are some of the most productive in the world. Huge numbers of anchovetta and sardine feed in the plankton-rich waters supporting a massive fishery as well as great numbers of seabirds and other wildlife.

Every 7 to 10 years or so, the trade winds cease to blow in their normal pattern from the east or south-east and warm equatorial water is blown in by abnormal winds from the west. In simple terms, pressure gradients in the west and east Pacific are reversed, causing a reversal of the trade winds and equatorial currents. A huge area of warm water is created and upwelling ceases thus reducing the supply of nutrients to the surface waters (Wuethrich, 1995). The increased water temperature kills many of the cold-water organisms normally present and, along with the lack of nutrients, leads to a dramatic reduction in primary production and a subsequent collapse of pelagic fish stocks. This has effects right up the food chain. Even the famous Galapagos iguanas suffer due to a reduction in the production of their seaweed food (Barber and Chavez, 1983).

The most extreme El Niño this century occurred in 1982–83. It started in July 1982 and continued until October 1983. Surface temperatures increased by 5°C and sea levels rose by as much as 22 cm in the Galapagos and 10 cm as far north as San Diego in California. Greatly increased rainfall caused extensive flooding in Peru and Ecuador. The exact mechanism is still not understood but two recent long-term programmes, TOGA, the Tropical Ocean Global Atmosphere, and WOCE (see Section 1.3.5) should one day allow scientists to model how the system works. This will help in the prediction of future El Niño southern oscillation events. Recently, there appears to have been an increase in the frequency of El Niños and this has led to considerable speculation on whether this could be due to global warming (see Chapter 10).

1.4 Biological features of the marine environment

Seawater is evidently an excellent medium for an abundance and variety of life. We know from geological findings that the seas have been well populated since the earliest time for which we have fossil records. It is widely thought that life originated in the sea, most likely in pools on the seashore where many different solutions of varying composition and concentration could accumulate in various conditions of temperature and illumination. Alternatively, some scientists are now postulating that life could have begun around deep-sea hydrothermal vents (see Section 6.4.4). The seas have now been populated for so long that it has been possible for marine life to evolve in great diversity.

Probably all natural elements are present in solution in the sea, and all the constituents needed for the formation of protoplasm are present in forms and concentrations suitable for direct utilization by plants (see Section 4.3.3). The transparency of the water and its high content of bicarbonates and other forms of carbon dioxide (see Section 4.3.2) provide an environment in the upper layers of the sea in which plants can form organic materials by photosynthesis, and in this way great quantities of food become available for the animal population. However, light penetrates only a short distance into the water. Marine plants must therefore be able to float close to the surface or, if attached to the bottom, are limited to shallow depths. Because water is relatively opaque to ultraviolet light, this property gives protection against the harmful effects of this part of the spectrum.

In an aquatic environment very simple and fragile forms of life can exist because the water affords them support, flotation, transport and protection, thereby permitting very simple reproductive processes, and minimizing the need for structural complications, such as locomotor organs, skeletons or protective coverings. In aquatic organisms there are several advantages in small size. For example, a large surface-to-volume ratio retards sinking, facilitates absorption of solutes at great dilution and favours light absorption. Also, small organisms can usually reproduce rapidly to take advantage of favourable conditions.

We shall discuss later (in Chapter 4) how organisms in the sea may be influenced by environmental conditions, notably the temperature, composition, specific gravity, pressure, illumination and movements of the water. However, we have already said enough about the circulation of the oceans to indicate that the water is kept well mixed, and this ensures a generally homogeneous environment. The composition of seawater (see Section 4.3) remains almost uniform throughout its extent despite considerable differences in the rates of evaporation and addition of fresh water in different localities. The composition of present-day seawater may differ in some respects from that of the remote past; but if so, marine organisms have been able to evolve and adjust to changing conditions. The body fluids of all the major groups of marine invertebrates are virtually isosmotic with seawater, and of a generally similar composition (see Section 4.3.1).

The high specific heat of water and the great volume of the oceans provides a huge thermal capacity, and the thorough mixing of the water ensures a fairly even distribution of heat. Consequently the temperature range of the oceans is relatively restricted and temperature changes occur slowly (see Section 4.2.1). The sinking of surface water due to cooling at high latitudes carries well-oxygenated water to the bottom, and thereby makes animal life possible at all depths. Despite biological activity, the buffer properties of the water (see Section 4.3.2, Carbon dioxide) are sufficient to keep the pH stable. It is therefore a notable feature of the marine environment that conditions are remarkably constant over great areas, and many marine plants and animals have correspondingly wide distributions. Such changes as do occur take place slowly, giving time for some organisms to acclimatize. However, stable conditions permit the evolution of a diversity of forms whose environmental requirements are very precise and whose range is limited by quite slight changes in their surroundings. It must therefore be evident that this combination of properties offers propitious conditions for a great variety of marine organisms of many types and sizes.

1.5 Elementary classification of the marine environment

Although the mixing effected by the oceanic circulation ensures that the major parameters vary but little throughout enormous volumes of water, there are nevertheless some strong contrasts between different parts of the sea. The cold, dark, slowly moving bottom layer of the deep ocean is obviously a very different

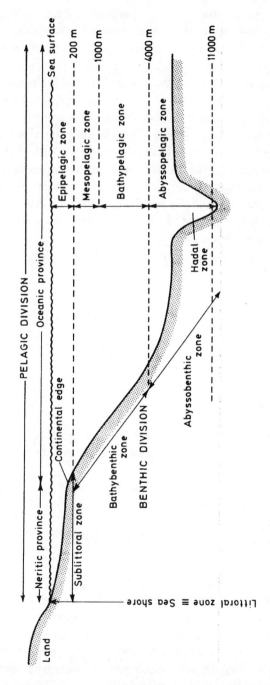

Figure 1.8   Main divisions of the marine environment.

environment from the well-illuminated, wave-tossed waters of the sea surface, or the strong currents and fluctuations of temperature and salinity that often occur near the coast. We therefore need a classification of subdivisions of the marine environment which takes account of different conditions of life in different parts of the oceans.

There are broadly two ways in which organisms live in the sea; they float or swim in the water, or they dwell on or within the sea bottom. We can correspondingly make two major divisions of the environment, the *Pelagic* and the *Benthic*, the Pelagic Division comprising the whole body of water forming the seas and oceans, and the Benthic Division the entire sea bottom (Figure 1.8).

In shallow water there is usually more movement and greater variations of composition and temperature than occur where the water is deep. We can therefore subdivide the Pelagic Division into (*a*), the *Neritic Province*, the shallow water over the continental shelf, and (*b*), the *Oceanic Province*, the deep water beyond the continental edge.

In deep water, conditions change with level and it is useful to distinguish four zones as follows:

(*a*)   The *Epipelagic Zone* from the surface to 200 m depth, in which there are sharp gradients of illumination, and often temperature, between the surface and the deeper levels; and also diurnal and seasonal changes of light intensity and temperature. In many areas the temperature gradient is irregular, involving discontinuities or thermoclines (see Section 4.2.2). Water movements may be relatively rapid.

(*b*)   The *Mesopelagic Zone* from 200 to 1000 m depth, where very little light penetrates, and the temperature gradient is more even and gradual without much seasonal variation. An oxygen-minimum layer (see Section 4.3.2) and the maximum concentrations of nitrate and phosphate (see Section 4.3.3) often occur within this zone.

(*c*)   The *Bathypelagic Zone* between 1000 m and 4000 m, where darkness is virtually complete except for bioluminescence, temperature is low and constant, and water pressure high.

(*d*)   The *Abyssopelagic Zone* below 4000 m; dark, cold, with the greatest pressures and very little food.

The sea bottom and the seashore together make up the Benthic Division which comprises three major zones, the *Littoral*, the *Sublittoral* and the *Deep Sea Zones*. The Littoral Zone includes the greater part of the seashore together with the wave-splashed region above high tide level (see Section 8.6). The Sublittoral Zone is the shallow sea bottom extending from the lower part of the shore to the continental edge. The Deep Sea Zone lies below the continental shelf, and can be subdivided into *Bathybenthic* and *Abyssobenthic Zones*. The Bathybenthic zone lies between the continental edge and a depth of about 4000 m, comprising mainly the continental slope. The Abyssobenthic Zone is the bottom below 4000 m, including the continental rise, abyssal plain and deeper parts of the sea floor.

The deepest parts of the ocean within the trenches below some 6000 to 7000 m are termed the *Hadopelagic* and *Hadobenthic Zones*.

Subdivisions of the marine environment with respect to temperature and light are mentioned later (see Sections 4.2.3 and 4.6).

Organisms of the Pelagic Division comprise two broad categories, *plankton* and *nekton*, differing in their powers of locomotion. The plankton consist of floating plants and animals which drift with the water, and whose swimming powers, if any, serve mainly to keep them afloat and adjust their depth level rather than to carry them from place to place. A brief account of the constituents of marine plankton appears in Chapter 2. The nekton comprises the more powerful swimming animals, vertebrates and cephalopods, which are capable of travelling from one place to another independently of the flow of the water.

The populations of the Benthic Division, the sessile and attached plants and animals and all the creeping and burrowing forms, are known collectively as *benthos*.

The term *benthopelagic* refers to animals, mainly fish, which live very close to, but not actually resting on, the bottom. Hovering slightly above the sea floor, they are well placed for taking food from the bottom.

References and further reading

Student texts
Bonatti, E. (1994). The Earth's mantle below the oceans. *Scient. Am.*, **270** (3), 26–33.

Duxbury, A.B. and Duxbury, A.C. (1996). *Fundamentals of Oceanography*. 2nd edition. Wm. C. Brown Publishers (WCB), USA. (A CD-ROM: Interactive Plate Tectonics by C.C. Plummer and D. McGeary can be obtained from WCB publishers.)

Freeman, W.H. (1976). *Continents adrift and aground, Readings from Scientific American*. San Francisco, W.H. Freeman, 230 pp.

Freeman, W.H. (1990). *Shaping the Earth – Tectonics of continents and oceans, Readings from Scientific American*. San Francisco, W.H. Freeman, 206 pp.

Graham, N.E. and White, W. B. (1988). The El Niño cycle: A natural oscillator of the Pacific Ocean-atmosphere system. *Science*, **240**, 1293–1302 (3 June 1988).

Ingmanson, D.E. and Wallace, W.J. (1995). *Oceanography: An Introduction*. 5th edition. Wadworth Publishing Co.

Macdonald, K.C. and Fox, P.J. (1990). The Mid-Ocean Ridge. *Scient. Am.*, **262** (6), 723–79.

Nance, R.D., Worsley, T.R. and Moody, J.B. (1988). The supercontinent cycle. *Scient. Am.*, **259** (1), 44–51.

National Geographic Society (1995). The *Earth's fractured surface*. National Geographic map, April 1995.

Summerhayes, C.P. and Thorpe, S.A. (eds) (1996). *Oceanography: An Illustrated Guide*. Southampton Oceanography Centre. Manson Publishing.

Vink, G.E., Morgan, W.J. and Vogt, P.R. (1985). The Earth's hot spots. *Scient. Am.*, **252** (**4**), 32–9.
York, D. (1993). The earliest history of the Earth. *Scient. Am.*, **268** (**1**), 82–8.

References

Anderson, C.H. (1992). *Changing the face of the Earth*. National Geographic map, June 1992.
Barber, R.T. and Chavez, P. (1983). Biological consequences of El Niño. *Science*, **222** (**4629**), 1203–10 (Dec. 16 1983).
Couper, A. (ed.) (1989). *The Times Atlas and Encyclopedia of the Sea*. London, Times Books.
Dietz, R.S. and Holden, J.C. (1970). The Breakup of Pangea. *Scient. Am.*, **223**, October.
Fofonoff, N.P. and Holliday, N.P. (1994). World Ocean Circulation Experiment (WOCE) four years on. *International Marine Science Newsletter* **69/70**, 1st semester 1994, UNESCO.
Ingmanson, D.E. and Wallace, W.J. (1995). *Oceanography: An Introduction*. 5th edition. Appendix VI, The Coriolis Effect. Wadworth Publishing Co.
McWilliams, J.C. (1977). On the Large Scale Circulation of the Ocean: A Discussion for the Unfamiliar. *Marine Science*, **5**, 723–47.
Perry, A.H. and Walker, J.M. (1979). *The Ocean-Atmosphere System*. London, Longman.
Richards, K.J. and Gould, W.J. (1996). Ocean Weather – Eddies in the Sea. In: *Oceanography; An Illustrated Guide*. C.P. Summerhayes and S.A. Thorpe, eds. Southampton Oceanographic Centre, Manson Publishing.
Robinson, A.R. (ed.) (1983). *Eddies in Marine Science*. Springer-Verlag, Berlin.
Whitmarsh, R.B., Bull, J.M., Rothwell, R.G. and Thomson, J. (1996). The Evolution and Structure of Ocean Basins. In: *Oceanography: An Illustrated Guide*. C.P. Summerhayes and S.A. Thorpe, eds. Southampton Oceanographic Centre, Manson Publishing.
Wuethrich, B. (1995). El Niño goes critical. *New Scientist*, **145**, No. 1963, 32–5.

2 Marine plankton

The word 'plankton' is taken from a Greek verb meaning to wander and is used to refer to those pelagic forms which are carried about by the movements of the water rather than by their own ability to swim. These organisms are called *planktonts*. The plants of the plankton are the *phytoplankton*, the animals the *zooplankton*. Early naturalists were enthralled by these tiny animals and plants and wrote some fascinating accounts of their observations (Fraser, 1962; Hardy, 1956 and 1967; Sverdrup *et al.*, 1946). This huge subject can be only introduced in this chapter.

The phytoplankton is responsible for most of the primary production in the sea (see Chapter 7). There would be virtually no life in the ocean without the photosynthesis carried out by these microscopic plants. On land the energy-fixing plants dominate the landscape in the form of grasses, shrubs and trees. In contrast the phytoplankton is only visible to us as a cloudiness or discolouration of the water when reproduction is rapid and a 'bloom' occurs.

Some planktonts can only float passively, unable to swim at all. Others are quite active swimmers but are so small that swimming does not move them far compared to the distance they are carried by the water. The swimming movements serve chiefly to keep them afloat, alter their level, obtain food, avoid capture, find a mate or set up water currents for respiration. Although the majority of planktonts are small, mainly of microscopic size, a few are quite large. For example, the tentacles of Portugese Man O'War (*Physalia*) sometimes extend 15 m through the water, and there are jellyfish (Scyphomedusae) which grow to over 2 m in diameter.

Organisms, whose entire lifespan is planktonic, are termed *holoplankton* or permanent plankton. The numbers of planktonts are swelled by the addition of organisms passing through a pelagic phase which is only part of the total life-span; for example, planktonic spores, eggs or larvae of nektonic (free-swimming) or benthic (bottom-living) organisms. These organisms, which spend only part of their lifespan in the plankton, usually the early stages, are termed *meroplankton* or temporary plankton. For fixed, sessile species, the planktonic larval stages provide an essential means of dispersal.

2.1   Definitions

Terms in wide use for referring to different components of the plankton include the following:

Size categories

There is no single agreed classification of plankton by size and the student will find the same terms applied to slightly different size categories in older texts. The definitions given below are based on those given in Omori and Ikeda (1984) which are both practical and easy to understand.

Megaloplankton – Gelatinous plankton such as medusae (jellyfish) and salps, >20 mm.

Micronekton – These fall into the same size category as megaloplankton but the term is usually applied to strongly swimming animals such as euphausids, mysids and fish larvae, 20–200 mm.

Macroplankton – Large planktonts visible to the unaided eye, such as pteropods, copepods, euphausiids and chaetognaths, 2–20 mm.

Mesoplankton – The principal components of zooplankton fall into this and the macroplankton category. Mesoplankton includes cladocerans, copepods, and larvaceans, 200 μm–2 mm.

Net plankton – The four larger size categories described above are collectively called net plankton because they can be effectively caught using nets.

Microplankton – Includes most phytoplankton species, foraminiferans, ciliates, rotifers and copepod nauplii, 20–200 μm.

Nanoplankton – Organisms such as fungi, small flagellates and small diatoms, 2–20 μm.

Ultraplankton or picoplankton – Mainly bacteria and cyanobacteria, < 2 μm. Water bottle plankton – The three smaller categories (micro-, nano- and ultra-plankton) which cannot effectively be caught by nets.

Habitat categories

Epiplankton – Plankton of the epipelagic zone, i.e. within the uppermost 200 m.

Pleuston – Passively floating organisms living at the air–sea interface, partially exposed to air and moved mainly by the wind (Cheng, 1975).

Neuston – Small swimming organisms inhabiting the surface water film, *epineuston* on the aerial side, *hyponeuston* on the aquatic side.

Bathyplankton – Plankton of deep levels.

Hypoplankton – Plankton living near the bottom.

Protoplankton – Pelagic bacteria and unicellular plants and animals.

Seston – Finely particulate suspended matter.

Tychopelagic – Organisms of normally benthic habit which occasionally become stirred up from the bottom and carried into the water.

2.2 Marine phytoplankton

Marine phytoplankton is made up of small plants, mostly microscopic in size and unicellular. Two orders of algae commonly predominate in the larger, net phytoplankton: Diatoms (Bacillariophyceae) and Dinoflagellates (Dinophyceae). The phytoplankton often also includes a numerous and diverse collection of extremely small, motile plants collectively termed microflagellates.

Floating masses of large algae are found living and growing in some areas, notably the Sargassum weed of the Sargasso area of the North Atlantic. Within very sheltered sea lochs on the Scottish coast, in areas where the salinity is reduced, there occurs a floating and proliferating form of the common littoral fucoid seaweed, *Ascophyllum nodosum*. These are not generally regarded as phytoplankton because they derive from the fragmentation of benthic plants growing on the sea bottom in shallow water.

Many of the phytoplankton species of the North Atlantic are described in detail in volumes by Lebour (1925, 1930) and Hendey (1964). There are many excellent photographs in Drebes (1974) although the text is in German. More recent, but less detailed accounts can be found in the student texts detailed at the end of this chapter.

2.2.1 Diatoms (class Bacillariophyceae)

The majority of diatoms are unicellular, uninucleate plants with a size range of about 15 μm to 400 μm in maximum dimension, although some smaller and a few considerably larger forms exist. The largest known diatom is a tropical species, *Ethmodiscus rex*, up to 2 mm in diameter. The diatom cell, known as a frustule, has a cell wall of unusual composition and structure. It is impregnated with siliceous material giving a glassy quality and consists of two parts, the valves. At its simplest, for example in *Coscinodiscus*, the cell wall is like a transparent pillbox (Figure 2.1), the larger valve or epitheca overlapping the smaller hypotheca much as the lid of a pillbox overlaps the base.

The valves are often very elaborately ornamented with an intricate sculpturing of minute depressions, perforations or tiny raised points which are sometimes arranged in beautiful symmetrical patterns of great variety. In some, the cell wall has larger projections forming spines, bristles and knobs. Ornamentation increases the surface area and also strengthens the cell wall, which in the majority of planktonic diatoms is very thin. In some species, growth occurs by elongation of the valves at their margins forming a number of intercalary bands, for example *Guinardia* (Figure 2.2a). Internal thickenings of these bands may form septa which partially divide the interior of the frustule.

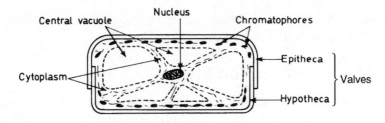

Figure 2.1   Diagrammatic section of a pillbox diatom.

The cytoplasm usually lines the cell wall and contains numerous small, brown chromatophores. There is a large central vacuole containing a cell sap. The nucleus with an enclosing film of cytoplasm is often suspended within the vacuole, supported by cytoplasmic threads extending from the peripheral layer. In planktonic diatoms the cell sap is probably lighter than seawater and may confer some buoyancy to support the heavier protoplasm and cell wall. In many diatoms the cytoplasm is not confined to the interior of the frustule, but exudes through small perforations to cover the surface or form long thin threads, and these may join the cells together in chains.

Planktonic diatoms present a considerable variety of shape, each in its way well adapted to provide a large surface/volume ratio which improves their photosynthetic efficiency. They may be grouped into four broad categories as follows:

(a)   *Pillbox shapes* – Usually circular and radially symmetrical when seen in top or bottom view, for example *Coscinodiscus* (Figure 2.2f), *Hyalodiscus*. Sometimes they are connected by protoplasmic strands to form chains, for example *Thalassiosira* (Figure 2.2h), *Coscinosira*.

(b)   *Rod or needle shapes* – The division between the valves may be at right-angles to the long axis of the cell, for example *Rhizosolenia* (Figure 2.2e), and these are often joined end to end to form straight chains. In others the division runs lengthways, for example *Thalassiothrix*, *Asterionella* (Figure 2.2d and g), and these may be joined to form starlike clusters or irregular zig-zag strands.

(c)   *Filamentous shapes* – Cells joined end to end by the valve surfaces to form stiff, cylindrical chains (*Guinardia*, Figure 2.2a) or flexible ribbons (*Fragilaria*, Figure 2.2c).

(d)   *Branched shapes* – Cells bearing various large spines or other projections, and sometimes united into chains by contact between spines, *Chaetoceros* (Figure 2.2i), or by sticky secretions, *Biddulphia* (Figure 2.2b).

In addition to the planktonic forms there are numerous benthic species of diatom occurring on the shore or in shallow water. These often grow on the surface of sediments, particularly in sheltered areas such as sea lochs. Here they

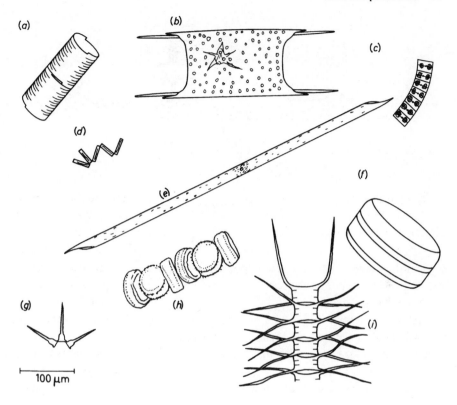

Figure 2.2 Some common diatoms from shallow seas around the British Isles. (*a*) Guinardia flaccida, (*b*) Biddulphia sinensis, (*c*) Fragilaria, (*d*) Thalassiothrix, (*e*) Rhizosolenia, (*f*) Coscinodiscus concinnus, (*g*) Asterionella japonica, (*h*) Thalassiosira, (*i*) Chaetoceros decipiens.

form a clearly visible, thin brown layer. On rocks and stones, they may form a slimy covering. Some project above the surface of the substrate on short stalks. Diatoms are also commonly found attached to the surface of other plants or animals. Benthic diatoms usually have appreciably thicker and heavier cell walls than the planktonic species. Certain benthic species living on sediments have some powers of motility, gliding within the interstices of the deposit so as to move to or from the surface with changing conditions (see page 292).

The usual method of reproduction is by simple asexual division. Under favourable conditions this may occur three or four times a day, so that rapid increase in numbers is possible. The protoplast enlarges and the nucleus and cytoplasm divide. The two valves become gradually separated, the daughter cells each retaining one valve of the parent cell. The retained valve becomes the epitheca

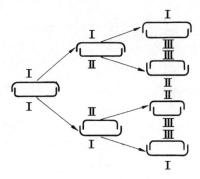

Figure 2.3   Reduction of mean cell size in diatoms following cell division.

of each daughter cell, and a new hypotheca is secreted, the margin of which fits inside the old valve. The new cell formed within the parent epitheca is therefore the same size as the parent cell, but the cell formed inside the original hypotheca is smaller. Because of this, it is a peculiarity of diatoms that the average size of the individuals in a population tends to decrease as division continues (Figure 2.3). This process of size reduction does not go on indefinitely. Eventually the valves of the smaller individuals separate, the protoplasm flows out and the valves are shed. The naked protoplasm, known as an auxospore, enlarges and grows new, larger valves.

Some diatoms can form resistant spores to carry them through unfavourable periods, for example, during the winter months in neritic water when the temperature falls and salinity may fluctuate appreciably. The cell vacuole disappears and the protoplasm becomes rounded, secreting a thick wall around itself. Probably many resistant spores sink to the bottom and are lost, but in shallow water some may be brought to the surface again later by wave action, currents and turbulence, and then germinate. In high latitudes, diatom spores become enclosed in sea ice during the winter months and germinate the following year when the ice melts.

Sexual reproduction has been observed in certain diatoms. In some species it precedes auxospore formation, the protoplasts of two diatoms fusing to form a single auxospore. In other cases, fusion of protoplasts appears to give rise to two or more auxospores. The formation of microspores has also been observed, the protoplast dividing numerous times to form minute biflagellate structures which are thought to act as gametes.

When planktonic diatoms die, fragments of their valves sink down to the sea-bed. In some areas, notably beneath the Southern Ocean and the northern part of the North Pacific, this accumulation of diatomaceous material gives rise to a siliceous ooze (see page 217).

2.2.2 Dinoflagellates (class Dinophyceae)

These are unicellular, biflagellate organisms with a range of size similar to diatoms but with a larger proportion of very small forms that escape through the mesh of fine plankton nets (nanoplankton). The arrangement of the flagella is characteristic of the group. Typically (Figure 2.4) the cell is divided into anterior and posterior parts by a superficial encircling groove termed the girdle, in which lies a transverse flagellum wrapped around the cell and often attached to it by a thin membranelle. Immediately behind the origin of the transverse flagellum, a whip-like longitudinal flagellum arises in a groove known as the sulcus, and projects behind the cell. The longitudinal flagellum performs vigorous flicking movements and the transverse flagellum vibrates gently, the combined effects driving the organism forwards along a spiral path. There are various departures from this characteristic form. For example, *Amphisolenia* has a thin, rod-like shape; *Polykrikos* (Figure 2.5f), has several nuclei and a series of girdles and sulci, usually eight, each provided with transverse and longitudinal flagella.

Many dinoflagellates have no cell wall. In these non-thecate forms the cytoplasm is covered only by a fine pellicle. Others are thecate and covered with a strong wall of interlocking cellulose plates. In certain species the cell wall is elaborated into spines, wings, or parachute-like extensions, and these are especially complex in some of the warm water forms (for example, *Dinophysis*, Figure 2.5g), perhaps assisting flotation. Dinoflagellates from the British Isles are described in Dodge (1982, 1985).

Dinoflagellates are mainly a marine planktonic group occurring in both oceanic and neritic water. They are most numerous in the warmer parts of the sea, where they sometimes outnumber diatoms, but are also found in cold areas. Around the

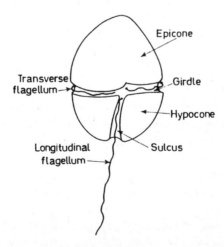

Figure 2.4   A simple non-thecate dinoflagellate.

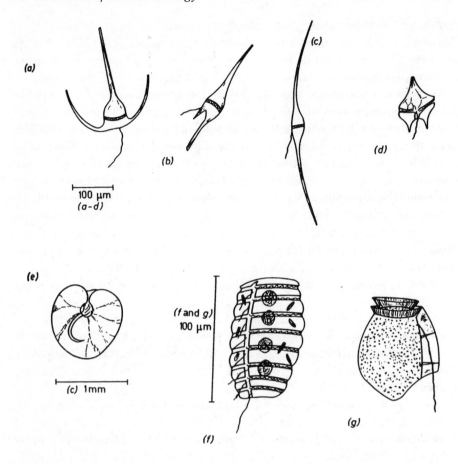

Figure 2.5 Some dinoflagellates of the north-east Atlantic. (a) Ceratium tripos, (b) Ceratium furca, (c) Ceratium fusus, (d) Peridinium depressum, (e) Noctiluca scintillans, (f) Polykrikos schwarzi, (g) Dinophysis.

British coasts they are scarce in the winter months, and reach their greatest abundance in the midsummer period when the low concentration of nutrients seems to have less effect in limiting the growth of dinoflagellates than of diatoms. Dinoflagellates also occur in fresh and brackish water, and are sometimes abundant in estuaries. Some are found in sand in the interstitial water between the particles. There are also many parasitic dinoflagellates infecting a variety of planktonic organisms including radiolaria, copepods, pteropods, larvaceans and fish eggs.

Reproduction is by asexual fission, but is not as rapid as in diatoms. In thecate dinoflagellates the process is somewhat similar to fission in diatoms, each daughter cell retaining part of the old cell wall and secreting the other part, but the old and new cell plates do not overlap, and there is consequently no size reduction as occurs in diatoms. The daughter cells do not always separate completely and repeated divisions then form a chain. Resistant spores may be produced during adverse periods.

Many dinoflagellates contain small chromatophores and perform photosynthesis. The group is certainly important as primary producers of food materials. Some of them are highly pigmented, and are sometimes so numerous that the water appears distinctly coloured, different species producing green, red or yellow tints. When this happens it is often described as a 'red tide' (see below). There are also many colourless dinoflagellates. Such thecate forms without chromatophores are presumably saprophytes, feeding off decaying material. However, some of the non-thecate, colourless forms are holozoic, feeding in an animal-like manner on various small organisms including other dinoflagellates, diatoms, microflagellates and bacteria. *Noctiluca* (Figure 2.5e), for example, devours copepod larvae and other small metazoa. Some of these holozoic dinoflagellates possess tentacles, amoeboid processes or stinging threads for capturing their food.

Some species are luminous, and can be the cause of remarkable displays of phosphorescence in seawater. *Noctiluca* is an example which sometimes occurs in swarms around the British Isles, visible at night as myriads of tiny flashes of light when stimulated by agitation of the water. This can be seen to dramatic effect when scuba diving. Various species of *Peridinium* flash spontaneously in undisturbed water.

Red tides

Certain species of dinoflagellates contain highly toxic substances which can cause the death of other marine creatures which eat them. Under certain conditions when an excess of nutrients is present and the sea is very calm, certain dinoflagellates multiply very rapidly and form a bloom. The water is often discoloured hence the name 'red tide'. Not all 'red tides' are red or toxic, but those that are occasionally cause the death of large numbers of fish. In recent years, many fish farms have suffered in this way. Examples of toxic dinoflagellates include *Gymnodinium brevis*, *Goniaulax polyedra*, and *Exuviella baltica*. In the Gulf of Mexico, tons of dead fish are periodically washed up along the Florida coastline. The fish are killed as they swim through blooms of the dinoflagellate *Gymnodinium breve*, rupturing the algae which release neurotoxins onto the gills of the fish (Anderson, 1994).

Sea birds, humans and marine mammals can also be poisoned indirectly when they eat filter-feeding shellfish that have accumulated the toxins. In the summer of 1968, cases of food poisoning in humans in the UK were traced to a dinoflagellate infection of mussels on the Northumberland coast, which also caused the death of seabirds, sand eels and flounders. In 1987, a large number of humpback whales died in Cape Cod Bay (USA). It was eventually shown that toxins from a dinoflagellate *Alexandrium tamarense* had killed the whales via their food web (Anderson, 1994). In the same month many fishermen, visitors and residents along the North Carolina coast, who had eaten local shellfish became ill with a variety of symptoms. In Canada, people who had eaten mussels from Prince Edward Island also became ill. This was traced to a toxin from a diatom, *Pseudonitzchia pungens*.

2.2.3   Some other planktonic plants

Many species of unicellular green algae are represented in the plankton. These are often motile or have a motile phase. *Halosphaera* (Figure 2.6) is one of the largest and belongs to the class Prasinophyceae. It consists of a single spherical cell with a tough elastic wall, and sometimes reaches nearly 1 mm in diameter. There is a large central vacuole in which a single nucleus is usually suspended, but fully grown cells may contain as many as eight nuclei. In the peripheral cytoplasm are numerous small, yellowish-green chromatophores giving the cell a vivid colour. Asexual reproduction involves a motile phase, repeated divisions of the protoplasm leading to the liberation of numerous four-flagellate spores. Sexual reproduction has not been recorded. *Halosphaera* is found throughout the North Atlantic but is commonest in tropical areas. It is sometimes carried into British waters in great numbers by the North Atlantic Drift and is often abundant in the northern North Sea in autumn.

Phaeocystis (Figure 2.7) belongs to a group of unicellular brown-coloured algae (class Haptophyceae). It is a minute biflagellate cell which develops into a colonial structure. The cells divide repeatedly, and extensive mucilaginous capsules form around them and bind them together in large gelatinous clumps up

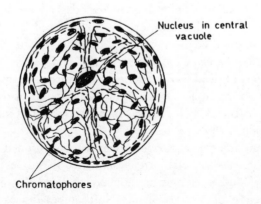

Nucleus in central vacuole

Chromatophores

Figure 2.6   Halosphaera.

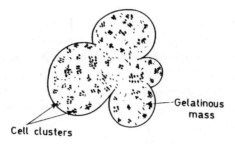

Figure 2.7   *Phaeocystis.*

to several centimetres in diameter. These are abundant in north temperate inshore waters in spring and summer. On occasion they are so numerous that they colour the water and give it a slightly slimy consistency. Fishermen call this 'weedy water' and it is apparently distasteful to some animals; for example, herring shoals seem to avoid this water and the catches are poor when *Phaeocystis* occurs in quantity on the fishing grounds. A *Phaeocystis* patch over 100 miles in extent was recorded from the North Sea in 1927.

Silico-flagellates are unicellular organisms belonging to the class Chryso-physeae which live predominantly in freshwater. The marine silico-flagellates are small, often about 100 μm in diameter, and have their protoplasm supported by an internal skeleton of interconnecting siliceous rods forming a capsular structure with outwardly radiating spines. There is usually a single flagellum. They are sometimes present in considerable numbers, mainly in the colder parts of the seas.

Blue-green algae (class Cyanophyceae) are also represented in the phyto-plankton. The commonest types form filaments of short chains of minute spherical

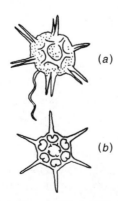

Figure 2.8   *A silicoflagellate showing* (a) *live animal;* (b) *part of skeleton.*

or oval cells with no definite chloroplasts, the pigments being diffused in the cytoplasm. In addition to chlorophyll, phycocyanin and phycoerythrin are present, giving the cells their bluish-green colour. Blue-green algae are different from all other classes of algae because, like bacteria, they have no organized nuclei. These plants have occasionally been found in large numbers at deep levels (see page 236) and may be able in some circumstances to feed saprophytically. Under certain conditions they form extensive blooms. With increasing eutrophication of enclosed fresh and brackish water areas such as estuaries, and reservoirs, such blooms are now causing amenity problems. Toxins released by the algae can cause rashes and illness in bathers and other water users.

2.2.4   Nanoplankton

The smallest planktonic organisms, less than 20 μm in diameter, are termed nanoplankton. These (and many microplanktonts) escape through fine-mesh plankton nets but may be collected by centrifuging, membrane filtration, or sedimentation (see page 67). This diverse group of minute organisms includes the smallest diatoms and dinoflagellates, coccoliths (see below) various other small flagellates known collectively as microflagellates, and also fungi and some bacteria.

It is only in fairly recent years that the nanoplankton has been much studied, but it now appears that the quantity of living material in the water in this form sometimes exceeds that present as diatoms and dinoflagellates. The nanoplankton is now thought to make a major contribution to the primary production of organic food in the sea, and is specially important as the chief food for many larvae. Modern analytical techniques have demonstrated that the shallow waters (epipelagic zones) of many temperate and tropical seas are dominated by nanoplankton, both in terms of numbers of individuals and amount of photosynthesis. In the tropics, nanoplankton may account for more than 80 per cent of photosynthetic activity in open ocean waters. In coastal (neritic) waters, nanoplankton play a less important role.

Coccolithophoridae

The Coccolithophoridae (Figure 2.9) are one of the more important groups of nanoplankton. They are minute, unicellular organisms, mostly some 5–20 μm in diameter and containing a few brown chromatophores (Thomas, 1993; Winter and Siesser, 1994). They are characterized by tiny calcareous plates which cover their outer surface. The plates are usually extremely finely and elaborately sculptured. At their simplest, they are oval discs, but in some species, the plates form long projections from the surface of the cell, often of bizarre design. They have complex life histories usually involving several morphologically different phases. Some have both a motile uni- or bi-flagellate phase and a non-motile

Figure 2.9   A coccolithophore.

pelagic phase, and some have a benthic filamentous phase. Coccoliths are widely distributed and are sometimes so numerous near the surface that they impart a slight colouration to the water. The condition which herring fishermen call 'white water', and which is regarded as indicating good fishing, is sometimes due to swarms of coccoliths. The calcareous plates of disintegrated coccoliths are a conspicuous component of the deep-sea sediment in some areas.

Coccoliths have occasionally been found in surprisingly large quantities far below the photosynthetic zone, sometimes very numerous between 200 and 400 m, and even in considerable abundance at depths of 1000–4000 m. This deep-water distribution suggests that some coccoliths can feed to some extent by methods other than photosynthesis, perhaps by absorption of organic solutes or even by ingestion of organic particles. In some areas they may be an important source of food for some of the animals at deep levels (see page 236).

Microflagellates

The seas contain many minute unicellular organisms which swim by means of one or more flagella, and are loosely termed microflagellates (Thomas, 1993). The majority are within the size range 1–20 μm in diameter, but there are a few larger species up to about 100 μm. Most of them contain chlorophyll, and there is no doubt that they are important as primary food producers. There are also many colourless saprophytic forms. The life histories are not well known. Some have been observed to reproduce by fission, while others are spores of larger algae. Non-motile cells, similar in general appearance to microflagellates but lacking flagella, are also known; and in some cases the life history includes both motile and non-motile pelagic stages.

2.2.5   Ultraplankton

Although the number of bacteria in seawater varies greatly with time and place, in coastal waters they sometimes contribute a significant part to the total biomass of the plankton. Bacterioplankton are often associated with floating organic debris known as 'marine snow' (see page 233). Bacteria and other micro-organisms are a constituent of much importance in marine ecosystems, functioning in many roles as decomposers, saprophytes and pathogens; regenerating nutrients, producing dissolved gases and ectocrine compounds, some of which may be essential for the normal growth of other organisms. They are a major source of food for protozoa and filter-feeding animals.

2.3   Marine zooplankton

The zooplankton includes a very wide variety of organisms. Every animal phylum represented in the sea, contributes at least to the meroplankton with eggs and larval stages, and it is beyond the scope of this book to attempt a comprehensive survey of the range of zooplankton organisms. Some excellent general accounts are now available, and several are listed at the end of this chapter. The purpose of this section is simply to mention the major groups of holoplanktonic animals with a few examples of species common in the north-east Atlantic.

Crustacea

These are the most conspicuous element in the permanent zooplankton, commonly amounting numerically to at least 70 per cent of the total. The predominant class is the Copepoda. These are present in incredible numbers and are represented by many species. There are about 1200 species in the British sea area, of which *Calanus* (Marshall and Orr, 1955), *Acartia*, *Centropages*, *Temora*, *Oithona*, *Pseudocalanus* and *Paracalanus* are especially common (Figure 2.10).

Another abundant group of planktonic Crustacea is the Euphausiacea. The best know of these is a large species, *Euphausia superba* (Figure 2.11a), which occurs in enormous numbers in the Southern Ocean south of the Antarctic Convergence. It constitutes the 'krill' upon which the giant baleen whales of the Antarctic feed, including the blue whale (*Balaena musculus*). In the north-east Atlantic, species of *Nyctiphanes*, *Meganyctiphanes* (Figure 2.11b) and *Thysanoessa* are common.

Other crustacean groups that are sometimes numerous in marine plankton are the Cladocera (water fleas), for example *Podon* and *Evadne* (Figure 2.12); the Ostracoda (seed shrimps), for example *Conchoecia*, *Philomedes*; and the Amphipoda, for example *Parathemisto* (Figure 2.13). Mysids or opossum shrimps (Mysidacea) mostly live close to the bottom but are sometimes found in coastal plankton, especially in estuarine regions. In addition to these holoplanktonic forms, great numbers of larvae are contributed to the temporary plankton by benthic crustacea. Larvae of crustaceans such as crabs, barnacles and lobsters are very different in form from the adults and often have striking adaptations of shape that prevent them from sinking.

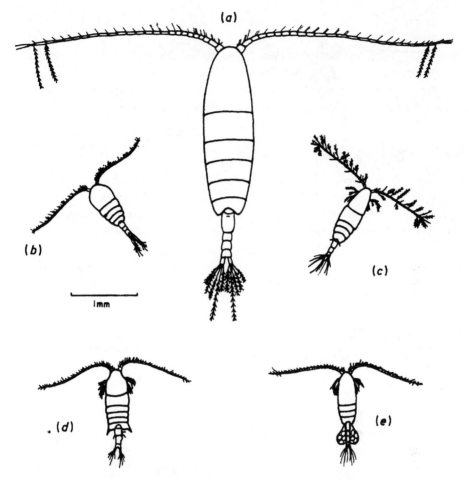

Figure 2.10 Planktonic copepods common around the British Isles. (a) Calanus finmarchicus, (b) Temora longicornis, (c) Acartia clausi, (d) Centropages typicus, (e) Pseudocalanus elongatus.

Chaetognatha

This phylum, the arrow worms, is widespread and well represented in most plankton samples. Around the British Isles there are several species of *Sagitta* (Figure 2.14d), and *Eukrohnia hamata* occasionally enters the North Sea from the Arctic. Chaetognaths are of special interest as 'indicator species' (see Section 4.7.1). They are an important group of planktonic predators.

Protozoa

The zooplankton includes a variety of Protozoa. We have mentioned earlier that some flagellates are holozoic. Foraminifera and Radiolaria are sufficiently

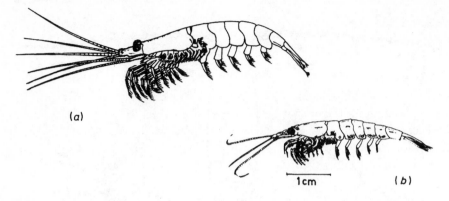

Figure 2.11   (*a*) Euphausia superba, *the 'krill' of Antarctic seas.* (*b*) Meganyctiphanes norvegica, *a common euphausid of the North Atlantic.*

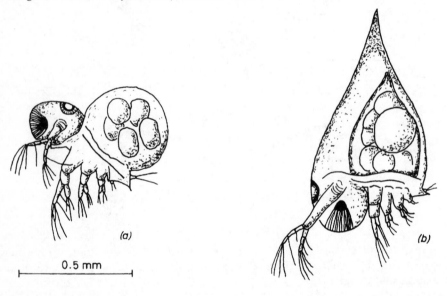

Figure 2.12   Two species of Cladocera *common in spring and summer around the British Isles.* (*a*) Podon polyphemoides, (*b*) Evadne nordmanni.

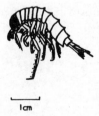

Figure 2.13   Parathemisto gaudichaudi, *a hyperiid amphipod which is epipelagic in colder parts of the north and south Atlantic.*

numerous for their skeletons to form a conspicuous part of some deep-water sediments (see Section 6.1.2). Ciliates are found mainly in coastal plankton.

Coelenterata

There are some holoplanktonic coelenterates, for example, the Trachymedusae and Siphonophora (Figure 2.14 a and b). However, pelagic jellyfish (Scyphozoa) such as *Aurelia* and *Cyanea*, mostly have an inconspicuous benthic stage, the scyphistoma, although this stage is omitted in some, for example *Pelagia*. The meroplankton includes great numbers of medusae set free by benthic hydroids.

Ctenophora

The comb jellies are holoplanktonic predators of widespread distribution. *Pleurobrachia* (Figure 2.14c) and *Beroe* are common around the British Isles. The movements of their rows of comb-like cilia result in waves of colourful irridescence, making them amongst the most beautiful of all planktonts.

Polychaetes

Several families of polychaete worms include holoplanktonic species, the most conspicuous being the family Tomopteridae. The genus *Tomopteris* (Figure 2.14e) occurs throughout the oceans. The majority of benthic annelids start life as planktonic larvae, and some also have adult pelagic phases, usually associated with mass spawning. A well-known example is the Pacific palolo worm. For just the few days of the full moon in October and November, the egg and sperm-filled hind ends of these normally benthic worms, break off and join the surface plankton.

Mollusca

A numerous and widespread group of planktonic molluscs is the Pteropoda, small opisthobranch gastropods. These are of two types, the tiny thecosomatous forms which have a very lightly built shell, for example *Spiratella* (*Limacina*) (Figure 2.14h) and the larger gymnosomatous forms which have no shell, for example the sea butterfly *Clione*. The latter is large enough to be seen regularly by divers. Various pelagic prosobranchs occur in warm oceans for example *Carinaria*, *Pterotrachea* and *Janthina*. These have light, fragile shells to assist floatation. *Janthina*, the violet snail, is holopelagic, drifting at the surface of the sea by means of a float of mucus mixed with air bubbles. It feeds on the pelagic coelenterates, *Porpita* and *Velella*, the by-the-wind sailor. Benthic molluscs produce innumerable planktonic larvae.

Urochordata

Urochordates are sometimes abundant. There are three planktonic orders; the Larvacea, for example *Fritillaria*, *Oikopleura* (Figure 2.14g) *Appendicularia*; the

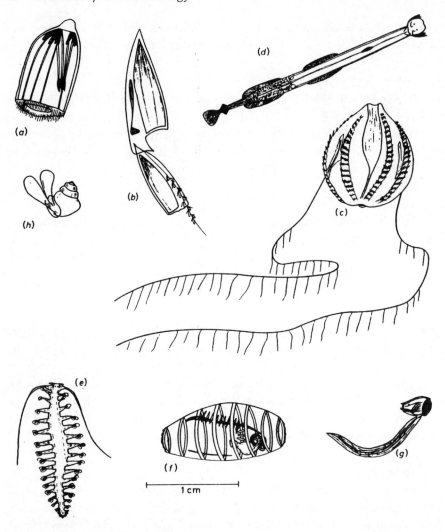

Figure 2.14 Various zooplanktonts from around the British Isles and the groups they represent. (*a*) Aglantha digitale (Trachymedusa), (*b*) Chelophyes appendiculata (Siphonophora), (*c*) Pleurobrachia pileus (Ctenophora), (*d*) Sagitta elegans (Chaetognatha), (*e*) Tomopteris septentrionalis (Polychaeta), (*f*) Doliolum (Dolioletta) gegenbauri (Thaliacea), (*g*) Oikopleura (Larvacea), (*h*) Spiratella retroversa (Pteropoda thecosomata.)

Thaliacea, for example *Doliolum* (Figure 2.14f), *Salpa*; and the Lucida, for example *Pyrosoma*. Species of the Larvacea are common all round the British Isles. The Thaliacea and Lucida in this area are mainly associated with the influx of warmer water from the south and west. On occasion, swarms of large salps are reported by divers off the west coast of Scotland, their rubbery bodies filling the surface layers.

Several other phyla, while not featuring prominently in the plankton, are sometimes represented. Small nematodes are occasionally seen in tow-net samples. Marine rotifers are sometimes collected close inshore. In deep water there are species of pelagic nemertines and holothurians.

The identification of zooplankton species is in many cases difficult, requiring detailed examination of minute structures, a considerable knowledge of the systematics of many groups, and access to a comprehensive library of taxonomic literature. The reference work of major importance for identification of North Atlantic zooplankton is the collection of *Fiches d'Identification du Zooplankton* published by the International Council for the Exploration of the Sea (ICES), Copenhagen (Jesperson and Russel). These 'fiches' are leaflets of line drawings of many planktonic groups with brief descriptions of recognition features, references to taxonomic literature and information on distribution. They are added to and revised from time to time. Several student books which are useful for provisional identification of the commoner planktonts are given in the references at the end of this chapter. The books, in addition to many line drawings, contain general information on plankton sampling, analysis of samples and the distribution of some species. Wimpenny (1966) gives an excellent general account of the plankton around the British Isles and has many clear drawings which are useful guides for identification by students.

The zooplankton includes both vegetarian and carnivorous feeders. The vegetarian forms feed upon phytoplankton, and are often referred to as 'herbivores' or 'grazers' because their position in the food chains of the sea is comparable with that of herbivorous animals on land. These animals have efficient filtration mechanisms for sieving microscopic food dispersed in large volumes of water. The planktonic herbivores are mainly copepods, euphausids, cladocera, mysids, thecosomatous pteropods and the urochordates. The Larvacea secrete peculiar 'houses' which contain an exceptionally fine filter for collection of nanoplankton and ultraplankton. These Urochordates are an important food for certain fish, especially in their young stages (see page 337), and thus form a significant link between the smallest plankton and the larger metazoa (Alldredge, 1977).

Planktonic carnivores include medusae, ctenophores, chaetognaths, polychaetes, hyperiid amphipods and gymnosomatous pteropods. Feeding habits differ between quite closely related forms; for example, although the majority of copepods feed chiefly on phytoplankton, some are carnivorous, particularly those that live at deep levels, and some of the common copepods of British coastal waters appear to be omnivorous. The euphausids are also predominantly vegetarian but there are carnivorous and omnivorous species. Some animals are highly specialized in their food requirements; e.g. *Clione limacina* apparently feeds only on species of the pteropod, *Spiratella*. In many cases, food requirements change with age. The majority of invertebrate larvae at first rely mainly upon phytoplankton for their food but become omnivorous or carnivorous later. Many fish larvae are at first omnivorous, but later take only animal food.

2.4 Other topics

Methods for sampling plankton (Omori and Ikeda, 1984), are discussed in Chapter 3 (Section 3.2.1); primary productivity (Raymont, 1980), in Chapter 5 (Sections 5.1 and 5.2) and Chapter 7 (Section 7.1); and plankton indicators in Chapter 4 (Section 4.7.1).

References and further reading

Student texts

Boney, A.D. (1989). *Phytoplankton*. 2nd edition. Studies in Biology. Cambridge University Press.

Newell, G.E. and Newell, R.C. (1977). *Marine Plankton. A Practical Guide*. Revised edition. London, Hutchinson.

Sykes, J.B. (1981). *An Illustrated Guide to the Diatoms of British Coastal Plankton*. Field Studies Council AIDGAP key. Reprinted from *Field Studies*, vol. 5 No. 3 (1981).

Todd, C.D. and Laverack, M.S. (1991). *Coastal Marine Zooplankton: a practical manual for students*. Cambridge University Press.

Wickstead, J.H. (1976). *Marine Zooplankton*. Studies in Biology, No. 62 London; Arnold. (Out of print but new edition under consideration by CUP.)

References

Alldredge, A.L. (1977). House Morphology of Appendicularians. *J. Zool. Lond.*, **181**, 175–88.

Anderson, D.M. (1994). Red tides. *Scientific American*, **271** (2), 52–8 (August).

Cheng, L. (1975). Marine Pleuston. Animals at the Sea–Air Interface. *Oceanogr. Mar. Biol. Ann. Rev.*, **13**, 181–212.

Dodge, J.D. (1982). *Marine dinoflagellates of the British Isles*. HMSO, London.

Dodge, J.D. (1985). *Atlas of dinoflagellates. A scanning electron microscope survey*. London, Farrand Press.

Drebes, G. (1974). *Marine phytoplankton*. Stuttgart.

Fraser, J.H. (1962). *Nature Adrift. The Story of Plankton*. London, Foulis.

Hardy, A. (1956). *The Open Sea, Part 1. The World of Plankton*. London, Collins.

Hardy, A.C. (1967). *Great Waters*. A voyage of natural history to study whales, plankton and the waters of the Southern Ocean in the old Royal Research Ship, *Discovery*, with the results brought up to date by the findings of RRS *Discovery 11*. London, Collins.

Hendey, N.I. (1964). An Introductory Account of the Smaller Algae of British Coastal Waters. Bacillariophyceae. *Fish. Investig. Lond. Series.*, **4**, Pt. 5.

Jesperson, P. and Russell, F. S. (eds.) (various). *Fiches d'Identification du Zooplankton*. Cons. perm. int. Explor. Mer. Copenhagen. (For identification of many North Atlantic zooplankton species.)

Lebour, M.V. (1925). *The Dinoflagellates of Northern Seas*. Marine Biological Assoc. UK.

Lebour, M.V. (1930). *The Planktonic Diatoms of Northern Seas*. London, Ray Society.

Marshall, S.M. and Orr, A.P. (1955). *The Biology of a Marine Copepod, Calanus finmarchicus*. Edinburgh and London, Oliver and Boyd.

Omori, M. and Ikeda, T. (1984). *Methods in Marine Zooplankton Ecology*. J. Wiley and Sons.

Raymont, J.E.G. (1980). *Plankton and productivity in the oceans*. 2nd edition. Vol. 1: *Phytoplankton*. Pergamon Press.

Raymont, J.E.G. (1983). *Plankton and productivity in the oceans*. 2nd edition. Vol. 2: *Zooplankton*. Pergamon Press.

Sverdrup, H.U., Johnson, M.W. and Fleming, R.H. (1946). *The Oceans*, Chap. 9. Populations of the Sea. New York, Prentice-Hall.

Thomas, C.R. (ed) (1993). *Marine Phytoplankton. A guide to naked flagellates and coccolithophorids*. London, Academic Press.

Wickstead, J.H. (1965). *An Introduction to the Study of Tropical Plankton*. London, Hutchinson Tropical Monographs.

Wimpenny, R.S. (1966). *The Plankton of the Sea*. London, Faber.

Winter, A. and Siesser, W.G. (1994). *Coccolithophores*. Cambridge University Press.

3 Measuring and sampling

In this chapter we will outline briefly some of the measuring and sampling techniques used at sea to obtain information of interest to ecologists. To evaluate the interactions between organisms and their environment, both oceanographic and biological data are required. Oceanographic data relate to the inorganic parameters of the environment, including measurements of water movement, temperature, composition, illumination and depth, and the nature of the sea bottom. Biological data relate to the distribution, numbers, activities and relationships of organisms in different parts of the sea. In order to collect this range of information, a great variety of apparatus has been devised. Techniques have changed greatly with the advent of modern electronics and computing capabilities and new methodologies are continually being devised. Only a few methods currently in wide use are described here. More detailed information can be found in the handbooks and references listed at the end of the chapter. Those interested in older techniques should refer to previous editions of this book.

3.1 Oceanographic data

3.1.1 Currents and water bodies

Ocean currents can be measured either by following floating objects or dyes within the water body or by measuring the speed and direction of the water as it passes a fixed point.

Drift bottles

The major surface currents of the oceans became known during the days of sailing ships when this knowledge was needed for successful ocean voyages. Information was accumulated by noting the course of drifting objects such as becalmed ships, drift-wood or pieces of wreckage. Oceanographers refined this technique by using specially designed drift bottles. These provided much of the original information about water movements around the British Isles. They are hardly ever used now but are described here for their historical interest and possible use in local inshore investigations.

A drift bottle has a long narrow neck and is usually ballasted to float with only the tip of the neck projecting above the surface, so that its course follows the movement of the water and is not much influenced by direct wind action on the bottle. The effect of the wind may be further reduced by attaching a small sea-anchor to the bottle. Inside the bottle is a postcard bearing an identification number and a request printed in several languages for the return of the card with details of time and place of finding, for which a small fee is paid.

Drift bottles can be used in several ways. They may simply be thrown overboard from an anchored vessel and their direction and speed of movement directly noted. This provides immediate information about the surface current in that locality at the time of observation. Usually they are set adrift in large numbers in the hope that some of them may eventually be found stranded, and their location reported. In this way, information regarding the general course of currents over wide areas and considerable periods has been obtained by a simple and inexpensive method. A variation is the use of ballasted drifters which hang below the surface or near the sea-bed. Strongly coloured dye can also be used although wind affects this more easily. In 1990 a lost cargo of about 40 000 pairs of Nike brand trainer shoes provided a free drift experiment in the North Pacific. Many of the shoes drifted ashore along the west coast of North America. An oceanographer, Curtis Ebbesmeyer from Seattle, realized their potential as 'drift bottles' and arranged to collate information on when and where the shoes landed up. Currently plastic yellow ducks, green frogs and blue turtles have reached the Bering Straits from a cargo of 29 000 bathtub toys lost in the Pacific in 1992!

Modern surface drifters

With the advent of modern satellite navigation, it has become possible to track floating buoys quickly and easily. A series of position fixes can be used to work out the speed and direction of the current. Modern surface buoys are deployed from ships and carry low-power satellite transmitters with a position accuracy of less than 1 km. They also carry instrument packages so that they can measure sea-surface temperature and barometric pressure whilst drifting. Most of the float lies below water to minimize wind influence, and they are generally fitted with net drogues which can be deployed at set depths down to several hundreds of metres. Currently the World Ocean Circulation Experiment (see Section 1.3.4) has deployed several thousand such drifters since 1991 to provide track charts for the upper ocean.

Sub-surface drifters

It is possible to detect and measure water movements at middle depths by using neutrally buoyant floats, the weight of which can be accurately adjusted to match the density of the water so that they sink to a predetermined depth and then drift

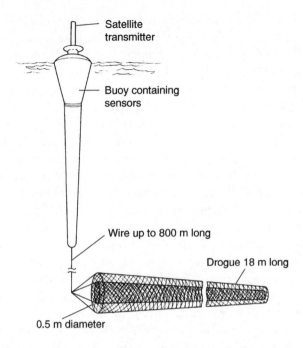

Satellite
transmitter

Buoy containing
sensors

Wire up to 800 m long

Drogue 18 m long

0.5 m diameter

Figure 3.1   An ARGOS satellite-tracked drifting buoy and its drogue.

with the current. Such floats are tracked acoustically or by satellite and are used extensively for open ocean studies. An early example is the 'Pinger' invented by Dr J.C. Swallow of the Institute of Oceanographic Sciences in the early 1950s. This is an aluminium tube containing a battery and acoustic transmitter. It can be ballasted to be neutrally buoyant at any required depth, and emits an intermittent 'ping' sound as it drifts. The principle on which this works is that the tube, whilst being negatively buoyant at the surface, gains buoyancy as it sinks because its compressibility is less than that of seawater.

There are now several modern versions of sub-surface floating buoys. One in present use is the RAFOS float. Like the original Swallow float, it is ballasted to be neutrally buoyant at pre-selected depths and can drift for several years. The float receives acoustic signals from moored sound sources, and by timing the arrival of these signals the float position can be determined and recorded. The data are transmitted back to base by satellite when the float surfaces following an appropriate acoustic signal. In other systems the floats emit the acoustic signals which are received by automatic listening stations up to 1500 km away. Recently similar floats have been used to track eddies in the NE Atlantic off the UK (see Section 1.3.3).

The WOCE experiment uses modern floats which go a step further and are

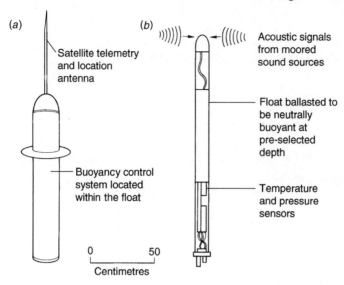

Figure 3.2 Sub-surface drifters: (a) an ALACE float; (b) a RAFOS float of the type used in the World Ocean Circulation Experiment.

completely independent of any tracking network. The Autonomous Langranian Circulation Explorer (ALACE) developed in the late 1980s, has been designed to rise to the surface at pre-set intervals. Once at the surface, it is located by satellite and transmits its data to the laboratory directly through the satellite system. This and other surface and sub-surface floats are described in more detail in Griffiths and Thorpe (1996).

Current meter moorings

The rate and direction of flow of water can be measured by various ingenious current meters placed in fixed positions. One of the first successful and widely used instruments for this purpose was the Ekman Current Meter. This was a mechanical instrument in which the rotations of a small propeller were counted and has now been superseded by more sophisticated electronic instruments.

The Aanderaa Recording and Telemetering Current Meter (Figure 3.3) was designed in the 1960s and has been so successful it is still widely used today. It can be deployed for up to a year and is reliable and relatively cheap. The design is a simple one and is rather like a weather vane and anemometer.

Magnets attached to a rotor turned by the current generate pulses at a frequency proportional to the current speed. A large vane aligns the apparatus with the current, the direction being sensed by a magnetic compass in the base of the container. Data on the speed and direction of the current are recorded within the instrument either on magnetic tape or in modern Aanderaa meters,

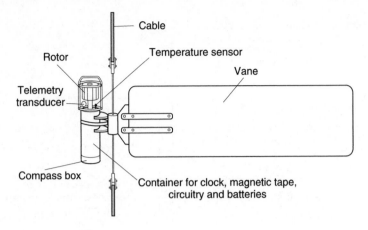

Figure 3.3   The Aanderaa™ recording current meter.

using solid-state electronics. Intervals at which readings are taken can be varied. Other sensors attached to the device measure water temperature, pressure and conductivity. As a recording instrument this meter can be used at virtually any depth and has been used for many years to measure deep ocean currents.

Other types of current meters are now available which avoid the use of rotors. Acoustic Doppler Current Profilers (ADCP) are used to look at current patterns in the upper parts of the water column and are either ship mounted or moored. They work on the basis of transmitting pulses of high frequency sound (mostly 150 or 300 kHz) and measuring the frequency shift caused by the back-scattering particles moving in the currents. The back-scatterers have to be half a wavelength across to be seen by the acoustics, which is about 1 cm for 150 kHz. A dense cloud of smaller particles can also reflect the sound. The 'particles' are, of course, often planktonic animals, and ADCP is now being used quite extensively by biologists to carry out synoptic surveys of pelagic communities. The technique is non-invasive and extremely rapid but has the disadvantage that you cannot tell what sort of animals the acoustics are seeing.

The direct measurement of water currents by meters suspended in the water involves difficulties connected with obtaining a fixed reference point, particularly in deep water. An anchored vessel is by no means stationary, and movements of the ship may falsify the measurements. However, recent developments using satellite position-fixing equipment allow ships to maintain their position relative to a fixed point on the sea-bed within a few metres. Alternatively a buoy attached to a taut anchor wire can be used as the reference point, movements of the ship relative to the buoy being allowed for in the interpretation of the meter readings. Or recording instruments may be attached to a submerged buoy system anchored to the sea bottom. The submerged floats with their attached meters can be retrieved via a special acoustical link that breaks away from the anchor when it receives the appropriate signal. Meters may also be attached to data-recording buoys (see page 55).

Anthropogenic tracers

Long-term mixing of water bodies is currently being studied using anthropogenic tracers (i.e. contaminants). During the past several decades, the surface ocean has been 'tagged' with substances that previously did not exist in the ocean in significant amounts. CFCs (chlorofluorocarbons) started to be commercially produced in the 1930s and 1940s for use as a refrigerant and in manufacturing processes. Their concentration in the atmosphere has gradually built up, and from here they have entered the oceans where they remain chemically stable. They can thus act as an indicator of the age of a water mass. Tritium, mostly from bomb testing, is used in a similar way. The substances are carried with surface water as it moves and changes into sub-surface water masses.

For example, North Atlantic Deep Water (NADW) is one of the major deep water masses of the world ocean. It has an important role to play in heat and CO_2 budgets and so affects global climate. It forms from surface and near surface waters, and so NADW formed in the last three to four decades will contain tritium and CFCs. Recent work following these anthropogenic tracers has shown the path taken by recently formed NADW in the Deep Western Boundary Current (Smethie, 1993).

3.1.2 Water samples and temperature/salinity measurements

Water samples from different depths are often required both by physical oceanographers and by biological oceanographers. Information from the analysis of such samples, along with temperature and salinity measurements, not only provides direct information about a water body, but has also been used to obtain indirect measurements of deep water circulation. Such ocean currents are difficult to measure directly, especially the slow movements of water at deep levels. The waters of different parts of the oceans are to some extent distinguishable by virtue of their physical and chemical characteristics, in particular their temperature, salinity, and content of oxygen, nitrate or phosphate. By studying the distribution of these quantities throughout the oceans, the movements of the water can be inferred. Temperature measurements and the analysis of water samples therefore provide much of the basic data of oceanography (Strickland and Parsons, 1968). The foundations of this science were laid during the voyage of HMS *Challenger*, 1872–76, when a large amount of this information was first collected from a series of depths at each of some 360 stations spread over the major oceans (Linklater, 1972).

Water bottles
on rosette
sampler

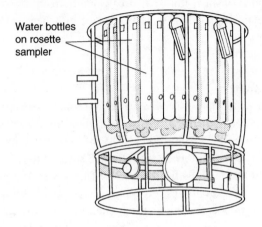

Figure 3.4   A hydrographic instrument package designed to take water samples with a rosette multisampler, along with a CTD instrument package to measure conductivity, temperature and depth.

CTD *instrument packages*

Temperature readings and water samples are usually taken together. Most water sampling from research vessels now utilizes modern electronic instruments known as CTD (conductivity-temperature-depth) probes developed in the 1970s. These are usually mounted on rosette water samplers. Temperature is measured through electrical resistance with an accuracy greater than that of mercury thermometers (0.002°C). Development of electrical resistance thermometers was made possible with the advent of the transistor and integrated circuit and such thermometers were first used in the early 1960s, although at this time they were not very accurate.

Salinity, to an accuracy of 0.005 units, is calculated from measurements of the electrical conductivity and temperature of the water using a *salinometer*. The data return to the ship as an electronic signal and are interpreted by on-board computers or connected to graphic or digital display units. However, the conductivity measurements need to be calibrated against water samples which is why CTDs are usually used as part of an instrument package which includes a water sampler. Rosette multisamplers can have as many as 24 bottles of 2, 10 or even 20 litre capacity which collect water at the required depths, when triggered electronically from the surface. Water samples are also necessary for chemical analysis of water. Small, hand-held salinometers on long cables are also available for use in shallow water from small boats.

CTDs can be lowered from stationary ships to provide continuous data from the surface down to about 6000 m. They can also be left in place if attached to a fixed buoy, to record changes in salinity and temperature over time. The data can be stored in the instrument or transmitted to a base via a satellite. Fluorometers (chlorophyll content) and oxygen electrodes are also routinely attached to the instrument package.

Oceanographic work requires extremely accurate thermometers because small variations in temperature produce considerable changes in water density. Modern CTD instruments have greatly increased the precision, speed and accuracy of measurements. As already mentioned, water samples are still needed to calibrate salinity measurements from these instruments. Precision thermometers are also used to check the accuracy of the CTD instruments. These usually have platinum resistance sensors and an LCD display rather than mercury in glass.

Water pumps

Water samples from the surface and the uppermost 60 m are frequently pumped aboard ship using hoses. Chemical analysis and temperature measurements can then be carried out. A continuous stream of water is sucked up and flushed over electrical thermo-salinometers. Surface temperatures and salinities along the cruise track of a vessel are often measured in this way. The water is then passed through autoanalysers and a range of minor constituents, notably ammonium, nitrite, nitrate, phosphate, silicate and organic carbon can be measured giving a complete record for each constituent as the vessel proceeds along its course. Modern autoanalysers can also measure fluorescence indicating chlorophyll content (a measure of primary productivity).

The use of hoses is cumbersome at greater depths, where water-bottles or rosette samplers still have advantages. Bottles are also more suitable than hoses where samples are to be examined for organisms which might be damaged by passage through pumps.

Water bottles

Before the advent of CTD probes and rosette samplers, most water sampling was done using simple water bottles such as the Nansen-Pettersson (Figure 3.5). Most such bottles take the form of an open-ended cylinder with spring-loaded valves for closing the ends. The sampler is lowered with the ends open so that water flows freely through it. When the required depth is reached, a release mechanism is operated by sending a slotted weight known as a messenger down the suspending wire, which causes the valves to snap shut closing the ends of the cylinder. The water bottle can then be hauled to the surface. Modern versions of such bottles are sometimes still used today and may be operated acoustically or electronically.

Before electrical resistance thermometers came into use, water samples and temperature measurements at deeper levels were often taken, using reversing bottles in conjunction with reversing thermometers. The reversing water-bottle is

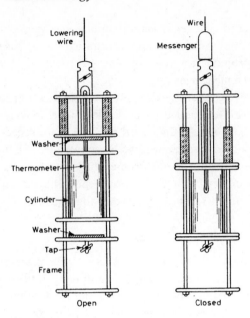

Figure 3.5 The Nansen-Pettersson.

a cylindrical container attached to the lowering wire by a hinged frame, the cylinder lying above the hinge when the bottle is in the open position. The bottle is closed when a messenger strikes a release mechanism which causes the cylinder to swing downwards through 180 degrees until it lies below the hinge. As this occurs, valves automatically close the ends of the cylinder (Figure 3.5). The purpose of this reversing mechanism is to allow the use of a reversing thermometer mounted on the side of the bottle. Such thermometers are rarely used today and are not described further here. Detailed descriptions are given in earlier editions of this book.

Towed undulators

Continuous records of temperature, depth and salinity can be obtained from moving ships throughout a range of depths from the surface to about 500 m by mounting electrical sensors such as CTDs on a towed undulator. These torpedo-like devices follow an undulant course controlled by movable vanes operated electrically from the ship. This saves considerable ship time since the ship no longer needs to stop and lower instruments. A popular model is the UK SeaSoar (Figure 3.6) which can be towed at 8 knots and can undulate from the surface to 500 m. As well as CTDs, the instrument package often includes pH, dissolved oxygen, and sound velocity meters. It can also be fitted with optical and acoustic sensors that measure colour and pigments in the water (indications of phytoplankton productivity), and zooplankton. Such packages can also be used

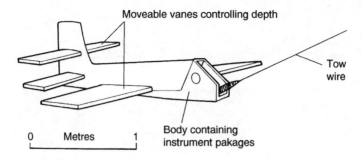

Moveable vanes controlling depth

Tow wire

Body containing instrument pakages

0   Metres   1

Figure 3.6   The SeaSoar towed undulator.

from stationary ships or attached to data-recording buoys (see below). Many smaller towed instruments have also been developed, some of which can be operated by non-specialists from merchant ships. One such is the Undulating Oceanographic Recorder which can operate down to 200 m.

Expendable probes

A relatively inexpensive method of data collection over a range of depths from the surface is the use of expendable probes. First developed as a means of quickly and cheaply plotting temperature profiles from a ship moving at full speed, these probes can now also be fitted with conductivity sensors for simultaneous salinity profiles.

An expendable probe is a small, bomb-shaped object carrying the sensors in the nose connected to a spool of fine electric cable contained within the tail. This cable is joined to a similar spool of wire in a launching tube which may be hand-held or deck-mounted on the ship. The ship end of the cable is connected to a signal-processing unit which energizes the probe and interprets and records the signal. Without the vessel needing to reduce speed the probe is allowed to drop from the launching tube into the water. The two spools of wire simultaneously run out freely. This double system of unspooling allows the probe to fall vertically through the water, virtually unaffected by the movement of the ship (Figure 3.7). As the probe descends it transmits electrical signals to the processing and recording unit on the ship, giving continuous readings of temperature and salinity until the probe falls away when the cable has been fully extended. As the rate of descent is known, the profiles can be directly related to depth on the vertical scale of a chart recorder. These devices are usually constructed to work to depths of 750 m but are obtainable for depths to 2000 m.

Data-recording buoys

The development of apparatus for data recording and telemetry has greatly extended the possibilities of long-term surveys by unattended instruments

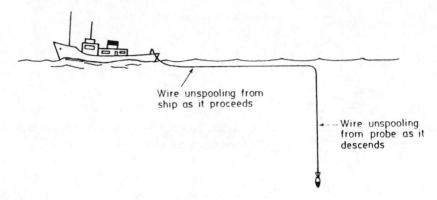

Figure 3.7   Expendable probe for obtaining temperature and salinity profiles from a moving ship.

suspended from moored buoys. This is much less expensive than the use of ships. Instruments can be attached at several depths along a cable below a buoy, including sensors for wave height, sea temperature, current speed and direction, salinity, pressure, pH and oxygen concentration. The surface float can be fitted with instruments for measuring atmospheric conditions such as air temperature, barometric pressure, wind speed and direction. All the collected measurements are recorded in a memory store within the buoy and can be transmitted periodically or on command by radio to shore stations.

'Pop-up' systems

Moored buoys floating on the surface have certain disadvantages for attachment of hydrographic instruments. Wave motion may cause misleading measurements for certain parameters, especially current speed. Heavy seas may lead to loss of the buoy and its instruments by breakage of the cable. Valuable equipment attached to a buoy may be stolen. These problems have prompted the development of 'pop-up' arrangements where the float is completely submerged, anchored below the surface to a weight on the bottom by a cable containing a release link. On receipt of a command signal, the release detaches from the anchor weight, allowing the float and attached instruments to float to the surface. Command releases operate on receiving a particular coded acoustic signal transmitted from the surface. The float is equipped with a radio beacon and light for easy retrieval.

3.1.3   Aerial and satellite surveys

Since the late 1970s, reasonably accurate remote sensing of the oceans from aircraft and satellites has been possible. Remote sensing depends on sensors mounted in the aircraft or satellite, picking up naturally occurring electromagnetic radiation, or generating electromagnetic energy and measuring it after reflection from the sea.

The advantage of these new systems is that large areas can be surveyed at one time. Considerable skill goes into the design of the sensors and the interpretation of the resulting data. The data that such systems can collect are restricted mainly to surface waters and concerns the temperature, colour, roughness and average slope of the sea surface. From these four basic measurements various characteristics can be derived. Perhaps the best known is an estimate of the chlorphyll content of the water which is related to primary production. This is derived from the colour of the surface waters and allows the production of monthly maps of primary productivity throughout the world's oceans. Other applications include investigating ocean currents from the surface slope; deriving wind speed from surface roughness; using temperature patterns to help predict events such as El Niño (see Section 1.3.6) and to provide information regarding boundaries of water masses and mixing processes; and determining the source and extent of sediment and pollution plumes from colour measurements (Robinson and Guymer, 1996). The latter is particularly useful for bodies such as the UK NRA (National Rivers Authority) charged with monitoring the quality of coastal waters.

3.1.4   Illumination

The rate of decrease of illumination with depth is expressed as the extinction coefficient, k_λ.

$$k_\lambda = \frac{2.30(\log I_{\lambda d_1} - \log I_{\lambda d_2})}{d_2 - d_1}$$

where $I_{\lambda d_1}$ and $I_{\lambda d_2}$ are the intensities of light of wavelength λ at depths d_1 and d_2 metres. For accurate measurements of illumination beneath the sea surface, photomultipliers and photoelectric photometers are used (Kampa, 1970). But for rough measurements, the secchi disc (Figure 3.8) provides a simple method which has often been used by biologists.

This is a white disc, 30 cm in diameter, which is lowered into the water to the depth at which it just disappears from sight. The extinction coefficient can then be roughly determined from the empirical relationship $k = 1.45/d$ where d is the maximum depth in metres at which the disc is visible (Walker, 1980).

3.1.5   Depth measurements

The classical method of measuring the depth of the sea was by means of sounding weights and lines. The weight was lowered from the vessel until it struck bottom and the length of line measured, usually by means of a meter attached to a sheave

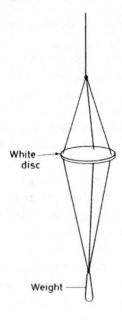

Figure 3.8   The secchi disc.

through which the line passed. This method has been superseded by sonic sounding.

Echo sounders and sonar systems

Sonar stands for 'sound and ranging' and is the detection of objects under water using sound. There are many applications including depth measurement, seabed profiling and location of objects such as wrecks.

Depth measurements are now predominantly made by echo-sounding. The depth of water is measured by timing the interval between the emission of a sound impulse at the surface and its return to the surface as an echo reflected from the sea-bed. The speed at which sound travels through water is known to be about 1500 m/s and so time can be converted to depth. The speed of sound through seawater varies slightly with temperature, salinity and pressure and so for really accurate sonic sounding, the actual speed needs to be measured. This can be estimated from hydrographic data where available, or direct measurements can be made by sensors lowered from a ship. These instruments contain solid-state circuitry generating acoustic pulses. The pulses are transmitted through the water to a reflector, usually a distance of 0.05 m, and back. When the reflected signal is picked up by the transducer this initiates another pulse. Thus the number of pulses generated in a standard time interval is a measure of the speed of sound in water.

The frequencies emitted by modern echo sounders lie above the audible range, usually between 15 and 50 kHz. The use of ultrasonic frequencies has several advantages. They can be focused into fairly narrow directional beams, giving a more precise echo than is obtainable from the audible part of the sound spectrum, and enabling a more detailed picture of the bottom profile to be drawn. There is also less interference from natural sounds.

A great variety of echo-sounders are now available to suit all types of vessels from inflatable boats to supertankers. One of the latest is a small hand-held version shaped like a torch. It can be used over the side of a small boat but as it is waterproof and pressure resistant, it can also be carried by a diver. The diver can thus measure the depth of seabed features below his operational limits.

Although the design varies, echo-sounders all work on the same principle. An acoustic transducer fitted on the underpart of the ship's hull or towed at a known depth, gives out short pulses of sound at a given frequency. The transducer also receives the reflected sound pulses on their return from the sea-bed. These transducers make use of the piezoelectric properties of quartz or the magnetostriction properties of nickel. In either case, rapid dimensional changes are produced by electrical excitation, causing brief pulses of vibration to be emitted into the water. The returning echo vibrates the transducer and sets up electrical signals which can be amplified and recorded as a trace on a cathode-ray tube or as a line drawn on a paper chart.

In the paper-recording instrument, which is not greatly used today, a strip of sensitized paper bearing a printed scale is drawn slowly through the recorder. A moving stylus in contact with the paper scans to and fro across the scale at a speed which can be adjusted in relation to the depth of water. A short pulse of sound is emitted from the ship's hull in a downward-directed beam as the stylus passes the zero mark of the scale. The stylus continues to move across the scale, and the echo signals are amplified and applied as an electric current to the stylus, marking the sensitized paper electrochemically. The position of this mark on the scale indicates the depth from which the echo is received. As the paper moves through the instrument, the repeated scanning of the stylus produces a series of marks which build up a line on the paper corresponding with the profile of the sea-bed (Figure 3.9).

Sonic techniques have further applications in marine biology. Fish shoals may be detected in this way, and different species may to some extent be identified by their characteristic echoes. The sonic scattering layers (see page 140) were discovered in the course of investigations with sonic equipment.

Side-scan sonar

Standard echo-sounders can only make measurements directly beneath the ship. This is adequate when depth is the main information required but where

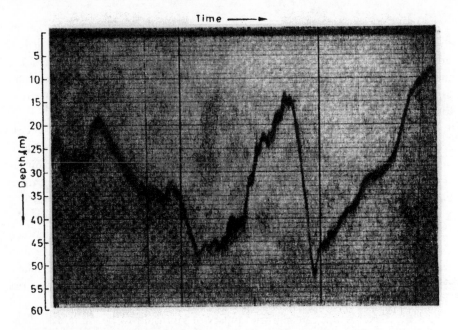

Figure 3.9   A typical sonic sounding trace over a widely varying sea-bed.

information on the type of sea-bed is also needed, the use of side-scan sonar is more appropriate. The technique of side-scan sonar makes use of a horizontally directed beam of sound to build up a series of images or sonographs of the sea-bed. Side-scan sonars are now extensively used for surveying broad swathes of the sea-floor. Different types of substrate reflect different patterns of echo trace and sonographs can provide information on the texture, orientation and composition of features. However, interpretation of complex sonographs requires considerable experience and expertise.

An instrument called GLORIA (Geological Long Range Inclined Asdic), is a sophisticated side-scan sonar about 8 m long developed at the UK Institute of Oceanographic Sciences. It is towed behind the vessel at a speed of 8–10 knots and at a depth of 40–80 m. In water 5000 m deep, it can survey a strip of sea-bed about 40–60 km wide. Features such as volcanoes, canyons, mud slides and nodule fields can be observed by overlapping several strips. Over the last 20 years or so, GLORIA has surveyed about 6 per cent of the ocean floor.

Side-scan sonars are now being put on to deep-towed instrument packages for carrying out fine-scale surveys of the sea-bed. These instruments are now revealing extraordinarily complex structures on the abyssal plains resulting from catastrophic failures of continental slopes. Structures such as meandering channels with levies and 'flood'-plains created by turbidity currents have been

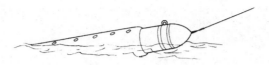

Figure 3.10   GLORIA 6.5 kHz side-scan sonar.

revealed, originating from events happening hundreds or even thousands of kilometres away on the continental margin.

A further use of side-scan sonar is in the detection of shipwrecks, and many archaeological finds and valuable cargoes have been recovered in this way. It is also being increasingly used for fish location. Echoes may be returned by objects such as fish, floating or swimming in the water in the path of the sound beam. Modern instruments are capable of detecting fish shoals, individual fish and very often, the species. The efficiency of this system has undoubtedly contributed to over-exploitation of some fish stocks (see Chapter 9). Sonar systems are becoming ever more sophisticated allowing us to 'see' the sea-bed in a way early oceanographers would have thought impossible (Riddy and Masson, 1996).

'Roxann' ultrasonic signal processor (USP)

Standard echo-sounders can provide only limited information concerning the nature of the sea-bed. This type of information is usually gained by use of side-scan sonar systems (see above). However, a new system for processing echo-sounder signals has recently been developed in the UK (Chivers *et al.*, 1990). It is designed to process information from a straightforward echo-sounder to provide simultaneous information on the nature of the sea-bed. The system is based on the electronic analysis of the first and second echoes visualized on the echo-sounder display. Straightforward desk-top computers and software are used to analyse and display the data.

The advantage of this system is its relative cheapness and its ability to interface easily with a wide variety of echo-sounders. In the UK it is finding application in initial low-cost 'look-see' surveys where mapping of bottom types and habitats is required in shallow water.

3.2   Biological sampling

3.2.1   Plankton

Plankton nets

Detailed methodologies for plankton sampling are given in a variety of texts such as UNESCO (1968) and Omori and Ikeda (1984). Samples of plankton are usually collected by plankton nets. These are of many designs, but all consist essentially

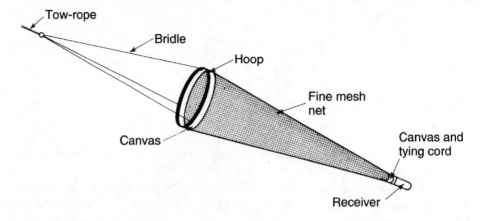

Figure 3.11   A simple plankton net.

of a long cone of fine-mesh net. The mouth of the net is usually some 50–100 cm in diameter, and is held open by a strong hoop to which the tow-rope is attached by three bridles (Figure 3.11). The narrow end of the net is firmly tied to a small metal or plastic vessel in which much of the filtered material collects. After hauling, the net is washed into a suitable receiver to collect any material left on the mesh.

To filter efficiently, plankton nets must be towed quite slowly, not faster than about 1–1.5 knots. Fine mesh presents high resistance to the flow of water through it, and if towed too fast, the net sets up so much turbulence in the water that floating objects are deflected away from the mouth. In many designs of plankton net the aperture is reduced by a tapering canvas sleeve as in the Hensen net (Figure 3.12) which cuts down the volume of water entering the net to give more effective filtering.

In the past, silk gauze was used for plankton nets. Modern nets are mostly made from nylon gauze. These can be constructed in a variety of ways but the best material is provided by monofilament heat-set plain weave (Omori and Ikeda, 1984). This has even meshes that do not deform under the strain of towing. The gauze comes in a variety of mesh sizes or grades identified by a rather confusing array of numbering systems. For example, nets with a mesh size of 330 μm are commonly used to capture meso- and macroplankton. This is grade 'bolting silk GG54' or 'nylon NGG52'.

Coarse mesh is more effective than fine mesh for catching the larger planktonts because it offers less resistance and allows a faster flow of water through the net and does not clog so easily. Omori and Ikeda (1984) recommend using a mesh size of about 75 per cent of the width of the smallest organism to be sampled, when towing at normal speeds (0.7–1.0 m sec^{-1}). For collecting organisms over a wide range of sizes several grades of net are often used together.

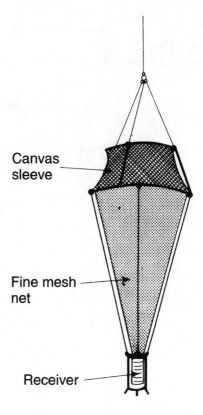

Figure 3.12 The Hensen net.

For collecting macroplankton such as euphasid shrimps, larger nets known as young fish trawls (YFTs) (Southward, 1970), are sometimes used, having a mesh of *ca*. 1 mm and an aperture of 1–2 m diameter. For collecting pleuston and neuston (see Section 2.1), a plankton net may be attached to a floating frame so as to skim the surface as it is towed (David, 1965).

Closing and opening-closing nets

Plankton nets can be towed behind a slowly moving vessel, or lowered from a stationary vessel and hauled vertically. When studying the vertical distribution and migration of plankton, it is necessary to have samples of plankton from particular levels. To avoid contamination of the samples by organisms entering the net while it is being lowered or raised, there must be some method of opening and closing the net at the required depth. The simplest method of closure is to encircle the mouth of the net with a noose which can be drawn tight, as in the Nansen closing net (Figure 3.13). This net is lowered vertically to the bottom of the zone to be sampled, not filtering on descent, and is then drawn up through

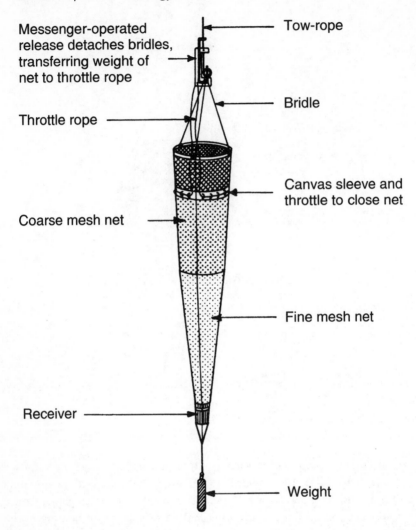

Messenger-operated release detaches bridles, transferring weight of net to throttle rope

Tow-rope

Throttle rope

Bridle

Canvas sleeve and throttle to close net

Coarse mesh net

Fine mesh net

Receiver

Weight

Figure 3.13   The Nansen closing net.

the sampling depths. A messenger sliding down the tow-rope then releases the bridles from the tow-rope, causing the throttle to draw tight and close the mouth of the net, which can then be hauled to the surface without further filtering.

Nets that can be both opened and closed under water are used mostly for horizontal and oblique tows. The net is kept closed whilst being lowered to the required depth, is then opened for the tow, and closed immediately afterwards. There are many different net types of varying efficiency. They can be opened and closed by messengers, electric, sonic, time or pressure releasing mechanisms.

The Leavitt system involves two messenger-operated throttles (Figure 3.14).

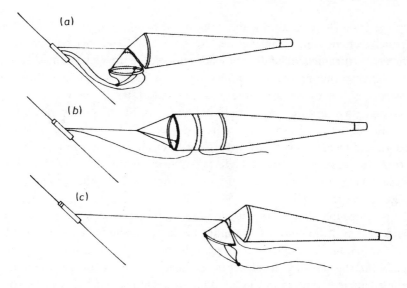

Figure 3.14 An opening and closing plankton net operated by throttles. (a) The net is lowered with first throttle closed. (b) The first messenger frees the first throttle and the net opens. (c) Second messenger releases bridles and the strain is taken by second throttle, closing the net before hauling.

The net is lowered with the mouth closed by a noose. The first messenger releases this noose and the mouth of the net opens. After towing, a second messenger releases the bridles and another noose closes the mouth before the net is hauled up. The Clarke–Bumpus net has messenger-operated valves which open and close the mouth. Large systems have several nets that can be used simultaneously at different depths.

Many planktonic organisms are very sensitive to temperature, and to keep a sample alive for any length of time it must be kept at an even temperature as close as possible to that of the water from which it was filtered. Vacuum flasks provide a means of doing this. For most purposes, however, preserved samples are needed. The addition to the sample of sufficient neutral formalin to produce a 4–5 per cent concentration will preserve the majority of planktonts satisfactorily.

Quantitative plankton studies

The aim of quantitative plankton studies is usually to estimate numbers or weights of organisms beneath unit area of sea surface or in unit volume of water. There are many difficulties in these studies. For instance, plankton is often very patchy in distribution, and it is difficult to obtain any clear picture of the amount and variety of plankton unless samples are taken at numerous stations spread over the area of investigation. Modern navigational aids enhance the precision of the sampling grid. There is also the difficulty of knowing how much of the plankton is actually retained in a plankton net, for some may be displaced from the path of the net by turbulence, small organisms may escape through the meshes and the larger active forms may avoid capture by swimming. A further difficulty is to know the volume of water filtered. If a net could filter all the water in its path, the volume passing through it would be $\pi r^2 d$, r being the radius of net aperture and d the distance of the tow, but in practice this formula can only give an approximate measurement. A net does not filter all the water in its path and the filtering rate reduces as material collects on the mesh and the resistance of the net increases. There is a further difficulty in knowing precisely the distance a net has moved during towing.

To measure the filtered volume more accurately, a flowmeter can be added to a plankton net. A flowmeter has a multi-bladed propeller which is rotated by the flow of water, and a simple counter records the number of revolutions. This can be placed in the aperture of the net to measure the volume of water entering, for example Currie–Foxton and Clarke–Bumpus nets; or the net may be surrounded by an open-ended cylinder and the flowmeter placed behind the net to measure the volume filtering through, for example the Gulf III Sampler (Figure 3.15).

Another method of collecting plankton is the *plankton pump*. This draws water up a hose and pumps it through nets or filters to trap the plankton. Simple pumps sampling shallow depths, can be operated from small boats. Large ships can operate submersible electric pumps capable of raising many litres of water a minute from depths down to about 100 m.

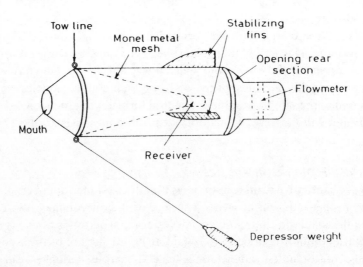

Figure 3.15   A high-speed plankton sampler of the Gulf III type.

With this method it is possible to measure quite accurately the volume of water filtered and the depth and to sample the actual water from which the plankton is filtered. After filtering, the water can be centrifuged or otherwise sampled for the smallest organisms that escape through nets. Despite these advantages the method has some drawbacks. Large creatures are prone to damage as they pass through the pump, the stronger-swimming planktonts may escape being sucked into the hose and there are difficulties in the use of pumps to obtain samples from deep levels.

Nanoplankton and ultraplankton

The smallest planktonts such as bacteria and microflagellates escape through the meshes of ordinary nets. Materials of finer mesh are now becoming available which retain much smaller organisms than hitherto, but if too fine, the water will not pass easily through them. So the sampling of nanoplankton and ultraplankton is usually done by collecting samples of seawater in sterile bottles or by pumping (see Section 3.1.2) and then concentrating the organisms by allowing them to settle, by centrifuging or by fine filtration.

Plankton counting

Plankton counters can be constructed for towing at sea and counting directly the number of small organisms present in the water which flows through the instrument. The apparatus is essentially a tube containing electrodes connected to circuitry which records the change of impedance when objects pass between the electrodes. This enables an estimate to be made of both number and size of organisms.

For most purposes quantitative investigations are done on plankton samples which have been filtered from a known volume of water, the method depending on the type of study. Sometimes an estimate is wanted of the gross quantity of plankton of all types. A rough volume estimate can be made very simply by allowing the sample to settle in a measuring cylinder and reading the volume directly from the scale. Measurements of displacement volume are probably rather more accurate, and estimates can also be made by weighing, either as a rough wet weight, or, better, by drying to constant weight.

The most detailed investigations are by direct counting. Large organisms are usually few in number and can be individually picked out and counted. Smaller organisms may be so numerous that the sample must be sub-sampled to reduce them to a number that it is practicable to count. The sub-sample can be spread out in a flat glass dish and examined, with a microscope if necessary, against a squared background. The count is then made, a square at a time. For very small organisms present in large numbers, the haemocytometer used by physiologists for counting blood cells can be used for plankton counting, or a Coulter particle counter can be adapted for this purpose. Other methods applicable to phytoplankton are mentioned on page 175.

Underway and continuous plankton samplers

We have previously pointed out that ordinary plankton nets must be towed slowly to be effective. They can therefore be used only from vessels operating for scientific purposes. Plankton samplers that can be used at higher speeds have some advantages; for example, they interfere little with the normal cruising of the ship and can therefore be towed behind commercial vessels proceeding on their normal routes, extending the scope of plankton studies. Also, they are probably more effective than slow-moving nets in capturing the more actively swimming creatures.

For high-speed sampling the net area must be very large in relation to the aperture to reduce the high back pressure developed when a fine-mesh net is towed rapidly through water. A simple sampler of this type is the Hardy Plankton Indicator consisting of a torpedo-shaped cylinder with stabilizing fins. Water enters a small opening at the front, filters through a disc or cone of bolting cloth and leaves through a rear aperture. A rather larger apparatus, the Gulf III Sampler (Southward, 1962) illustrated in Figure 3.15, incorporates a flowmeter. The Jet Net is similar in appearance and designed to reduce the speed of water flow through the mesh, thus reducing damage to organisms.

The Hardy Continuous Plankton Recorder (Figure 3.16) is a high-speed sampler which provides a continuous record of the plankton collected over a long-distance haul. The recorders are towed at a depth of 10 metres on a standard length of wire rope and are deployed at monthly intervals as far as possible. Modifications can be made so that the angle of the depressor plate alters during towing, causing the apparatus to take an undulating course and sample over a range of depths. Electronic sensors for depth, salinity and temperature are also often fitted.

Water enters through a small aperture (1.25 cm) in the nose-cone. Once inside, the water is slowed to about a thirtieth of its original speed by the enlargement of the tunnel. It is then filtered through a slowly moving band of filtering gauze. Organisms trapped on the cloth are covered by a second band of gauze to form a 'sandwich' which is wound onto a take-up spool in a formalin reservoir where the plankton is preserved undisturbed.

At the end of a voyage, the instrument is sent to the headquarters of the CPR survey (Sir Alister Hardy Foundation for Ocean Science) in Plymouth, UK. The spool of gauze is unwound and each 10 cm length, representing 10 nautical miles of tow, is subjected to a standard routine analysis. The plankton is identified and counted and the data fed into a computer. By comparing the sample number with the ship's log, a general picture of the distribution of plankton can be built up.

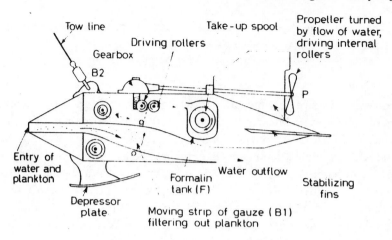

Figure 3.16 Diagrammatic section of a Hardy Continuous Plankton Recorder.

These instruments have been in regular use for over 60 years, towed by commercial vessels on many routes around the British Isles. Between 1931 and 1990, recorders have been towed for 3 733 746 nautical miles by merchant ships and ocean weather ships operating in the North Atlantic and North Sea (Colebrook *et al.*, 1991). The survey is on-going and continues to provide a means of mapping plankton populations and pin-pointing changes. This is of particular interest in terms of possible changes resulting from global warming. The results of these investigations are published at intervals in the *Bulletin of Marine Ecology*. The design of the recorder has not been changed in any way that would affect the sampling characteristics.

In the Longhurst–Hardy Plankton Recorder the principle of collecting the catch on a long strip of mesh is applied to a unit attached to the apex of a conventional plankton net in place of the usual receiver. Instead of continuously winding on, the filtering strip moves intermittently at 30 s intervals. It therefore carries a series of catches each representing filtering for a 30 s period. This mechanism is electrically driven, powered by batteries carried within the apparatus. It also contains a flowmeter and sensors for salinity and depth, these data being recorded on a miniature chart-recorder contained in the unit.

Diver collection and observation

Gelatinous plankton such as scyphozoans and thaliaceans are best captured by divers using wide-mouthed containers. Divers have also contributed much useful information on the behaviour and mode of life of these animals (Hamner, 1974; Hamner, 1975; Hamner *et al.*, 1975). Plankton for use in the laboratory study of living animals can also be collected using special buckets instead of the normal cod-end collectors on plankton nets. The buckets are generally much larger than normal and have mesh windows in the top part.

3.2.2   Nekton

Various types and sizes of midwater trawl have been designed to attempt the capture of nekton at middle depths down to 1000 m or more (Harrisson, 1967). The Isaacs–Kidd net is an elongate conical bag, usually with a mouth aperture of 8 m, and with an angled depressor plate to keep the net below the surface while towing. It can be fitted with a depth recorder, and its depth during towing can be monitored on hydrophones by attaching to the net a pressure-sensitive sound-emitter with a pulse frequency which varies with the depth. A modification is a double-ended net with the opening to the two receivers controlled by a pressure-operated valve. Above a preselected depth the captured material goes into one container, while below this level a flap deflects the catch into the other container, thereby providing a deep-level sample separate from material collected during lowering and hauling.

A midwater trawl (Baker *et al.*, 1973; Clarke, 1969) developed for use from the RRS *Discovery*, and operated to depths of over 1500 m, has a mouth which can be opened and closed by remote control (Figure 3.17). With the mouth closed the net is lowered to the required depth, as indicated by the pulse frequency of a pressure-sensitive sound-emitter on the net. The mouth is then opened by a release mechanism activated by an acoustic signal from the ship. At the end of fishing a second acoustic signal from the ship causes the mouth of the net to close before hauling. An extension of this system contains two acoustically operated nets in one framework, a small upper net with a 0.32 mm mesh above a larger coarse-mesh net of 8 m² rectangular mouth and 4.5 mm mesh. There is also a monitor on the frame which measures depth, flow and temperature, telemetering the data acoustically to the ship.

Large active bathypelagic fish and cephalopods have proved extremely difficult to catch, and little is known about their distribution. Probably very large pelagic trawls are needed for their capture. Some abyssal fish have been taken by line, either laid on the bottom or simply suspended from the surface, and some success has also been achieved with deep-water fish traps.

3.2.3   Benthos

There are several branches of science which seek information about the sea bottom; for example, oceanography, geology and palaeontology as well as marine biology. Each makes use of apparatus designed primarily to collect information needed in that particular field of study, but there is so much overlap between these various aspects of marine science that data relevant to one may well be of interest to another, and there is consequently a variety of instruments for studying the sea

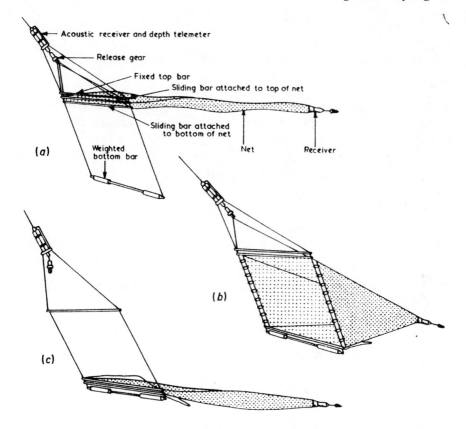

Figure 3.17 Midwater trawl for mesopelagic organisms, used on RRS Discovery. The net can be opened and closed by sonic signals sent from the ship, operating the release mechanism. The depth of the net is indicated by the pulse frequency of sonic signals generated by the depth telemeter. (a) Net closed for lowering; (b) net open; (c) net closed for hauling.

bottom which produce information relating to marine biology. We shall refer to only a few. Those selected for mention here are of two general types; instruments which are intended mainly for collecting samples of sediment, and those for collecting benthic organisms. The distinction is not a firm one because sediment samplers are likely to include small organisms in the material brought up, and apparatus designed to catch bottom-dwelling creatures may also retain some of the deposit.

Sediment samplers

Various small spring-loaded, snapper grabs have been devised which take a shallow bite out of the sea floor (Figure 3.18). Some investigations seek

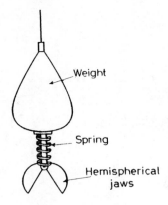

Figure 3.18   A spring-loaded snapper grab.

information about the deeper layers of deposit, and instruments known as corers are used for this purpose.

A corer is a long tube which can be driven down into the sea floor, and then withdrawn enclosing a core of sediment. The coring tube contains a separate liner to facilitate removal of the core. Considerable force is required to drive a corer far into the deposit, and several methods are used for this purpose.

The corer may be heavily weighted and allowed to descend at speed, penetrating the sediment under its own momentum. An explosive charge can be detonated to drive the tube downwards. A *vibrocorer* is driven into the sea-bed by a vibrating motor at the top of the tube, operated by electricity or compressed air.

The Kullenberg piston corer makes use of hydrostatic pressure to assist deep penetration. This corer consists of a weighted coring tube with brass liners, inside which fits a sliding piston attached to the lowering cable. The corer is lowered with the piston at the lower end of the tube, and the apparatus is slung from a release mechanism held in the closed position by counter-weights suspended below the nose of the coring tube. When these counter-weights touch bottom, the release mechanism opens to let the coring tube fall under its own weight. At this moment, the reduced strain on the lowering cable is indicated on the vessel by a dynamometer, and the cable winch is stopped immediately so that the piston attached to the cable is held stationary as the coring tube plunges downwards. This creates a tremendous suction inside the tube which helps to overcome the resistance of the substrate to penetration. Undisturbed cores over 20 m long have been obtained from very deep water with this device.

A type of piston corer has been used in conjunction with the drilling tube of the drilling ship *Glomar Challenger*. This ship makes drill borings in the deep ocean floor as part of a research project, the International Programme of Ocean

Drilling (formerly the Deep Sea Drilling Project), studying the structure of the earth's crust beneath the sea. The drilling bits used for boring hard rock disrupt the soft uppermost sediments, but by first dropping piston corers down the drill pipe it has been possible to obtain cores of undisturbed sediment up to about 200 m long. These cores contain the remains of planktonic organisms deposited on the sea-bed over a period of several hundred thousand years, and something may be learnt of oceanic conditions in the past by studying the variations in composition at different levels of the core.

For taking short cores up to about 1 m long from deep water, a 'free fall corer' may be used. This is a weighted coring tube which has no cable for lowering and hauling, but is simply thrown over the side of the vessel to sink freely. When the corer has sunk into the bottom an automatic release frees the tube, and a pair of glass floats carries the tube and enclosed sample to the surface. Retrieval is aided by a flashing light on the instrument.

Sediment traps

In the deep sea, the amount of particulate organic material reaching the deeper layers and the sea-bed affects the composition of animal communities. The amount and nature of the biogenic particles that rain down throughout the year can be measured using sediment traps. The traps can be moored at various depths above the sea-bed and left in place for a year or more. Modern traps work automatically using rotary collectors (Figure 3.19). Each collector opens for a set number of days, commonly seven, then closes (Lampitt, 1996). These techniques and time-lapse photographic techniques have been widely used in the recent Joint Global Ocean Flux Study (JGOFS) (Ducklow and Harris, 1993).

Collecting organisms from the bottom

Methods of sampling the benthic population vary with the types of organisms under study, and the type of bottom. Demersal fish and many other creatures that live on, rather than within, the sea bottom can be captured by the trawls and seines used by commercial fisheries, described later (see page 315). For research purposes, the nets are usually of smaller mesh than is permitted for commercial fishing. Animals living within the sediment must be dug out using grabs and corers. Rock-living species can usually only be sampled using divers or photographed using remote cameras or submersibles. Practical details of all these methods are given in Holme and McIntyre (1984).

Trawls and dredges (qualitative)

Trawl nets are designed to skim over the bottom and as they cover a large area, they have a good chance of collecting widely dispersed species. They are, however, often rather selective. A net much used for biological work is the Agassiz trawl

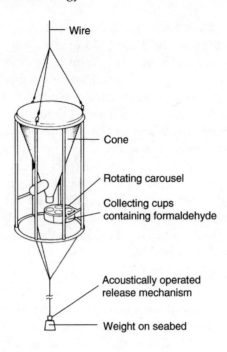

Wire

Cone

Rotating carousel

Collecting cups
containing formaldehyde

Acoustically operated
release mechanism

Weight on seabed

*Figure 3.19   A time series sediment trap. Falling material enters the cone and falls down
into a collecting cup filled with formalin. After a set number of days, the next collecting
cup moves round into position.*

(Figure 3.20), which has the advantage of very easy handling because it does not
matter which side up it reaches the bottom. The mouth of this net is held open
by a metal frame, and it can be fitted with fine-mesh net to retain small creatures.
It is simply dragged along the bottom.

To capture animals that live beneath the surface, the sampling device must be
capable of digging into the deposit. The naturalist's dredge is a simple device
which can be operated from a small boat. It consists of a bag of strong sacking
or wire mesh held open by a heavy, rectangular metal frame. This can bite a few
inches into a soft sediment as it is hauled along, but tends to fill mainly with
material lying on the bottom. The leading edges of the frame can be angled and
sharpened to increase the tendency to dig rather than to ride along the surface,
but it does not catch the deeper-burrowing creatures.

An example of an instrument that takes a considerably deeper bite is the Forster
anchor dredge (Forster, 1953) (Figure 3.21). This requires a sizeable vessel for its
operation. The net is attached to a strong rectangular metal frame with a long,
forward-projecting upper arm and a lower, downward-sloping digging-plate. The
dredge is lowered to the bottom and remains stationary as the ship moves slowly

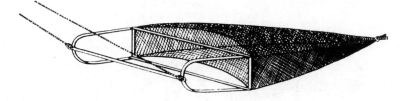

Figure 3.20 The Agassiz trawl.

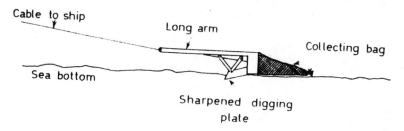

Figure 3.21 The Forster anchor dredge.
(Based on Forster, 1953.)

astern paying out a long length of cable, three to five times the depth of the water. The winch brakes are then applied to the cable, and the strain exerted as the ship is stopped causes the dredge to tilt and bite deeply into the substrate. So instead of sliding along the bottom the dredge digs in like an anchor to a depth of about 25 cm. Finally, the cable is winched in until the dredge eventually breaks out, the contents being retained within the net. For ease of use in deep water this type of dredge can be made with digging-plates on both sides of the arm, so that it will bite whichever way up it lands on the bottom.

Epibenthic sledges

Epibenthic sledges are widely used for sampling smaller seabed animals, especially in the deep sea. The simplest sledge consists of a mesh bag or bags mounted in a metal frame on runners. The collecting bag is protected inside a steel mesh cage. The angle and height of the cutting plates on the mouth of the frame can be adjusted and the sledge can operate either way up.

More sophisticated sledges (Gage and Tyler, 1991) have various instruments mounted on the top and so must operate the right way up. Cameras photograph the bottom area ahead of the sledge. Acoustic devices indicate the position of the sledge relative to the bottom, and an odometer wheel coupled to a potentiometer measures the distance the sledge has travelled over the sea-bed.

Plankton living near the bottom (hypoplankton) can be collected in a plankton net attached to a sledge (Figure 3.22) and dragged over the sea floor.

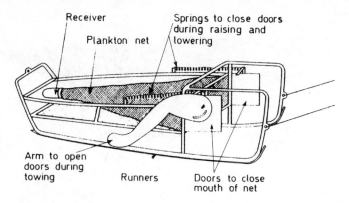

Receiver

Plankton net

Springs to close doors
during raising and
lowering

Arm to open
doors during
towing

Runners

Doors to close
mouth of net

Figure 3.22   The Bossanyi hypoplankton net.

Quantitative bottom sampling

Quantitative studies of benthic populations require samplers which take a standard bite of known area and depth. A large number of remote samplers have been designed for use in a variety of depths and conditions. For small organisms (micro- and meiobenthos), most of which live close to the surface, short coring tubes can provide satisfactory samples from soft deposits. Capturing larger creatures presents more difficulty because some can escape the sampling gear by crawling away or moving deeper down their burrows. The main types of grabs and corers in current use have been reviewed by Eleftheriou and Holme (1984) and aspects of their use are described in Hartley and Dicks (1987). Only those most commonly used are described here.

On soft sediments the Petersen grab (Figure 3.23) has been much used. It consists of a pair of heavy metal jaws which are locked wide apart while lowering to the sea-bottom. The grab sinks into the deposit under its own weight, and as the cable goes slack the lock holding the jaws apart is automatically released. On hauling, before the grab lifts off the bottom, the tightening cable first draws the jaws together enclosing a bite of the substrate of approximately 0.1 m² surface area. The grab bites fairly well into soft mud, but on sand or gravel it digs only to a depth of some 3–4 cm and many creatures escape. Stones or pieces of shell may wedge between the jaws, preventing complete closure, and much of the catch may then be lost during hauling.

The Petersen grab is less used nowadays than several other samplers which work on similar principles. Three grabs in common use because of their simplicity and ease of handling are the Smith–McIntyre, Day and Van Veen grabs.

The much used Smith–McIntyre grab has jaws which are spring-loaded to drive them into the sediment. Hauling on the cable then closes the buckets before lifting the grab off the bottom. The Day grab is based on the Smith–McIntyre design but

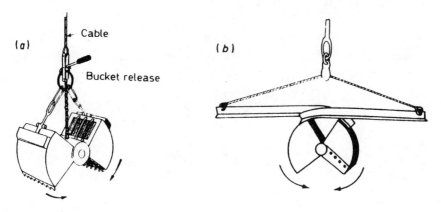

Figure 3.23   (a) The Petersen grab. (b) The Van Veen grab.

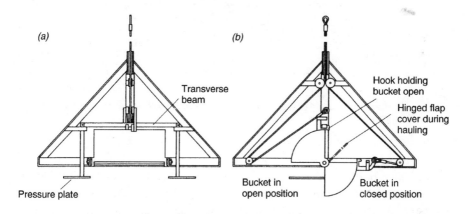

Figure 3.24   The Day grab. (a) End view open for lowering; (b) Side view, one bucket open the other closed. On reaching the sea-bed, the two pressure plates are pushed upwards releasing the transverse beam so that the hooks holding the buckets open are released. The buckets are closed by tension on the two cables.
(From: Day (unpublished manuscript) in Holme and McIntyre, 1984.)

is not spring-loaded, and is therefore possibly somewhat safer in operation. It has a strong pyramid-shaped frame and can be almost guaranteed to land, and stay the right way up (Figure 3.24). The simple design of the Van Veen grab makes it especially suitable for small boat use (Figure 3.23b). It has arms attached to the jaws to give greater leverage for forcing the jaws together. Its lightness means it works best in soft sediments and calm weather.

An example of a sampler which can be used on rather coarser deposits is the Holme scoop sampler, used in studies of the biomass of the English Channel. This digs by means of semi-circular scoops. Two models have been designed, one having a single scoop and the other a pair of counter-rotating scoops (Figure

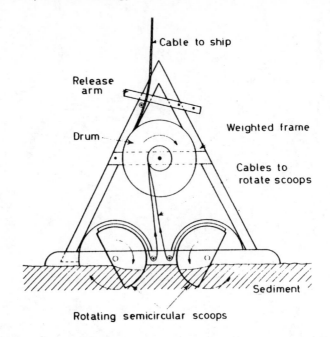

Figure 3.25   Diagrammatic representation of the Holme double scoop sampler.
(Based on Holme, 1953.)

3.25). The apparatus is lowered with the scoops in a fully open position. On reaching the bottom, a release mechanism operates so that, when hauling commences, the strain on the cable is first applied to the scoops, turning them through 180° so that they dig into the substrate. Each scoop samples a rectangular area of approximately 0.05 m² although a later model has scoops twice the width. In favourable conditions each bite is semi-circular in vertical section with a maximum depth of 15 cm.

In deep-sea investigations, the USNEL box corer (Hessler and Jumars, 1974), is now the standard gear for quantitative sampling. It consists of a square, open-ended steel core box fixed to a weighted column, mounted on a support frame. After the corer has sunk into the sediment, a spade swings down to close the bottom and the top is closed by flaps (Figure 3.26). When operated carefully the sample retains an undisturbed sediment surface.

One of the problems with both grabs and box corers is the 'bow wave' generated as the apparatus nears the sea-bed. This can 'blow away' small surface-dwelling animals such as amphipods. The USNEL box corer has been designed to minimize this effect. Its operation is described in detail in Gage and Tyler (1991).

Another type of bottom sampler for quantitative work is the Knudsen Suction sampler. This is a short corer of wide bore with a suction pump in the upper part

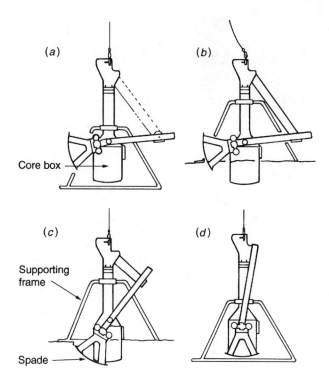

Figure 3.26 USNEL box corer. (a) Gear in ready position before reaching sea-bed; (b) on sea-bed, heavy corer enters sediment; (c) the spade swings down to seal the box as the slack wire is winched in; (d) gear and sample being hauled to surface.
(From Gage and Tyler (1991), by courtesy of Cambridge University Press.)

of the instrument. After reaching the bottom, tension on the hauling cable first turns the pump, thereby generating a suction inside the tube to assist penetration. When the corer breaks out of the substrate it automatically turns upside down to avoid loss of contents.

Rough quantitative comparisons can also be made using the anchor dredge (see page 75), which takes a fairly uniform bite.

Diver-operated samplers

In shallow water (less than about 25 m), diver-operated suction samplers provide accurate quantitative samples of soft-bottom benthos. All the designs utilize the lift generated by compressed air to suck up the sediment. They are most effective in sand and least in mud. Their use is described in detail in Hiscock (1987).

Early designs are based on the Barnett-Hardy suction corer. The diver positions the corer and presses it into the sediment (Figure 3.27). A compressed air line from the surface, or more usually a diving cylinder, then generates an air-lift in a suction pipe connected to the coring tube, whereby water is sucked out of the corer to

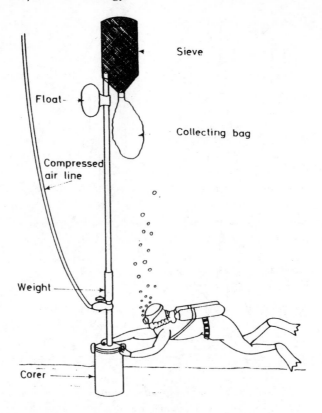

Figure 3.27   The Barnett-Hardy diver-operated suction corer.

force it down into the mud. When it has penetrated fully, the top of the coring tube is opened and the air-lift sucks the contents of the tube up into a sieve which separates the sample into a collecting bag. There are many variations on this theme, all utilizing compressed air to create a lift. Similar devices are also extensively used in underwater archaeology to remove sediment from around artifacts.

In soft sediments, small quantitative samples can be collected by divers using something as unsophisticated as plastic drain pipe. The volume sampled can be adjusted by using pipes of different diameter and sampling to a known depth. Use of a 'milk crate' to hold the tubes and rubber bungs to retain the samples allows for replicate samples and for sampling along a transect line.

3.3   Underwater observations

It would simplify many problems in marine biology if the range of direct observation could be extended. The only marine populations which are easily accessible to close inspection are those of the seashore, and then only for a part of each tidal cycle. Our knowledge of the rest of marine life comes almost entirely from the incomplete samples obtained by nets, dredges, grabs and similar devices. Recently, new techniques for visual underwater exploration have been developed, and have already provided much new information on marine organisms.

3.3.1 Diving

Diving by means of air pumped down a tube from the surface to a man enclosed in a special helmet and diving suit was first introduced in 1819 by Auguste Siebe. This was the prototype of hard hat diving. Apparatus of this type is still used by commercial divers working in connection with underwater constructions or salvage operations, but has found little application in biological work. Apart from its unwieldiness and expense, it has the drawback that it does not permit free movement of the diver over a wide area because he is limited by the length of his breathing tube and the need to keep it free from snags.

This problem has disappeared with the development of the aqualung or SCUBA gear (Self-Contained Underwater Breathing Apparatus). This provides a diver with a means of moving freely underwater, unencumbered by an airpipe, his air supply being carried on his back in compressed air cylinders. In 1942, Jaques Cousteau and Emile Gagnan developed the first fully automatic regulator or 'demand valve' that provided air from the cylinder only 'on demand'. Modern equipment is based on their original design.

In recent years, the design and amount of diving equipment available have increased tremendously. Large-capacity diving cylinders and warm dry-suits have increased the efficiency and time under water of divers. Diving computers allow for easier decompression planning. Underwater tape recorders, communication systems and cameras enable more data to be collected. Towed and self-propelled sledges allow divers to survey large areas.

Diving carries with it certain inherent risks and proper training and certification are essential. In most instances, training with organizations such as the BSAC (British Sub Aqua Club) or PADI (Professional Association of Diving Instructors) is adequate. However, in Great Britain, since 1981, paid work involving diving has brought biologists and scientists under the HSE (Health and Safety Executive) regulations and additional training and certification may be required.

Physiological hazards

The physiology and problems associated with diving are described in detail in a wide variety of publications such as the British Sub-Aqua Club diving manual (BSAC, 1985; Bennett and Elliott, 1993; Moon and Bennett, 1995). Only a short summary of the major problems is given here.

The special problems of breathing underwater are due to the pressure of water surrounding the body. On the surface at sea level, the normal air pressure is 1 atmosphere (atm) or 1 bar absolute. The pressure increases approximately 1 atm for every 10 m of depth. For a diver to be able to expand his lungs against the water pressure, he must be supplied with air at a pressure equal to that of the water. Whereas at the surface we breathe air at atmospheric pressure (1 atm), at a depth of 10 m the diver must have air at double this pressure, that is at 2 atm; at 20 m, 3 atm and so on. The aqualung cylinder contains air at very high pressure (200–300 bars). Modern demand valves or regulators reduce this pressure in two stages to match the ambient pressure of the surrounding water. The first stage (the reducing valve) is attached to the valve of the diving cylinder and reduces the pressure to about 8–10 bars above ambient. The second stage (the demand valve) is held in the diver's mouth and reduces the pressure to ambient i.e. equal to the pressure of the surrounding water.

Diving using air is without serious physiological hazards down to about 9 m depth. Below this, precautions must be taken to avoid the dangerous condition of 'decompression sickness' during ascent. The 'bends', as it is often called, occurs when bubbles of gas (mainly nitrogen in air-breathing divers) are liberated into the tissues or the blood. As the diver ascends, the ambient pressure falls, and gas dissolved under pressure in the blood can come out of solution too rapidly – a process similar to the fizzing of soda-water when the bottle is unstoppered. Depending on how large the bubbles are and where they lodge, decompression sickness can produce a variety of severe symptoms including intense joint pains (bends), paralysis or, in extreme cases, death.

Decompression sickness is avoided by following a decompression table which tells the diver how long he can safely stay at a particular depth and still come straight to the surface. Beyond this time limit, it is necessary to make a gradual ascent involving a series of pauses, or 'decompression stops', to give ample time for the excess nitrogen dissolved in the blood and tissues to be eliminated in the expired breath. Nowadays, modern diving computers are often used in place of tables.

The normal safe depth limit for compressed air diving is considered to be 50 m and, in Britain, divers who are covered by the HSE safety regulations are limited to this depth. Below about 30 m, air-breathing divers face another problem known as nitrogen narcosis. Large quantities of nitrogen dissolved under pressure in the blood have an effect on the brain producing a condition of rapturous inebriation in which the diver loses control of his actions, with possibly fatal results. Commercial divers overcome this problem by using 'heliox' – a mixture of oxygen and helium. This can be used down to about 500 m and is much safer than compressed air but has the drawback of causing distortion of the voice, making speech communication difficult. Experiments are now underway in which oxygen is mixed with hydrogen to produce 'hydrox' which can be used at even greater depths. Use of such equipment is expensive, requires special training and is not much used in scientific work. However, recent developments in the field of 'nitrox' diving are now allowing safer, deeper diving with an acceptable level of extra cost and training. Nitrox gas consists of an oxygen/nitrogen mix in different proportions to atmospheric air. Closed-circuit, re-breathing technology also looks set to extend diving limits for scientific divers in the future.

Saturation diving from underwater chambers in which divers can live at pressure has obvious advantages in time and costs and is much used by the offshore oil industry. At any particular pressure the body can absorb only a certain amount of gas before becoming saturated. Once the saturation point is reached, no more gas will be absorbed and the time required for decompression will then be the same, however long the duration of the dive. Divers return to their submerged 'house' to feed or sleep and replenish their gas supplies, allowing them to have relatively long working periods at depth without time wasted on frequent decompression and resurfacing. A diver who spends a week at 150 m requires only the same decompression period as one who descends to that depth for only one hour.

Attempts to dive much deeper than 200 m encounter additional dangers. Modern medicine is suggesting that the breathing of oxygen and inert gases such as helium at high pressure causes causes long-term changes in the diver's nervous system and physiology. Another serious hazard of deep saturation diving is bone necrosis, the death of areas of bone apparently caused by blockage of blood vessels, sometimes leading to severe arthritis. Recent predictions put the lowest limit to which divers exposed to pressure may be able to work at between 500 and 1000 m. At the time of writing, the record for successful descent and return using aqualung equipment appears to be held by six French divers operating from a diving bell and breathing an oxygen-helium mixture. At a depth of 460 m they each worked for periods up to 2 hours 20 minutes over four days. Two of them then descended for ten minutes to 501 m, after which about ten days was spent on decompression.

The aqualung in ecological research

The aqualung is a tool that has many applications in marine biological investigations in shallow water. It makes possible many quantitative studies on distribution and growth of marine organisms by direct observation with minimal disturbance of their natural environment. The behaviour of marine animals can be recorded in their normal surroundings. Photographs can be taken of precisely selected areas and events and changes can be monitored. Divers can operate many types of underwater equipment which would otherwise have to be remotely controlled from the surface or might not be usable at all in particular localities (Kritzler and Eidemuller, 1972; Potts, 1976).

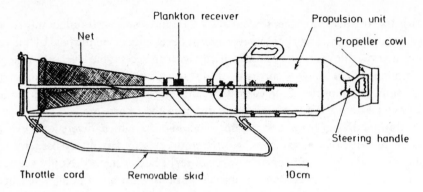

Figure 3.28   A diver-controlled plankton net.
(From Potts (1976) by courtesy of Cambridge University Press.)

Diver-controlled nets (Figure 3.28) can be used in rock gullies and around submerged reefs where it would not be possible to tow conventional nets from the surface. Diver-controlled dredges can be opened and closed so as to sample only selected parts of the sea floor, and can be raised or steered to avoid snags and obstacles. Grabs and corers can be exactly positioned to take only the material needed (Figure 3.27). In sediment areas, divers have been used to study crab and fish burrows by taking resin casts – rather like animal track casts on land.

Diving is particularly useful when studying the shallow, rocky areas just below the shore – the sublittoral. Remote sampling is extremely difficult in such areas and it is only in the past 15 to 20 years that an accurate picture of such areas has been obtained. A similar comment applies to the study of coral reefs. Baseline ecological surveys of marine habitats and species are vitally important in terms of marine conservation. It is essential to know what is there before sensible decisions can be taken regarding development (e.g. oil exploration, marina development, etc.) and response to pollution incidents. With modern equipment, divers are able to carry out research even in such adverse places as under the Antarctic ice.

In Britain, divers are playing an important part in the Marine Nature Conservation Review, a project started in 1987 to survey and assess coastal marine habitats (see page 231). Standard recording methods for the survey of shallow sublittoral areas using divers have been developed. The massive increase in sport diving in recent years has also led to the development of a breed of 'underwater naturalists'. There is now a wide range of marine projects, expeditions and surveys in which amateur divers can participate and the results are providing many useful data.

3.3.2 Submersibles
Below the depth that can be safely reached by divers, exploration is possible in submersibles where air can be breathed at normal pressure. Such vehicles must

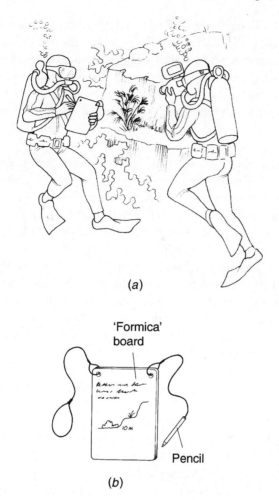

(a)

'Formica'
board

Pencil

(b)

Figure 3.29 (a) *Divers using slates and cameras to record habitats and species;* (b) *diver's slate.*

be of great strength to withstand the enormous water pressures encountered in the deeps and must be provided with lighting equipment to illuminate the surroundings. In recent years, the development of the offshore oil industry has led to the design of a number of manned submersibles and unmanned remotely operated vehicles (ROVs). In addition to the vehicle itself, surface support vessels, handling gear, logistic and maintenance support are needed.

Early deep-sea exploration had of necessity to concentrate primarily on the logistics of the exercise and only limited observations were possible. Between

1930 and 1934, William Beebe and Otis Barton broke all previous records for descent into the deep sea. Their bathyshere was a spherical observation cabin, 1.5 m in diameter, which was lowered and raised on a cable from a winch on a surface vessel. This device was the first to reach the deep-sea bottom at a depth of nearly 1000 m, the limit of length of cable then available. They were the first to see many deep-sea fish previously only known from trawled-up, damaged specimens. In 1950, in a later version known as the Benthoscope, Barton made a deeper descent to 1300 m.

The next development after the bathysphere was the bathyscape, invented by Professor Auguste Piccard and able to operate without cables. The design combined an observation cabin or gondola with an underwater float, which may be likened to an underwater balloon. The small spherical, pressure-resisting cabin with portholes was suspended beneath the large, lightly built float filled not with gas but with aviation fuel. Being much lighter than water and virtually incompressible, the fuel provided adequate buoyancy to support the heavy cabin, and the float did not need to be constructed to withstand great pressure. The bathyscape carried iron-shot ballast and sank freely under its own weight. To ascend, sufficient ballast was shed for the vehicle to float up to the surface. In some models, electrically driven propellers provided a limited amount of horizontal movement when submerged.

Piccard called his bathyscape the *Trieste* and made the first manned dives in it in 1953 funded largely by the Italian government. In 1958 the *Trieste* was bought by the the US Navy and on 23 January 1960, Auguste Piccard's son Jaques and Donald Walsh embarked on one of the most perilous underwater journeys ever undertaken. On that day, they made a successful return voyage from nearly 11 000 m in the Challenger Deep of the Marianas Trench, only about 122 m short of the deepest known part of the ocean floor.

At the moment there is only one vessel, the Japanese 'Kaiko', able to reach the bottom of the Marianas Trench. It is unmanned and operated via a 12-km-long cable. An American prototype manned 'Deep Flight' submarine is currently being developed to enable it to reach 11 000 m. The one-person submarine will 'fly' on inverted wings rather than simply sinking under its own weight (Mullins, 1995).

A variety of small, manoeuvrable submersibles, both manned and unmanned, are now in use, mainly for underwater engineering projects or geophysical research. Unmanned submersibles (remote-operated vehicles or ROVs) can reach 6000 m and carry a wide array of instruments and cameras. An example is the ROV Jason operated by the US Woods Hole Oceanographic Institution. It carries cameras and sonar and is connected to the support vessel by a 10-km-long fibre-optic cable which allows the operator to 'see' and steer the vehicle. Untethered ROVs that operate under their own power without connecting cables are also coming into increasing use.

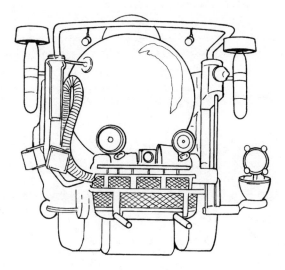

Figure 3.30   Deep-sea submersible, Johnson Sealink.

Manned research vessels have greater depth restrictions than ROVs and there are presently only six in the world, capable of reaching depths of between 4000 m and 6000 m or so. The Alvin, commissioned by the US Navy in 1964, is one of the most widely used. Only 7.6 m long, it has now made more than 1700 dives. It was from this submersible that new forms of life around deep-sea vents were first discovered (see Section 6.4.4). The Alvin was also used in 1986 to view the remains of the infamous ocean liner, the *Titanic*, which sank to a depth of 3810 m after a collision with an iceberg on its maiden voyage in 1912. The United States also has *Sea Cliff*. Other well-known submersibles include the Russian *Mir I* and *Mir II* carrying two crew and a scientist; the French *Nautile*; and the Japanese *Shinkai 6500*.

Other manned submersibles operate down to only 1000 m or so. An example is the US *Johnson Sealink* which has an acrylic dome allowing excellent viewing, and carries two crew and two scientists (Figure 3.30).

The chief advantage of manned submersibles is in enabling direct visual observations to be made at great depths. Submersibles have various applications in biological work, and have been used for benthic surveys and photography of the sea bottom and of mesopelagic and bathypelagic organisms. Actual collection of small organisms by submersible can be done with a 'slurp gun', a form of suction pump with a flexible hose gripped by the submersible's manipulator arm. Organisms can be gently sucked into the nozzle of the hose and thence into a cannister. Fragile animals such as radiolaria, medusae, siphonophores, ctenophores, amphiphods, mysids and small fish have been collected in this way

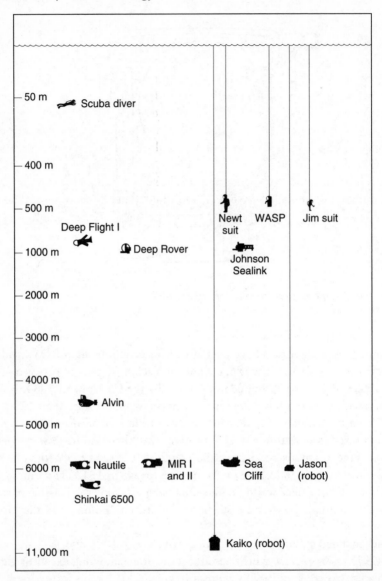

Figure 3.31   Diagram illustrating depth capabilities of various submersible craft and suits.

in excellent condition, which when collected by deep-water nets are usually severely damaged.

Unmanned submersibles do not need a pressure sphere to accommodate crew and since they draw power from the surface via a tether, are not limited by battery power. These systems continue to be developed, and in the future unmanned submersibles with sophisticated robotic capabilities will undoubtedly allow extensive sampling, visualization and experimentation in the deep sea. Already sophisticated camera equipment allows stereo viewing and virtual reality technology can be used to control manipulator arms with great accuracy.

Recent years have seen the development of a variety of one- and two-man submersibles for use in shallow water down to several hundred metres, and mainly developed in connection with the offshore oil and gas industry. Armoured diving suits operate in a similar manner to submersibles but are 'submarines you can wear'. The diver's limbs are encased in tubes, with hands of claws or claspers appropriate to the work to be undertaken. The suit contains a life support system, with a duration of about 70–90 hours, and can operate down to several hundred metres. Examples are the Jim, Wasp, Newt and Spider suits (Sisman, 1982). In general they are too clumsy and expensive for scientific use.

Currently, simple, easy-to-use and cheap submersibles are being developed for use by scientists and even sports divers, to depths of about 100 m. Further development of these will enable scientists and laymen alike to observe and record without the restrictions imposed by diving.

3.3.3 Underwater photography

In recent years, there have been rapid advances in the development of underwater cameras, television and video. The current interest in sport diving has led to the development of a wide variety of underwater camera equipment designed for easy use by scuba divers. Further impetus has come from the need of the oil and gas industry to inspect their underwater hardware. Photography is widely used by marine biologists to support other sampling methods. It can also be used as a recording tool on its own. For example, stereophotography, using two cameras mounted in a frame, provides three-dimensional images allowing size measurements to be taken. Repeat visits to marked areas can be made to record changes in species composition, growth rates and seasonal changes.

Diver-operated cameras

Underwater photography has become an increasingly important tool for diving marine scientists over the past 15 years or so. Diver-held underwater cameras fall into two main categories: standard 35mm land cameras and flashguns enclosed in pressure-proof housings; and amphibious, waterproof cameras with dedicated flash guns of which the Nikonas system is the most widely used (Figure 3.32). Both systems can be fitted with a variety of wide-angle and close-up lenses. Housed cameras tend to be bulky but have the inherent capabilities of any single lens reflex camera allowing automatic exposure and focusing. Several amphibious cameras now have automatic exposure capability with through-the-lens (TTL) metering but to date, only one, the Nikonas VI, has automatic focusing. The high price of this model puts it beyond the reach of many divers.

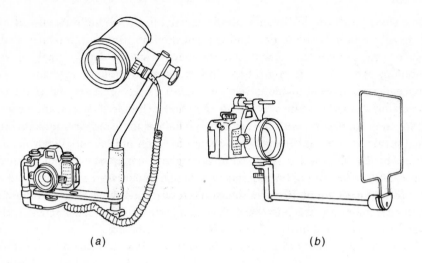

(a)   (b)

Figure 3.32   Nikonas V amphibious camera equipment; (a) camera with dedicated flashgun allowing TTL metering; (b) camera with close-up lens framer.

The main problems with underwater photography stem from the lack of available light. Except in very shallow depths, a flashgun is always necessary, both to provide enough light, and to restore colours at the red end of the spectrum (see page 138). Backscatter from particles in the water reflected by the flash makes distance photography difficult. This can be partially overcome by use of wide-angle lenses which allow a close approach to the subject whilst covering a reasonable area. A wide variety of books, videos and specialist courses on underwater photography are now available.

Underwater housings have also been developed for the lightweight, compact video cameras now available on the market. These are easy and cheap to use and are especially useful at deeper diving depths where diver time is limited. Underwater video is now being increasingly used in biological survey and monitoring work.

Remote cameras

Remote, automatically operating underwater cameras deployed on wires from research vessels have been in use since about the 1970s. Nowadays they are more often mounted on sledges, submersibles or on towed arrays of multi-instruments. They can also be mounted above baited traps. Modern cameras and lighting units provide high-resolution photographs. Cameras operating in the deep sea do not necessarily need a shutter due to the lack of light at such depths.

The camera systems described here are those most commonly in use today for studying the deep-sea bed.

Flat plan pictures of the sea-bed are taken using stereo cameras pointing straight down (Southward *et al.*, 1976). The instrument package in its supporting frame is lowered from the ship until a trip-weight hanging beneath the apparatus touches bottom. This fires the electronic flash so that a photograph is taken, and the film is wound on. It also causes an acoustic signal to be transmitted to the ship. As soon as this is heard on the ship's hydrophone the apparatus is raised slightly above the sea-bed and then lowered again. As the weight touches bottom a further picture is taken, and the process is repeated for each photograph. Alternatively, either bottom or midwater cameras can be triggered by levers carrying bait, each photograph being taken at the instant an animal grasps the bait.

A camera can be mounted on a sledge and towed across the sea-bed by the ship. In this case the camera is usually obliquely mounted and programmed to take photographs at set intervals. It is often easier to identify small animals from oblique, rather than straight down pictures.

A recent exciting development is in free vehicle cameras that can be deployed on the sea-bed and left there, with no connection to the surface, for many months at a time. Their use in time-lapse photography at abyssal depths has been revolutionizing concepts of life on the deep-sea bed. The camera takes photographs at set intervals and other instruments record current speed and direction at the same times. The instrument package, including the camera, is recovered by sending an acoustic signal which releases it from its expendable

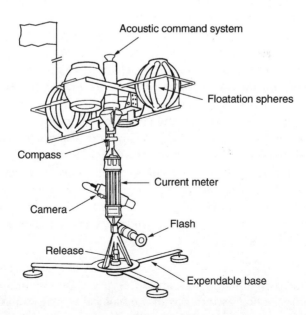

Figure 3.33 '*Bathysnap*', *a free-vehicle time-lapse camera system.*
Reproduced from Lampitt and Burnham (1983), with permission from Elsevier Science Ltd.

ballast base. It then floats to the surface and can be tracked and retrieved by the research vessel. A system called 'Bathysnap' (Figure 3.33), developed in the UK by the Institute of Oceanographic Sciences (Lampitt and Burnham, 1983), has recently been used to follow seasonal changes in the amount of phytodetritus at abyssal seabed sites (Rice *et al.*, 1994), especially in relation to the spring plankton bloom in the North Atlantic (see page 195).

These cameras can be used in combination with baited traps allowing photographs to be taken of otherwise motile and well-dispersed species. Giant amphipods and various fish have been photographed in this way.

The appearance of the sea-bed shown in photographs provides information about the nature of the sediment and the speed of movement of the bottom water. Organisms can be seen undisturbed in their natural environment, or their presence known from their tracks or burrows. Where creatures can be easily recognized, quantitative information from photography is probably more reliable than that obtained by grab samples.

3.3.4   Underwater television and video

Underwater closed-circuit television has been in use for some years. It first achieved a notable success in locating a sunken submarine on the bottom of the English Channel in 1951. Underwater television has found some applications in biological work, having advantages over photography in allowing immediate, continuous observation. The apparatus is more complex and costly than photographic cameras, and there are greater difficulties in its use at very deep levels (Barnes, 1963). As yet, it is not often used in the deep sea because of the very heavy power drain from the cables connecting the camera to the surface ship. Fibre optics may help to resolve this problem in the future.

Television cameras can be mounted on towed underwater sledges along with still cameras. The television signals give a continuous record of the strip of sea bottom traversed by the sledge, and colour photographs show greater detail of particular areas (Holme and Barrett, 1977). If the distance travelled by the sledge is measured, then quantitative estimates of fauna can be made.

In recent years, high-resolution video cameras have been developed and used attached to submersibles or free vehicles. This system was used to survey the remains of the *Titanic* after she was discovered in 1985. In biology, such video footage has proved of great use in studying the behaviour of deep-sea animals such as those around deep-sea vents (see Section 6.4.4).

3.3.5   Position fixing

Whatever the method used to obtain a biological sample from the sea-bed, it is important to know exactly where the sample came from. Most modern research vessels are equipped with Decca Navigator or Loran C systems which utilize

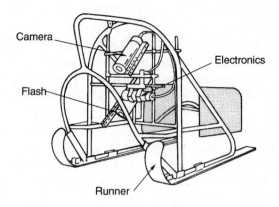

Figure 3.34   Towed camera sled used by A.J. Southward at MBA, Plymouth.
Reproduced from Gage and Tyler (1991) from photographs and drawings in Southward *et al.* (1976).

medium-frequency radio direction-finding transmitters. Satellite navigation systems utilizing transit satellites are also in common use.

The global positioning system (GPS) is a recent development from the USA that provides precise positioning to within only a few metres. The system works using an array of up to 24 satellites at least 4 of which will always be above the horizon and therefore within range of the ship. Small, hand-held units are now available so that the system can even be operated from small inflatable boats. However, the available accuracy of the system for most users is only about 50–100 m because the system has been degraded by modifying the signals as a national security precaution. Authorized users such as the US military can be given full accuracy by decoding.

References and further reading

Student texts
Baker, J.M. and Wolff, W.J. (eds) (1987). *Biological surveys of estuaries and coasts.* Estuarine and Brackish-water Sciences Association Handbook. Cambridge University Press.
Elliott, J.M. and Tullett, P.A. (1978). *A bibliography of samplers for benthic invertebrates.* Freshwater Biological Association, Occasional Publications No. 4.
Summerhayes, C.P. and Thorpe, S.A. (eds) (1996). *Oceanography: An Illustrated Guide.* Southampton Oceanography Centre, Manson Publishing.

References
Baker, A. de C., Clarke, M.R. and Harris, M.J. (1973). Combination midwater trawls. *J. Mar. Biol. Ass. UK*, **53**, 167–84.
Barnes, H. (1963). Underwater television. *Oceanogr. Mar. Biol. Ann. Rev.*, **1**, 115.

Bennett, P.B. and Elliott, D.H. (eds) (1993). *The Physiology and Medicine of Diving*. 4th edition. W.B. Saunders.

BSAC (1985). *Sport Diving*. The British Sub-Aqua Club Diving Manual. Stanley Paul.

Chivers, R.C., Emerson, N. and Burns, D.R. (1990). New acoustic processing for underway surveying. *The Hydrographic Journal*, **56**, 9–17.

Clarke, M.R. (1969). A new midwater trawl. *J. Mar. Biol. Ass. UK*, **49**, 945–60.

Colebrook, J.M, Warner, A.J, Proctor, C.A., Hunt, H.G., Pritchard, P., John, A.W.G., Joyce, D. and Barnard, R. (1991). *60 years of the continuous plankton recorder survey: a celebration*. The Sir Alister Hardy Foundation for Ocean Science.

David, P.M. (1965). The neuston net. *J. Mar. Biol. Ass. UK*, **45**, 313.

Ducklow, H.W. and Harris, R.P. (eds) (1993). JGOFS: The North Atlantic bloom experiment. *Deep-Sea Research II*, **40**, Nos. 1 & 2. Pergamon Press.

Eleftheriou, A. and Holme, N.A. (1984). Macrofaunal techniques. In: *Methods for the Study of Marine Benthos*. N.A. Holme and A.D. McIntyre, eds., pp. 140–216. Oxford, Blackwell Scientific Publications.

Forster, G.R. (1953). A new dredge for collecting burrowing animals. *J. Mar. Biol. Ass. UK*, **32**, 193.

Gage, J.D. and Tyler, P.A. (1991). *Deep-sea biology: A natural history of organisms at the deep-sea floor*. Cambridge University Press. (Chapter 3, Methods of study of the organisms of the deep-sea floor.)

Griffiths, G. and Thorpe, S.A. (1996). Marine instrumentation. In: *Oceanography: An Illustrated Guide*, C.P. Summerhayes and S.A. Thorpe, eds. Southampton Oceanographic Centre, Manson Publishing.

Hamner, W.M. (1974). Blue water plankton. *National Geographic Magazine*, **146**, 530–45.

Hamner, W.M. (1975). Underwater observations of blue-water plankton. *Limnology and Oceanography*, **20**, 1045–51.

Hamner, W.M. *et al.* (1975). Underwater observation of gelatinous zooplankton. *Limnology and Oceanography*, **20**, 907–17.

Harrisson, C.M.H. (1967). On methods for sampling mesopelagic fishes. *Symp. Zool. Soc. Lond.*, **19**, 71–126.

Hartley, J.P. and Dicks, B. (1987). Macrofauna of subtidal sediments using remote sampling. In: *Biological surveys of estuaries and coasts*. J.M. Baker and W.J. Wolff, eds. Cambridge University Press.

Hessler, R.R. and Jumars, P.A. (1974). Abyssal community analysis from replicate box cores in the central North Pacific. *Deep-sea Research*, **21**, 185–209.

Hill, M.N. (ed.) (1963). *The Sea. Ideas and Observations in the Study of the Seas*. Vol. 2. New York and London, Interscience.

Hiscock, K. (1987). Subtidal rock and shallow sediments using diving. In: *Biological surveys of estuaries and coasts*. J.M. Baker and W.J. Wolff, eds. Cambridge University Press.

Holme, N.A. (1953). The biomass of the bottom fauna of the English Channel. Part II. *J. Mar. Biol. Ass. UK*, **32**, 1.

Holme, N.A. and Barrett, R.L. (1977). Sea-bed photography. *J. Mar. Biol. Ass. UK*, **57**, 391–403.

Holme, N.A. and McIntyre, A.D. (1984). *Methods for the Study of Marine Benthos*. 2nd edition. IBP handbook No. 16. Blackwell Scientific Publications.

Kampa, E.M. (1970). Underwater daylight and moonlight measurements in the eastern North Atlantic. *J. Mar. Biol. Ass. UK*, **50**, 397–420.

Kritzier, H. and Eidemuller, A. (1972). A diver-monitored dredge for sampling motile epibenthos. *J. Mar. Biol. Ass. UK*, 52, 553–6.

Lampitt, R.S. (1996). Snow falls in the open ocean. In: *Oceanography: An Illustrated Guide*. C.P. Summerhayes and S.A. Thorpe, eds. Southampton Oceanographic Centre, Manson Publishing.

Lampitt, R.S. and Burnham, M.P. (1983). A free fall time lapse camera and current meter system, 'Bathysnap' with notes on the foraging behaviour of a bathyal decapod shrimp. *Deep-Sea Research*, 31, 329–52.

Linklater, E. (1972). *The Voyage of the Challenger*. London, Murray.

Moon, R.E. and Bennett, P.B. (1995). *Scientific American*, 273 (2).

Mullins, N. (1995). Voyage to the bottom of the sea. *New Scientist*, 145 (No. 1966), 26–9.

Omori, M. and Ikeda, T. (1984). *Methods in Marine Zooplankton Ecology*. J. Wiley and Sons.

Potts, G.W. (1976). A diver-controlled plankton net. *J. Mar. Biol. Ass. UK*, 56, 959–62.

Rice, A.L., Thurston, M.H. and Bett, B.J. (1994). The IOSDL deepseas programme: Introduction and photographic evidence for the presence and absence of a seasonal input of phytodetritus at contrasting abyssal sites in the northeastern Atlantic. *Deep-Sea Research I*, 41 (9), 1305–20.

Riddy, P. and Masson, D.G. (1996). The sea floor – exploring a hidden world. In: *Oceanography: An Illustrated Guide*. C.P. Summerhayes and S.A. Thorpe, Southampton Oceanographic Centre, Manson Publishing.

Robinson, I.S. and Guymer, T. (1996). Observing oceans from space. In: *Oceanography: An Illustrated Guide*. C.P. Summerhayes and S.A. Thorpe, Southampton Oceanographic Centre, Manson Publishing.

Sisman, D. (ed.) (1982). *The professional diver's handbook*. Submex.

Smethie, W.M. Jr. (1993). Tracing the thermohaline circulation in the western North Atlantic using chlorofluorocarbons. *Progress in Oceanography*, 31 (1), 51–99.

Southward, A.J. (1962). The distribution of some plankton animals in the English Channel and Western Approaches. II. Surveys with the Gulf III High-Speed Sampler 1958–60. *J. Mar. Biol. Ass. UK*, 42, 275–375.

Southward, A.J. (1970). Improved methods of sampling post-larval young fish and macroplankton. *J. Mar. Biol. Ass. UK*, 50, 689–712.

Southward, A.J. *et al.* (1976). An improved stereocamera and control system for close-up photography of the fauna of the continental slope and outer shelf. *J. Mar. Biol. Ass. UK*, 56, 247–57.

Strickland, J.D.H. and Parsons, T.R. (1968). *Practical Handbook of Seawater Analysis*. Bulletin 167. Ottawa, Fisheries Research Board of Canada.

UNESCO (1968). *Zooplankton Sampling*. Paris, UNESCO.

Walker, T.A. (1980). Correction to the Secchi Disc light-attenuation formula. *J. Mar. Biol. Ass. UK*, 60, 769–71.

4 The seawater habitat – physical and chemical conditions

4.1 Introduction

Although many features of the marine environment are virtually uniform over wide areas, different parts of the sea are populated by different communities of organisms. The aim of marine ecological studies is to discover what these differences are and why they exist and to evaluate the factors responsible for them. These investigations encounter many difficulties.

There are the obvious problems of working in an environment to which we have no easy access. Observations and measurements have mostly to be made with remotely controlled instruments. Some of the physical and chemical conditions can be measured with precision; but biological measurements involve many uncertainties because sampling apparatus such as nets, dredges and grabs are not instruments of high accuracy. Measurements of the activities of marine organisms in their natural surroundings are limited to diver observations in shallow water and remote videos and submersibles in deep water. Organisms can be brought into the laboratory and kept alive for a time, but here their behaviour may not be the same as in natural surroundings because it is obviously impossible to simulate closely in a tank all the conditions of the open sea.

Because several properties of the marine environment usually vary together, the effects of variation in single factors are seldom evident in natural conditions. There are two major zonations of distribution in the sea – between the tropics and the poles, and between the surface and the depths. Both are associated with differences of penetration and absorption of solar radiation, and therefore with gradients of temperature, illumination, and to a lesser extent salinity. Vertical distribution is also influenced by pressure. The distribution of a species is consequently associated with a complex of variables and it is not easy to assess the role of each factor independently.

The effects of variation in single factors can be studied to some extent in controlled conditions in the laboratory but in this unnatural environment the responses may be abnormal. There is also the complication that several factors often interact in their effects; for example, in some species the tolerance to salinity change is modified by temperature, and temperature tolerance may itself vary with salinity. Furthermore, observations on specimens from one locality may not hold for an entire population of wide distribution because each species exhibits a range of variation for each character, and these may be related to the geographical situation due to selection or acclimatization.

Apart from the effects of the inorganic environment, there are also many ways in which organisms influence each other. Even where physical and chemical conditions seem suitable, a species may not flourish if the presence or absence of other species has an unfavourable effect. Predation may be too severe. Other competing forms may be more successful in the particular circumstances. The environment may be lacking in some essential resources contributed by other species, such as food, protection, an attachment surface or some other requirement. These biological factors are obviously of great importance, but their evaluation is extremely difficult.

Generally, the distribution of a species is an equilibrium involving many complex interactions between population and environment which are at present very incompletely understood. Nevertheless, a start can be made in tracing the complicated web of influences which control the lives of marine organisms by first studying the individual environmental variables, noting the extent to which each can be correlated with the distribution and activity of different species, and observing the effects of change both in natural conditions and in the laboratory. The variable conditions of obvious biological importance which we shall refer to in this chapter are temperature, the composition of the water, specific gravity and hydrostatic pressure, viscosity, illumination and water movements.

For detailed descriptions of the composition, properties and conditions for life in seawater, the student should refer to one of the many general texts available on oceanography. A selection of these is listed in the references at the end of this chapter and Chapter 1.

4.2 Temperature

4.2.1 Geographical and depth variations

The continual circulation of the oceans and their enormous heat capacity ensure that the extent of temperature variation in the sea is small despite great geographical and seasonal differences in absorption and radiation of heat. Except in the shallowest water, the temperature range in the sea is less than that which occurs in most freshwater and terrestrial habitats, and the relative stability of sea temperature has a profoundly moderating effect on atmospheric temperature change.

The highest sea surface temperatures are found in low latitudes where much of the oceanic surface water is between 26 and 30°C. In shallow or partly enclosed areas like the Arabian Gulf, the surface temperature may rise to as high as 35°C during the summer, and conditions are extreme on the shore where intertidal pools sometimes exceed 50°C. At the other extreme, the freezing point of seawater varies with the salinity, and is depressed below 0°C by the dissolved salts. At a salinity of 35‰ (see page 109), seawater freezes at approximately −1.91°C.

Excluding the shore and shallow water, the extreme temperature range between the hottest and coldest parts of the marine environment is therefore in the order of 30–35°C, but in any one place the range of temperature variation is always much less than this. In high and low latitudes, sea temperature remains fairly constant throughout the year. In middle latitudes, surface temperature varies with season in association with climatic changes. The range of seasonal temperature change depends upon locality, but is commonly about 10°C. Off the south-west coast of the British Isles, the temperature usually varies between about 7°C in winter and 16°C in summer, while off the north coast of Scotland the range is 4°C in winter to about 13°C in summer. The greatest seasonal variations of sea temperature are about 18–20°C, this range being recorded in the China Sea and Black Sea. Inland seas such as the Caspian also exhibit large ranges.

While surface water varies in temperature from place to place and time to time, the deep layers throughout the major ocean basins remain fairly constantly cold. The coldest water is at deep levels of the Arctic where the temperature is between 0 and −1.9°C. In the Atlantic, Pacific, Indian and Southern Oceans, the temperature of the bottom water lies between 0°C near Antarctica and 2–3°C at lower latitudes. Quite exceptional conditions are found in small pockets of deep water in zones of submarine volcanism along tectonic plate boundaries. Water at temperatures of between 200 and 360°C, spouts out from deep-sea hydrothermal vents (see page 237) warming the surrounding water to around 10–17°C. Within pits of the Red Sea floor remarkably high temperatures up to 56°C have been recorded in water of abnormally high salinity (up to nearly 300 parts per thousand) and unusual composition, rich in trace metals.

4.2.2 Thermoclines

In high latitudes, heat passes from the sea to the atmosphere. Surface cooling of the water produces convectional mixing, and there is, therefore, little difference in temperature between the surface and the deep layers. Through the whole depth of water the temperature range is usually within the limits of −1.8 to 1.8°C. There is often an irregular temperature gradient within the top 1000 m because the surface is diluted by fresh water from precipitation or melting ice. This forms a low-density layer of colder water above slightly warmer, but denser, water of higher salinity entering from middle latitudes (Figure 4.1a). Below 1000 m the temperature is almost uniform to the bottom, decreasing only slightly with depth.

At low latitudes, heat absorption at the sea surface produces a warm, light surface layer overlying the cold, denser, deep layers. Here the temperature gradient does not descend steadily but shows a distinct step, or *thermocline*, usually between about 100 and 500 m (Figure 4.1b), where temperature falls quite sharply with depth. This zone is termed a *discontinuity layer*. Above it, surface mixing maintains a fairly even warm temperature, a stratum referred to as the *thermosphere*. Below the thermocline is the *psychrosphere* where the water is cold, and there is only a slight further decrease of temperature towards the bottom. To a considerable extent the thermocline acts as a boundary between a warm-water population above and a cold-water population below.

In middle latitudes, the surface water becomes warm during the summer months and this leads to the formation of *temporary, seasonal thermoclines* near the surface, commonly around 15–40 m depth (Figure 4.1c). In winter, when the surface water cools, these temporary thermoclines disappear and convectional mixing may then extend to a depth of several hundred metres. Below the level to which convectional movements mix the water, there is usually a permanent but relatively slight thermocline between about 500 and 1500 m.

4.2.3 Temperature tolerances and biogeography

Water temperature exerts a major control over the distribution and activities of marine organisms (Kinne, 1963). Temperature tolerances differ widely between species, but each is restricted in distribution within its particular temperature range. Some species can only withstand a very small variation of temperature, and are described as *stenothermal*. *Eurythermal* species are those of wide temperature tolerance. Strict stenotherms are chiefly oceanic forms, and their distribution may alter seasonally with changes of water temperature. Eurytherms are typical of the more fluctuating conditions of shallow water. Sessile organisms have generally a rather wider temperature tolerance than free-living creatures of the same region.

Because water temperature has so great an effect on distribution, the extent of marine biogeographical regions can be related more closely to the course of the isotherms than to any other factor. The definition of biogeographic subdivisions of the sea is inevitably somewhat vague because the marine environment contains few firm ecological boundaries. Land barriers account for some differences between oceanic populations, and wide expanses of deep water prevent the spread of some littoral and neritic species; but for the most part the transition between one fauna and another is gradual, with a broad overlap of populations. However,

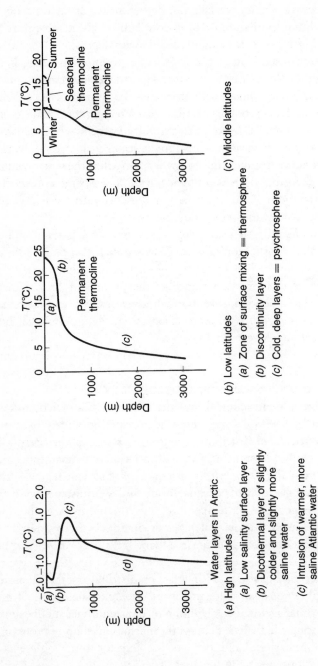

Figure 4.1 *Temperature profiles in the deep ocean.*

(a) High latitudes

Water layers in Arctic

(a) Low salinity surface layer
(b) Dicothermal layer of slightly colder and slightly more saline water
(c) Intrusion of warmer, more saline Atlantic water
(d) Arctic deep water

(b) Low latitudes

(a) Zone of surface mixing ≡ thermosphere
(b) Discontinuity layer
(c) Cold, deep layers ≡ psychrosphere

(c) Middle latitudes

in a general way the populations of the surface waters fall into three main groups associated with differences of water temperature: namely, the warm-water populations, the cold-water populations, and populations which inhabit waters of intermediate temperature where the temperature of the surface layers fluctuates seasonally, i.e. temperate waters. These major divisions of the marine population may be almost endlessly subdivided to take account of local conditions.

Warm-water populations are mainly to be found in the surface layers of the tropical belt where the surface temperature is above about 18–20°C (Figure 4.2). This warm-water zone corresponds roughly with, but is rather more extensive than, the zone of corals which have their main abundance in clear shallow water where the winter temperature does not fall below 20°C. Within the warm-water regions of the oceans there is little seasonal variation of temperature. At the Equator, the temperature of the surface water in most areas is between 26 and 27°C, and does not change appreciably throughout the year.

Cold-water populations are found in the Arctic and Southern Oceans where the surface temperature lies between about 5°C and a little below 0°C. In the Southern Ocean the cold water has a well-defined northern boundary at the Antarctic Convergence (see page 17) where it sinks below the warmer sub-Antarctic water. The sharp temperature gradient at this convergence effectively separates many species of plant and animal, and forms a distinct northern limit to the Antarctic faunal and floral zones. The southern boundary of the Arctic zone is less distinct except at the convergences of the Labrador Current and Gulf Stream in the Atlantic, and of the Oyo-Shiwo and Kuro-Shiwo currents in the Pacific. Broadly, the Arctic zone comprises the Arctic Ocean and those parts of the Atlantic and Pacific Oceans into which Arctic surface water spreads, the limiting temperature being a summer maximum of about 5°C.

The temperate sea areas lie between the 5 and 18°C mean annual surface isotherms, and here the surface water undergoes seasonal changes of temperature. The colder part of the temperate regions between the 5 and 10°C isotherms are termed the Boreal zone in the Northern hemisphere and the Antiboreal zone in the Southern hemisphere.

The course of the surface isotherms is determined largely by the surface circulation. On the western sides of the oceans the warmest water reaches higher latitudes, and the coldest water lower latitudes, than on the eastern sides. The temperate zones are therefore narrow in the west and much wider in the east, where they extend further to both north and south. On the basis mainly of water temperature we can designate some of the chief biogeographic subdivisions of the littoral and epipelagic zones as follows, their positions being shown in Figure 4.2.

1 Arctic and Subarctic regions.
2 East Asian Boreal region.

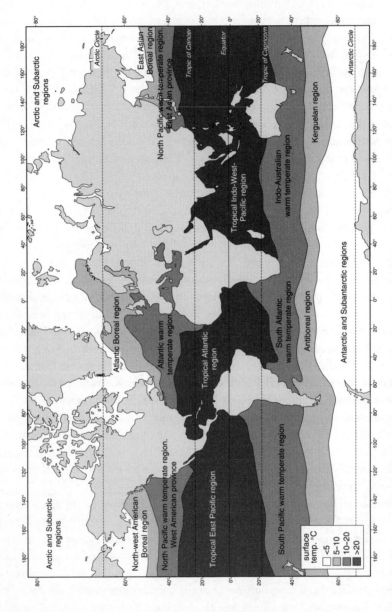

Figure 4.2   Approximate positions of mean annual isotherms and marine biogeographic areas.

3   North-west American Boreal region.
4   Atlantic Boreal region.
5   North Pacific warm temperate region. East Asian province.
6   North Pacific warm temperate region. West American province.
7   Atlantic warm temperate region (Lusitanian).
8   Tropical Indo-West-Pacific region.
9   Tropical East Pacific region.
10 Tropical Atlantic region.
11 South Pacific warm temperate region.
12 South Atlantic warm temperate region.
13 Indo-Australian warm temperate region.
14 Antiboreal region.
15 Kerguelen region.
16 Antarctic and Subantarctic regions.

There are some cases of the same species, or very closely related forms, occupying zones of similar temperature in middle or high latitudes in both northern and southern hemispheres, although absent from the intervening warm-water belt. Such a pattern of distribution is termed *bipolar*. The bipolar distribution of a pelagic amphipod *Parathemisto gaudichaudi* is shown in Figure 4.3, approximating to the distribution of surface water between 5 and 10°C. Among numerous examples of bipolarity, Ekman (1953) mentions the following inhabitants of the North-East Atlantic, the barnacle, *Semibalanus balanoides* (North Atlantic, Tierra del Fuego and New Zealand); the tunicates *Botryllus schlosseri* and *Didemnum albidum* (both North Atlantic and New Zealand); the genus *Engraulis* (anchovies) and the entire order Lucernariida. In some cases apparent bipolarity is really a continuous distribution through the colder layers of water underlying the warm surface layers of the tropics, i.e. tropical submergence. *Eukrohnia hamata* (Alvarino, 1965) (Figure 4.4), *Parathemisto abyssorum* and *Dimophyes arctica* are examples of amphipod species found at the surface in both Arctic and Antarctic waters, and present at deeper levels at low latitudes.

Two major biogeographical provinces meet around Great Britain: the Lusitanian to the south and the boreal, which is centred on the British Isles. The British Isles lie across the 10°C mean annual surface isotherm, and in winter the 5°C isotherm moves south along these coasts. It is therefore possible here, to distinguish certain species as belonging to a northern group of Arctic and Boreal forms, and others as a southern Lusitanian group of Mediterranean and temperate water species. There are seasonal changes in distribution and a broad overlap of populations, but the 10°C isotherm lies approximately between the two groups. Among the fishes of the area, the northern group includes cod (*Gadus morhua*), haddock (*Melanogrammus aeglefinus*), ling (*Molva molva*), plaice (*Pleuronectes platessa*), halibut (*Hippoglossus hippoglossus*), and herring (*Clupea*

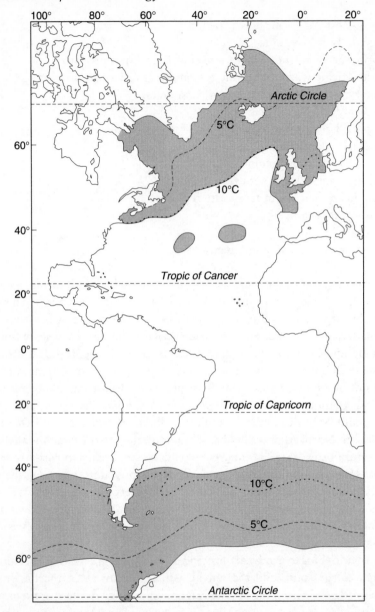

Figure 4.3   Approximate known distribution of Parathemisto gaudichaudi *in the Atlantic, and mean annual isotherms for 5°C and 10°C.*
(Based on a map of the world by courtesy of G. Philip and Son Ltd.)

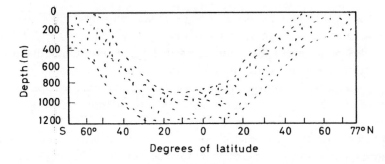

Figure 4.4   Distribution in depth of Eukrohnia hamata *in the Pacific, from the Bering Sea to MacMurdo Sound.*
(From Alvarino (1965) by courtesy of Allen & Unwin.)

harengus). Examples of southern forms are pollack (*Pollachius pollachius*), European hake (*Merluccius merluccius*), Dover sole (*Solea solea*), turbot (*Scophthalmus maximus*), pilchard (*Sardina pilchardus*), anchovy (*Engraulis encrasicolus*), mackerel (*Scomber scombrus*) and tunny (*Thunnus thynnus* and *T. alalunga*).

On the seashore almost all boreal species can occur all round the British coast but a few (e.g. the barnacle *Semibalanus balanoides*, the limpet *Acmaea tessulata*, the blenny *Zoarces viviparus* and the spindle shell *Neptunea antiqua*), which are common in the north and east, become scarce or absent towards the south-west. There are a larger number of species which are abundant in the south-west and often extend up the west coast, but are absent in the north and east, the British Isles being the northernmost limit of their range. These southern forms include the barnacles *Chthamalus montagui*, *C. stellatus*, and *Balanus perforatus*, the top shells *Monodonta lineata*, and *Gibbula umbilicalis*, the limpet *Patella depressa*, the snakelocks anemone *Anemonia sulcata*, the prawn *Leander serratus* and the cushionstar *Asterina gibbosa*.

Long-term changes in mean annual sea temperatures and short-term weather extremes can alter local patterns of distribution. For example, the distribution of *Monodonta lineata* eastwards along the coast of the English Channel appears to be affected by events such as the unusually cold winter of 1962–3. After about 1961, mean annual sea temperatures around the British Isles fell slightly compared with those of the previous 25 years. The distribution of many marine organisms correspondingly shifted slightly southwards. In the western part of the English Channel cold-water species such as cod, Norway pout, ling and herring became more numerous, whereas warm-water species, notably pilchards and hake, declined in numbers over the same period and tended to spawn later. Effects could also be seen on the relative abundance of some species of barnacles (see page 288). Around south-western shores, numbers of the southern *Chthamalus montagui* and *C. stellatus* steadily increased, along with sea temperature, until around 1960 (although the variation in sea temperature was only about 0.5°C in 50 years). After 1961 or so, mean sea temperatures fell again and the relative proportion of the boreal *Semibalanus balanoides* started to increase. In *S. balanoides*, high summer temperatures prevent final maturation of gametes and may also affect adult survival. Now sea temperatures in the North Atlantic seem to be rising again and we may see a further shift in these proportions.

Over the past 15 years or so, there has been a great increase in the number of sublittoral surveys using divers, around the coastlines of the UK and Ireland (see page 231). This has resulted in many new records of the northern and southern distributional limits of species as well as new UK records (Earll and Farnham, 1983; Erwin *et al.*, 1990; JNCC, 1995). During the summer in the UK, there are regular reports of warm-water fish species such as seahorses (*Hippocampus*) and trigger fish (*Balistes*) extending northwards into the English Channel. Sometimes these species may even overwinter here, but they cannot successfully breed. The effect of water temperature on breeding is one of the key factors influencing the distribution of marine organisms (see Section 4.2.4). The exceptionally hot summer of 1995 resulted in the appearance of many exotic species in UK waters. In Cornish waters, a big-eyed thresher shark (*Alopias superciliosus*) commonly found off Florida, and a sailfin dory were landed by fishing boats – the first records of these species in UK waters. Other visitors included mako and hammerhead sharks in the Irish Sea, bluefin tuna in Shetland and sunfish in the North Sea. Sea temperatures in the North Atlantic are now rising and such tropical species could become regular visitors if the warming continues. The Marine Biological Association in Plymouth (UK) has established a British Marine Fishes Database. It is currently being used as a tool for monitoring the distribution and abundance of uncommon species in response to environmental change.

Whereas the distribution of many littoral, sublittoral and epipelagic species is fairly fully recorded, knowledge of the distribution of species at deep levels is very incomplete. It is becoming apparent that there is greater diversity of abyssal species than was earlier thought, and that some are relatively restricted in distribution. However, the limits of abyssal zoogeographic regions are even less clearly defined than those of the surface layers. The most distinct deep-level boundary is probably the system of submarine ridges separating the Arctic basin from the North Atlantic which, together with the shallow water of the Bering Strait, form a barrier which some abyssal species do not cross. North of the North Atlantic Ridge much of the bottom water of the Arctic is colder than 0°C, while throughout the rest of the abyss almost all the water lies between 0 and 4°C. Relatively few species appear to be common to the bottom of both Arctic and other deep oceans. The deep levels of Atlantic, Indian and Pacific Oceans are all connected by the deep water of the Southern Ocean, and throughout this vast extent of abyss many species are widely distributed.

4.2.4 Physiology

Except for sea birds and mammals, marine organisms are poikilothermic, i.e. their body temperature is always close to that of the surrounding water and varies accordingly. In the coldest parts of the sea, where temperatures are close to −2.0°C, the blood of fishes would be below freezing point but for the antifreeze action of relatively high concentrations of glycoprotein in the plasma. However, some large and active forms have body temperatures higher than water temperature due to the release of heat by metabolism. In fast-swimming sharks and tuna the temperature in the swimming muscles is sometimes 10°C above water temperature.

The physiological effects of temperature change are complex but, in simple terms, rates of metabolic processes increase with rising temperature, usually about 10 per cent per 1°C rise over a range of temperatures up to a maximum (Figure 4.5) beyond which they fall off rapidly. Death occurs above and below certain limiting temperatures, probably because of disturbances of enzyme activity, water balance and other aspects of cellular chemistry. Marine creatures usually succumb more rapidly to overheating than to overcooling. The limits of distribution of a species in the sea do not coincide closely with the normal occurrences of rapidly lethal temperatures but are much more restricted. In freak climatic conditions, extremes of heat or cold may have rapid and devastating effects on marine populations, especially those of the shore, but in normal circumstances temperature probably controls distribution in subtle and gradual ways through its influence on several major processes including feeding, respiration, osmo-regulation, growth and reproduction, especially the latter.

Temperature regulates reproduction in several ways. It controls the maturation of gonads and the release of sperms and ova, and in many cases the temperature

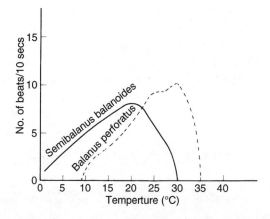

Figure 4.5   Relation of rate of ciliary beat to temperature in two species of barnacle.

tolerance of embryonic and larval stages is less than that of the adults. Temperature has therefore a major influence on the breeding range and period, and on mortality rates during early stages of development and larval life. Along the fringes of distribution there are usually non-breeding zones where the adults can survive but cannot reproduce, the population being maintained by spread from the main area of distribution within which breeding is possible. Several Lusitanian species which are quite common along south and west coasts of the British Isles, e.g. the crawfish *Palinurus vulgaris*, probably seldom if ever breed in these waters. They are carried into the area from the south as larvae which successfully metamorphose and complete their development here.

In temperate seas many species virtually cease feeding during the winter. In some cases reduced feeding is simply the result of shortage of food, but many creatures definitely stop eating below a certain temperature. Food requirements are reduced during cold periods because the respiration rate is low and growth ceases. These interruptions of growth may produce periodic markings in growing structures; for instance, the annual winter rings on fish scales (see page 363).

Despite the depressing effects of cold on growth, it is nevertheless generally observed that where the distribution of a marine species covers a wide range of temperature the individuals living in colder areas attain larger adult sizes than those in the warmer parts of the distribution. This trend is associated with a longer growing period, later sexual maturation and a longer life in cold water. There are exceptions to this trend, and some species reach larger sizes in warmer water e.g. the gastropod *Urosalpinx cinerea* and the sea urchin *Echinus esculentus*.

Apart from direct physiological effects, changes of temperature have certain indirect effects by altering some of the physical properties of the water, notably density, viscosity and the solubility of gases, which in turn influence buoyancy, locomotion and respiration. There are instances e.g. the summer and winter forms of the diatom *Rhizosolenia hebetata* (Figure 4.6), where the morphology of a species appears to vary with changes of temperature, possibly because of alterations in viscosity and buoyancy. The viscosity of water falls considerably with increasing temperature, which may partly account for the increased setation of the appendages of many warm-water planktonts as compared with cold-water forms. The greater surface area of finely divided appendages increases their floating ability.

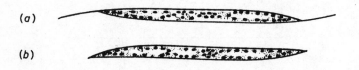

Figure 4.6 Rhizosolenia hebetata *in* (a) *summer and* (b) *winter forms.*

4.3 The composition of seawater

Seawater is an extremely complex solution, its composition being determined by an equilibrium between rates of addition and loss of solutes, evaporation and the addition of fresh water. The original source of seawater is uncertain, but was probably by condensation of water vapour and solutes released into the atmosphere from hot rocks and volcanic action at an early stage of the Earth's history. At the present time many constituents of seawater are continually added from various sources; for instance, in 'juvenile water' released from basalts which flow into the sea floor along the separating boundaries of the Earth's crustal plates (see page 7), in volcanic gases escaping into both oceans and atmosphere and in processes of weathering and erosion of the Earth's surface. Loss of solutes from the water occurs by precipitation on the bottom (Glasby, 1973; MacIntyre, 1970). Short-term, minor fluctuations of composition occur through biological processes involving absorption and release of solutes by organisms and detritus. There are also interchanges of gases between sea and atmosphere.

4.3.1 Major constituents and salinity

It is uncertain to what extent the composition of seawater may have changed during geological time, but it is not thought to have varied very widely over the period that life has existed. At present, the principal cations are sodium, magnesium, calcium, potassium and strontium, and the chief anions are chloride, sulphate, bromide and bicarbonate. These make up over 99.9 per cent of the dissolved material, forming approximately a 3.5 per cent solution. The amount of inorganic material dissolved in seawater expressed as weight in grams per kilogram of seawater is termed the salinity (S), and usually amounts to about 35 g/kg, i.e. S = 35 parts per thousand (generally written 35‰). The quantities of the major constituents of a typical sample of ocean water are shown in Table 4.1.

The relative proportions of the major ionic constituents in ocean water remain virtually constant despite some variation in total salinity. Estimation of the concentration of any of these ions therefore enables the total salinity to be calculated. Salinity determinations may be made by titrating seawater with silver nitrate solution. This precipitates the halides, mainly chloride with a trace of bromide, and their total weight in grams per kilogram of seawater is termed the chlorinity, *Cl*. The salinity is then determined from the empirical relationship known as Knudsen's formula:

$$S‰ = 0.030 + (1.805 \times Cl‰)$$

A convenient method is to titrate 10 ml of seawater with silver nitrate solution containing 27.25 g/ℓ, using a chromate or fluorescein indicator. The added volume of silver nitrate in millilitres is approximately equal to the salinity in grams per

Table 4.1 Major constituents of an ocean water. $S = 35.00‰$

Constituent	g/kg
Sodium	10.77
Magnesium	1.30
Calcium	0.412
Potassium	0.399
Strontium	0.008
Chloride	19.34
Sulphate as SO_4	2.71
Bromide	0.067
Carbon, present as bicarbonate, carbonate and molecular carbon dioxide	0.023 at pH 8.4 to 0.027 at pH 7.8

kilogram, and a small correction is made from tables to allow for the slight differences in weight of unit volume of seawater at different salinities.

Greater accuracy is obtained by using 'standard seawater' for comparison. This is water of very accurately known chlorinity available from specialized suppliers. By comparing the titrations of silver nitrate against both sample and standard seawater the calculation of chlorinity becomes independent of the concentration of silver nitrate solution, and all measurements are made to the same standard.

There are various objections to using chlorinity measurements as a basis for all determinations of salinity. It assumes a constant ratio between chlorinity and total amount of dissolved material, which obviously cannot be true for all dilutions of seawater with other waters of differing compositions. Also, silver nitrate is an expensive reagent, and titration is a relatively time-consuming technique. Consequently, attention has turned to other methods of salinity measurement. Several physical properties of seawater vary with the amount of dissolved salts and can be used for salinity determination (see Section 3.1.2); for example, electrical conductivity, density, vapour pressure, freezing point, refractive index and sound conductivity (Grasshof, 1976; Johnston, 1969).

The electrical conductivity of seawater increases the more ions it contains. This makes it possible quickly and directly to measure salinity using an instrument called a salinometer. An electrical probe can be lowered into the water to the required depth and a direct reading taken. Since it is actually the conductivity that is being measured, the instrument must be calibrated for temperature if it is to read out directly in salinity units. Salinometers are of great value in areas such as enclosed sea lochs where there may be a halocline – a sudden change in salinity with depth. These instruments should be periodically tested and calibrated by comparing the readout with samples of the same water for which the salinity has been determined chemically.

The salinity of most ocean water is within the range 34–36‰. There are slight seasonal variations of salinity, and average positions for the surface isohalines during the northern summer are shown in Figure 4.7. High salinities are associated with low rainfall and rapid evaporation, especially where the circulation of the water is relatively poor. Such conditions are found in the Sargasso area of the North Atlantic and in the South Atlantic off the east coast of Brazil, where the surface salinities are about 37‰. In high latitudes, the melting of ice, heavy precipitation and land drainage together with low evaporation reduce the salinity of the surface water. In the Arctic, the surface salinity fluctuates between 28 and 33.5‰ with alternate melting and freezing of ice.

In land-locked areas there are appreciable departures from the normal oceanic range of salinities. For instance, in the Baltic, dilution by fresh water reduces the salinity from 29‰ in the Kattegat region to below 5‰ in the Gulf of Bothnia. In the Black Sea, rainfall and the outflow of the Danube, Dnieper and Dniester lower the surface salinity to 18‰ or below. This low-salinity water forms a low-density layer overlying the more saline, deep layers with little mixing between them, and cuts off the depths of the Black Sea from the air, producing the peculiar hydrographic conditions mentioned later (see page 114). In hot regions, high surface salinities are found in enclosed seas due to rapid evaporation. Throughout most of the Mediterranean surface salinities are above 37‰, increasing from west to east and reaching about 39‰ in the eastern part. In the Red Sea, surface salinities may exceed 40‰. On the shore the salinity of evaporating pools is sometimes greater than 100‰. Peculiar salinities occur in deep-sea pits at tectonic plate boundaries (see page 98).

The salinity of neritic water is subject to fluctuation due to changes in the rate of dilution by fresh water from the land. River water often contains ions in very different proportions to those of normal seawater, and this may produce appreciable changes in the composition of seawater near a river mouth.

Except for the teleosts and higher vertebrates the majority of marine creatures are in osmotic equilibrium with the surrounding water. The ionic composition of their internal fluids has, in most cases, a close similarity to that of seawater, containing relatively high concentrations of sodium and chloride and con-siderably lower concentrations of potassium, magnesium and sulphate. There is commonly, though not invariably, a rather higher proportion of potassium to sodium in body fluids than that which occurs in seawater, and somewhat less magnesium and sulphate (see Table 4.2).

External salinity changes usually produce corresponding changes in the concentration of internal fluids by passage of water into, or out of, the body (osmotic adjustment) to preserve the osmotic equilibrium, and these changes are often accompanied by alterations in the proportions of the constituent ions of the internal fluids. Beyond limits, which differ for different species, departures from the normal concentration and composition of the internal medium cause metabolic disturbances and eventual death.

Figure 4.7 Approximate positions of mean annual isohalines. Salinities in ‰.

Table 4.2 Concentrations of Ions in Body Fluids of some Marine Invertebrates (g/kg)

	Na	K	Ca	Mg	Cl	SO$_4$
Seawater (*S‰* = 34.3)	10.6	0.38	0.40	1.27	19.0	2.65
Aurelia aurita	10.2	0.41	0.39	1.23	19.6	1.46
Arenicola marina	10.6	0.39	0.40	1.27	18.9	2.44
Carcinus maenas	11.8	0.47	0.52	0.45	19.0	1.52
Mytilus edulis	11.5	0.49	0.50	1.35	20.8	2.94
Phallusia mammillata	10.7	0.40	0.38	1.28	20.2	1.42

The majority of organisms of the open sea have very limited tolerance of salinity change, i.e. they are *stenohaline*. *Euryhaline* forms which can withstand wider fluctuations of salinity are typical of the less stable conditions of coastal water (Kinne, 1963, 1964). Extreme euryhalinity characterizes estuarine species.

Organisms which remain in osmotic balance with their surroundings when the salinity varies are termed *poikilosmotic,* and these include some widely euryhaline creatures. The lugworm *Arenicola marina* is a familiar example from the British coastline, where it is widely distributed in marine, brackish and estuarine muddy sands and able to survive salinities down to about 18‰. In other parts of its range, for example the Baltic, it is found at even lower salinities. Other examples from the British fauna which are poikilosmotic and moderately euryhaline are the bivalves *Mytilus edulis, Cerastoderma (Cardium) edule,* and *Mya areanaria,* the barnacles *Semibalanus balanoides* and *Balanus improvisus,* the polychaete worms *Hediste (Nereis) pelagica* and *Perinereis cultrifera* and many other common shore forms.

Some animals are able to control within limits the concentration of their internal fluids independently of salinity changes in the water. This process is known as osmoregulation, and organisms which maintain this stability of internal environment are described as *homoiosmotic.* The shore crab *Carcinus maenas* is a very euryhaline osmoregulator which extends up estuaries to levels where it encounters immersion in fresh water. Some powers of osmoregulation are also present in the ragworm *Nereis diversicolor,* the prawn *Palaemon serratus,* and the amphipods *Gammarus locusta, G. duebeni* and *Marinogammarus* (= *Chaetogammarus) marinus.* The ability to osmoregulate is influenced by temperature and fails above and below certain limiting temperatures.

In marine teleost fish the concentration of salts in their internal fluids is lower than in seawater, so water tends to pass out of their tissues by osmosis. To counteract this water loss and maintain a correct water balance the fish swallow seawater and absorb it through the gut. The excess salts, and much of their excretory nitrogenous products, are eliminated by special secretory cells in the gill membranes. The kidneys of many marine teleosts have a much reduced number of glomeruli, or glomeruli may be absent. Urine is produced in small quantity and is nearly isotonic with the blood. Waste nitrogen in the urine of teleosts is excreted mainly as trimethylamine oxide in substitution for ammonia, which is the chief nitrogenous end-product of the majority of aquatic organisms. Excretion of ammonia, which is highly toxic compared with trimethylamine oxide, requires a copious, very dilute urine. Replacement of ammonia by trimethylamine oxide in the urine of marine teleosts is a useful adaptation for conserving water.

Vascular plants growing on the seashore are exposed to a very different environment to that of other terrestrial plants. Compared with normal soil water, the concentration of salts is much higher and the ionic composition of the water quite different. Almost all halophytes have adapted to these conditions by increasing the intracellular concentration of their tissues sufficiently to be able to take in water by osmosis, and by selective control of ion absorption.

Change of salinity alters the specific gravity of the water, and this influences pelagic organisms indirectly through its effects on buoyancy.

4.3.2   Dissolved gases
All atmospheric gases, including the inert gases, are present in solution in seawater.

Oxygen
The oxygen content of seawater varies between 0 and 8.5 ml/ℓ, mainly within the range 1–6 ml/ℓ. High values occur at the surface, where dissolved oxygen tends to equilibrate with atmospheric oxygen. Rapid photosynthesis may sometimes produce supersaturation. Because oxygen is more soluble in cold water than in warm, the oxygen content of surface water is usually greater at high latitudes than nearer the Equator, and the sinking of cold surface water in polar seas carries oxygen-rich water to the bottom of the deep ocean basins.

Although the deep layers of water are mostly well oxygenated, oxygen is by no means uniformly distributed with depth, and in some areas there is an *oxygen-minimum layer* at a depth somewhere between 100 and 1000 m. This is most evident in low latitudes, where the water at 100–500 m has sometimes been found to be almost completely lacking in oxygen. The reasons for this are uncertain, but the oxygen-minimum zone often appears to be well populated, and one cause of the deficiency of oxygen may be depletion by a large amount of animal and bacterial respiration in water where relatively little circulation is taking place.

The exceptional conditions in the Black Sea were mentioned earlier. Cut off from the Mediterranean by the shallow water of the Bosphorus, there is little mixing between the low density surface water (see page 111) and the denser, more saline deep water. The deep levels are virtually stagnant and have become completely depleted of oxygen. Animal life is impossible below some 150–200 m, but anaerobic bacteria flourish in the deep layers, mainly sulphur bacteria which metabolize sulphate to sulphide and produce large quantities of H_2S, giving the deep water a very objectionable smell. Comparable conditions sometimes arise in other land-locked areas of deficient circulation, such as deep lochs, fiords and lagoons.

Carbon dioxide

The preponderance in seawater of the strongly basic ions, sodium, potassium and calcium, imparts a slight alkalinity and enables a considerable amount of carbon dioxide to be contained in solution. This is of great biological importance because carbon dioxide is a raw material for photosynthesis. Under natural conditions, plant growth in the sea is probably never limited by shortage of carbon dioxide.

Carbon dioxide is present in seawater mainly as bicarbonate ions, but there are also some dissolved CO_2, undissociated H_2CO_3 and carbonate ions. At the surface, dissolved CO_2 tends towards equilibrium with atmospheric CO_2, the oceans acting as a regulator of the amount of CO_2 in the atmosphere. The overall equilibrium can be represented as follows:

$$\text{Atmospheric } CO_2$$
$$\Updownarrow$$
$$\text{Dissolved } CO_2 \rightleftharpoons H_2CO_3 \rightleftharpoons H^+ + HCO_3^- \rightleftharpoons H^+ + CO_3^{2-}$$

The role of the oceans and especially of coral reefs in reducing the amount of atmospheric carbon dioxide and hence helping to control global warming is discussed in Chapter 10.

The pH of seawater normally lies within the range 7.5–8.4, the higher values occurring in the surface layer where CO_2 is withdrawn by photosynthesis. The presence of strong bases together with the weak acids H_2CO_3 and H_3BO_3 confers an appreciable buffer capacity. Addition of acid to seawater depresses the dissociation of H_2CO_3 and H_3BO_3, and there is not much change of pH while reserves of CO_3^{2-}, HCO_3^- and $H_2CO_3^-$ ions remain. Addition of alkali increases the dissociation of H_2CO_3 and H_3BO_3, and the pH remains fairly stable so long as undissociated acid is still present.

The dissociation constants of the equilibrium are influenced by temperature, pressure and salinity. Increase of temperature or pressure causes a slight decrease of pH. At great depths the lowering of pH due to pressure may be sufficient to cause solution of some forms of calcium carbonate, which is not a conspicuous component of sediments below about 6000 m (Pytkowicz, 1968).

Nitrogen

Amounts of uncombined nitrogen in seawater vary between 8.4 and 14.5 ml/ℓ.

Nitrogen-fixing bacteria are known to occur in the sea, but the quantity of nitrates formed by their activity is probably very small. There is also some return of nitrogen from the oceans to the atmosphere by the nitrogen-freeing activity of denitrifying bacteria and blue-green algae. With increasing quantities of atmospheric nitrogen being fixed by industrial processes for fertilizers, the biological freeing of nitrogen from nitrate becomes of increasing importance in maintaining the equilibrium of the nitrogen cycle.

4.3.3 Minor constituents

In addition to the major constituents listed in Table 4.1 there are many other elements present in seawater in very small amounts (Table 4.3). The most abundant of the ionized minor constituents are silicate ions at concentrations up to 6 mg/kg, and fluoride ions up to 1.4 mg/kg. The combined weights of all the other minor constituents, numbering nearly fifty, total less than 2 mg/kg, and at this dilution the estimation of many of them is very difficult. Probably all natural elements occur in seawater, though some at infinitesimal concentrations. Several are known to be present mainly because they are concentrated in the bodies of marine organisms. Details of many of the laboratory procedures for determining biologically significant constituents in seawater can be found in references at the end of this chapter (Grasshof, 1976).

Few marine organisms survive for long in an artificial seawater which contains only the major constituents in correct proportion. The minor constituents are evidently of biological importance although in many cases their role is uncertain. Some are known to be essential for the normal growth of plants; for example nitrate, phosphate, iron, manganese, zinc, copper and cobalt. Silicon is an ingredient of diatoms, and some marine algae require molybdenum and vanadium. Many of the minor constituents are also necessary for animal life. Silicon is included in the spicules of most radiolaria and some sponges. Iron is required by all animals. Copper is present in the prosthetic group of the blood pigment haemocyanin which occurs in some molluscs and crustacea. Vanadium and niobium occur in the blood pigment of ascidians. The vertebrate hormone thyroxin is an iodine compound.

Certain organisms concentrate the minor constituents to a remarkable extent. Vanadium in ascidians is an outstanding example, occurring in some species at concentrations approximately a million times greater than in seawater. Iodine, nickel, molybdenum, arsenic, zinc, vanadium, titanium, chromium and strontium are concentrated in the tissues of various marine algae, and some fish concentrate silver, chromium, nickel, tin or zinc. Certain of the heavy metals appear to be essential for normal enzyme activity, notably copper, though toxic at abnormal concentrations.

Whereas the major constituents of seawater, and some of the minor constituents, remain virtually constant in proportion (conservative constituents), certain minor constituents fluctuate in amount due to selective absorption by organisms (non-conservative constituents). The latter include nitrate, phosphate, silicate, iron and manganese, and the list will probably increase as our knowledge of the requirements of marine organisms grows.

Nitrate and phosphate

Nitrogen in combined form is present in seawater as nitrate, nitrite, ammonium ions and traces of nitrogen-containing organic compounds. Nitrate ions predominate, but in the uppermost 100 m and also close to the bottom there are sometimes appreciable amounts of ammonium and nitrite formed by biological activity.

Phosphorus is present almost entirely as orthophosphate ions $H_2PO_4^-$ and HPO_4^{2-} with traces of organic phosphorus. The concentrations of these combined forms of nitrogen and phosphorus generally fall within the following wide ranges:

$NO_3\text{-}N = 1\text{--}600\,\mu g/\ell$ $(0.1\text{--}43\,\mu g$ atoms $N/\ell)$
$NO_2\text{-}N = 0\text{--}15\,\mu g/\ell$
$NH_3\text{-}N = 0.4\text{--}50\,\mu g/\ell$
Organic-N $= 30\text{--}200\,\mu g/\ell$
Phosphate-P $= <1\text{--}100\,\mu g/\ell$ $(0.01\text{--}3.5\,\mu g$ atoms $P/\ell)$
Organic-P $= <1\text{--}30\,\mu g/\ell$

Nitrogen and phosphorus are important requirements for plants. Together with other essential trace constituents they are commonly referred to as 'nutrients', and the amount of plant growth in the sea is largely controlled by the availability of nutrients in the surface layers.

Quantities of nitrate and phosphate vary greatly with depth. They are generally low and variable at the surface, reflecting the uptake of these ions by plants. Surface values usually fall within the range $1\text{--}120\,\mu g/\ell$ $NO_3\text{-}N$ and $0\text{--}20\,\mu g/\ell$ phosphate-P, with highest values in winter and lowest in summer. At deeper levels the concentration of nutrients is considerably greater, and where there are high values in the surface layers they are generally due to admixture with water from below by convectional mixing or upwelling. Near the coast, high values often occur due to the stirring up of bottom sediments or to large amounts of nitrate and phosphate present in some river water.

In deep water below the level of surface mixing there is usually a gradient of increasing concentration with depth, and a zone of maximum concentration between about 500 and 1500 m, with quantities in the range $200\text{--}550\,\mu g/\ell$ $NO_3\text{-}N$ and $40\text{--}80\,\mu g/\ell$ phosphate-P. Below this there may be a slight decrease but values remain high and fairly uniform. There may be some increase close to

Table 4.3 Geochemical Parameters of Seawater. (From Hill, 1963 by courtesy of Interscience)

Element	Abundance (mg/ℓ)	Principal species	Residence time* (years)
H	108 000	H_2O	
He	0.000005	He(g)	
Li	0.17	Li^+	2.0×10^7
Be	0.0000006		1.5×10^2
B	4.6	$B(OH)_3$; $B(OH)_2O^-$	
C	28	HCO_3^-; H_2CO_3; CO_3^{2-}; organic compounds	
N	0.5	NO_3^-; NO_2^-; NH_4^+; N_2(g); organic compounds	
O	857 000	H_2O; O_2(g);SO_4^{2-} and other anions	
F	1.3	F^-	
Ne	0.0001	Ne(g)	
Na	10 500	Na^+	2.6×10^8
Mg	1 350	Mg^{2+}; $MgSO_4$	4.5×10^7
Al	0.01		1.0×10^2
Si	3	$Si(OH)_4$; $Si(OH)_3O^-$	8.0×10^3
P	0.07	HPO_4^{2-}; $H_2PO_4^-$; PO_4^{3-}; H_3PO_4	
S	885	SO_4^{2-}	
Cl	19 000	Cl^-	
A	0.6	A(g)	
K	380	K^+	1.1×10^7
Ca	400	Ca^{2+}; $CaSO_4$	8.0×10^6
Sc	0.00004		5.6×10^3
Ti	0.001		1.6×10^2
V	0.002	$VO_2(OH)_3^{2-}$	1.0×10^4
Cr	0.00005		3.5×10^2
Mn	0.002	Mn^{2+}; $MnSO_4$	1.4×10^3
Fe	0.01	$Fe(OH)_3$(s)	1.4×10^2
Co	0.0005	Co^{2+}; $CoSO_4$	1.8×10^4
Ni	0.002	Ni^{2+}; $NiSO_4$	1.8×10^4
Cu	0.003	Cu^{2+}; $CuSO_4$	5.0×10^4
Zn	0.01	Zn^{2+}; $ZnSO_4$	1.8×10^5
Ga	0.00003		1.4×10^3
Ge	0.00007	$Ge(OH)_4$; $Ge(OH)_3O^-$	7.0×10^3
As	0.003	$HAsO_4^{2-}$; $H_2AsO_4^-$; H_3AsO_4; H_3AsO_3	
Se	0.004	SeO_4^{2-}	
Br	65	Br^-	
Kr	0.0003	Kr(g)	
Rb	0.12	Rb^+	2.7×10^5
Sr	8	Sr^{2+}; $SrSO_4$	1.9×10^7
Y	0.0003		7.5×10^3
Nb	0.00001		3.0×10^2
Mo	0.01	MoO_4^{2-}	5.0×10^5
Ag	0.0003	$AgCl_2^-$; $AgCl_3^{2-}$	2.1×10^6
Cd	0.00011	Cd^{2+}; $CdSO_4$	5.0×10^5
In	<0.02		
Sn	0.003		5.0×10^5
Sb	0.0005		3.5×10^5
I	0.06	IO_3^-; I^-	

Table 4.3. (*cont.*)

Element	Abundance (mg/ℓ)	Principal species	Residence time* (years)
Xe	0.0001	Xe(g)	
Cs	0.0005	Cs$^+$	4.0×10^4
Ba	0.03	Ba^{2+}; BaSO$_4$	8.4×10^4
La	0.0003		1.1×10^4
Ce	0.0004		6.1×10^3
W	0.0001	WO$_4{}^{2-}$	1.0×10^3
Au	0.000004	AuC$_{l4}^-$	5.6×10^5
Hg	0.00003	HgCl$_3^-$; HgCl$_4{}^{2-}$	4.2×10^4
Tl	<0.00001	Tl$^+$	
Pb	0.00003	Pb^{2+}; PbSO$_4$	2.0×10^3
Bi	0.00002		4.5×10^5
Rn	0.6×10^{-15}	Rn(g)	
Ra	1.0×10^{-10}	Ra^{2+}; RaSO$_4$	
Th	0.00005		3.5×10^2
Pa	2.0×10^{-9}		
U	0.003	UO$_2$(CO$_3$)$_3{}^{4-}$	5.0×10^5

*Assuming a steady state where the concentration of a constituent is constant in the long term, i.e. that the rates of addition and removal are equal, the *residence time* of a constituent is its total quantity in the oceans divided by the rate of either addition or loss.

the sea-bed due to release of nutrients by bacterial decomposition of organic matter deposited on the bottom.

Despite fluctuations in total amount, the relative concentrations of nitrate and phosphate remain fairly constant, the nitrate:phosphate ratio usually being about 7:1 by weight and 15:1 by ions. This close relationship indicates that the two ions are probably absorbed by living organisms, and subsequently released, in much the same proportions as they are present in the water.

Silicate

Silicon is present in seawater chiefly as silicate ions and possibly sometimes minute traces of colloidal silica. It is a constituent of the diatom cell wall, some radiolarian skeletons and some sponge spicules.

The concentration of silicate at the surface is usually low, but increases with depth to between 1 and 5 mg Si/ℓ in deep ocean water, the highest values being in the deep Pacific. In the English Channel, surface concentrations of 200–400 µg Si/ℓ have been recorded during the winter, falling rapidly to as low as 10 µg/ℓ during the spring diatom peak. During the summer months, surface concentrations often show rapid and considerable fluctuations.

Although much of the silicate incorporated in the diatom cell wall is probably returned to the water fairly quickly after death, siliceous deposits of planktonic origin cover large areas of the sea-bed (see page 216).

Iron and manganese

Ferric hydroxide is almost insoluble within the pH range of seawater. The amount of iron in true solution is probably not more than $2\ \mu g$ Fe/ℓ, but there are appreciable quantities of iron in particulate form as colloidal micelles, mainly ferric hydroxide and traces of ferric phosphate, ferricitrate or haematin. This particulate iron can be removed from seawater by ultrafiltration.

Estimates of total iron vary considerably with time, place, and different techniques of measurement, generally within the limits of 3–$70\ \mu g$ Fe/ℓ. Inshore water usually contains much more iron than oceanic water, especially in the vicinity of estuaries where river water often transports relatively large quantities of both dissolved and particulate iron. In the English Channel, fluctuations from about 3 to $150\ \mu g$ Fe/ℓ have been recorded, high values being obtained at the surface and near the bottom. Surface values usually show marked seasonal reductions following peak periods of diatom growth.

There is probably a continual loss of iron from seawater, and accumulation at the bottom, due to adsorption on sinking detritus and sedimentation. Iron is an essential plant nutrient, and also has various roles in animal physiology. It is a component of the cytochrome enzyme system and the blood pigments haemoglobin (vertebrates, some annelids and some molluscs), haemerythrin (some molluscs and crustacea) and chlorocruorin (some annelids). The amount of iron in solution seems inadequate to support rapid plant growth, and it is possible that marine plants can utilize particulate iron in some way, perhaps by gradual solution of particles adsorbed on the cell wall, or even by actual ingestion by certain plants which have exposed protoplasm.

Manganese is a plant nutrient which, like iron, is probably present mainly in particulate form as oxide micelles, in amounts between 0.3 and $10\ \mu g$ Mn/ℓ. In deep water the surface layers may become depleted of particulate and adsorbed iron and manganese by losses through sinking, and this may limit the amount of plant growth that can be supported. There is often more iron and manganese in neritic water due to replenishment by land drainage. In experimental cultures, enrichment of samples of ocean water by addition of iron and manganese sometimes results in a considerably increased growth. Recent experiments in which iron sulphate was added directly to areas in the Pacific where phytoplankton biomass was low, demonstrated greatly increased phytoplankton growth (see Section 9.6.4).

Dissolved organic matter (DOM)

Particulate organic matter is always present in seawater, but in addition, varying quantities of organic compounds are present in solution. This is referred to as dissolved organic matter (DOM) or dissolved organic carbon (DOC) (Jorgensen, 1976; Williams, 1975). The estimation of minute quantities of organic solutes is difficult, but it appears that ocean water commonly contains about 2 mg carbon per litre in dissolved organic forms, in some of which nitrogen, phosphorus, sulphur, iron or cobalt are included. However, recent analyses suggest that the amounts may be higher than this (Toggweiler, 1988). Although these concentrations may seem small, the total quantity in the ocean is very large, and it has been calculated that there are on average about 15 kg of DOM beneath each square metre of the ocean surface. This greatly exceeds the amount of organic matter present in particulate form as either living material or organic debris. The yellow colour sometimes detectable in seawater derives mainly from various organic compounds in solution.

These solutes evidently come from a variety of sources. They are predominantly of biological origin. Some enter the seas via land drainage, some on airborne particles or vapours. The major contribution must certainly derive from the activities of marine organisms in metabolic processes of photosynthesis, feeding, excretion, breakdown of dead tissues, etc. Possibly traces of organic matter are leached from living tissues, and some organisms are known to liberate organic secretions into the water as part of their normal metabolism. Such external metabolites are termed exocrines. In recent years, man-made organic compounds have been produced in increasing amounts, and many of these eventually find their way into the sea (see Chapter 10).

In temperate waters the total quantity of DOM shows seasonal changes with a marked increase following the spring phytoplankton bloom (see page 195), giving high summer values declining to a minimum in later winter. During summer, greater amounts of DOM are found above the thermocline than below it. It is now certain that a significant fraction of primary production is quickly released via various routes as DOM. Some of this material may leach from plant cells, some may be lost during cell division, and much may be routed through the feeding, excretion and egestion of animals. Some zooplanktonts grazing on phytoplankton liberate into the water appreciable quantities of DOM from plant cells by breaking up the frustules prior to ingestion, so-called 'sloppy feeding'.

There is little precise information regarding the quantities and chemical nature of these compounds, but a growing list of substances identified in seawater includes various hydrocarbons, carbohydrates, urea, aminoacids, organic pigments, lipids, alcohols, and vitamins such as ascorbic acid and components of the vitamin B complex, e.g. thiamin and cobalamin.

The way in which organic solutes influence the biological properties of seawater is a matter for speculation, but it is clear that at least some are absorbed by organisms and may have various effects. Artificial seawater made from all known inorganic constituents including the trace elements does not have the same biological quality as natural seawater, and organisms do not thrive in it. If, however, small quantities of organic material are added in such forms as soil extracts, urine, extracts of various tissues or even small volumes of natural seawater, the quality of the solution as a satisfactory medium for marine life is greatly enhanced. The fact that the concentrations of organic solutes in natural seawater are low suggests that they may be continuously absorbed from the water by organisms. The importance of DOM in the organic food cycle in the sea, has only recently been appreciated. The role of DOM in the so-called 'microbial loop' is described in Section 5.1.2.

It is now clear that certain phytoplankton species readily utilize dissolved organic nitrogen (DON) and phosphorus (DOP) compounds, indeed some species may even thrive better on organic solutes than inorganic. Although in temperate waters the concentrations of nitrate and phosphate in the surface layers become markedly depleted following the spring growth of diatoms, with generally low amounts continuing through the summer, the total dissolved nitrogen and phosphorus in both inorganic and organic forms undergoes no corresponding reduction. Evidently many of the inorganic nutrients taken in by the phytoplankton reappear as DOM, and are thus available to maintain production of species which absorb nitrogen and phosphorus in organic form. The seasonal succession of phytoplankton species which occurs during the productive period may partly reflect the seasonal changes in DOM, with spring species flourishing on inorganic nutrients and later dominants utilizing organic solutes which occur in the water following the spring bloom. This may also explain why measurements of production during the summer sometimes indicate higher rates than would seem to be consistent with the low concentrations of organic nutrients if these were limiting factors.

There is also evidence that certain forms of DOM in seawater may become aggregated into particles at air–water interfaces, especially on the surface of bubbles, thereby adding to particulate food supplies in the surface layers.

Much of the dissolved organic matter in seawater is fairly resistant to bacterial degradation, constituting a sort of stable marine humus. The easily degradable small molecules of glucose, aminoacids or fatty acids are found mainly in the euphotic zone or in sediments, where they are presumably produced and quickly consumed. The amount of DOM used as a source of nutrition is therefore probably greater at the surface and bottom than at middle depths. However, occasional reports of large numbers of pelagic saprophytes at deep levels around 3000 to 4000 m suggest that DOM may also be a significant food source for abyssopelagic organisms. Although less in quantity here than at the surface or bottom, its relative importance may be great because other food is lacking.

The role of DOM as a source of food to benthic animals is discussed in Section 6.4.

On land, many animals communicate by means of airborne chemicals such as pheromones. It is now known that various interactions between marine organisms are brought about by external secretions that dissolve in, and are carried around by, the water. The olfactory sense is highly developed in many marine creatures and probably serves several useful functions; for example, the recognition and location of other individuals, the detection of food, and in certain species it is important in connection with navigation and migration.

Various micro-organisms, particularly flagellates and bacteria, are known to produce substances toxic or repellent to other organisms. It has been demonstrated that bacterial respiration is sometimes depressed in the presence of diatoms performing photosynthesis, presumably due to antibiotic substances set free by the plants. *Phaeocystis* when abundant gives off acrylic acid which depresses the growth of certain other phytoplankton species. This may even be an important aspect of competition, and possibly also of defence. It has also been suggested that zooplankton avoids water containing large numbers of phytoplankton because of distasteful external metabolites produced by the plants (see page 189). Some dinoflagellates contain toxic substances and may cause the death of other marine creatures when present as a bloom or 'red tide' (see Section 2.2.2).

In some cases, organic solutes have a growth-promoting effect, and seem to have a role in nutrition comparable to that of vitamins, auxins or hormones. Experiments with cultures of marine algae indicate that some have specific organic requirements for normal growth, for example cobalamin. Chelating agents in seawater may also be of some importance, favouring plant growth by bringing into solution essential trace metals which occur in particulate form. The enhanced fertility of coastal water for phytoplankton growth may be partly attributable to the large amounts of DOM derived from land drainage. The group of coloured organic solutes broadly termed 'humic substances' are also thought to fulfil a nutrient function for some marine plants.

Early experiments by Wilson and Armstrong in the 1950s on the rearing of invertebrate larvae, for example the sea urchin *Echinus* and the polychaete *Ophelia*, demonstrated differences in survival rates of larvae reared in samples of water collected from different areas although there were no obvious chemical or physical features of the water to which the differences could be clearly attributed. The factors responsible for these variations in biological quality of the water could not be determined, but some growth-promoting or growth-inhibiting substances were evidently present. The result of mixing seawaters of different quality suggested that the differences were probably due to the presence of beneficial substances in some samples, rather than to harmful substances in others. It has been suggested that differences in biological properties between seawaters of different areas may depend partly upon their biological histories through the effect of metabolites produced by preceding generations of organisms.

Dimethylsulphide (DMS)
During the 1980s, the importance of a constituent of DOM called dimethylsulphide $(CH_3)_2S$ was realized. DMS is a gas found in solution near the sea surface.

The gas escapes from the ocean surface and it has been estimated that it accounts for as much as 25 per cent of the sulphur that enters the atmosphere. The exact process by which it is liberated remains unknown but it originates from dimethylsulphoniopropionate (DMSP) found in many marine algae, both seaweeds and phytoplankton, where it performs various roles in metabolism. DMSP released directly from plants into seawater is somehow rapidly converted by free-living bacteria into DMS. Digestion of plant material by animals such as copepods, also converts DMSP to DMS via bacteria living in the gut.

DMS released into the atmosphere is oxidized to sulphur dioxide, which may result in the acidification of rain. Airborne sulphate ions ($SO_{4=}$) are also produced and these, often in combination with atmospheric ammonia, are a major consituent of the tiny particles known as cloud condensation nuclei. Water condenses on these particles and clouds are built up. Thus the metabolism of marine phytoplankton, by producing DMS in the ocean surface, is thought to be linked with cloud formation which in turn affects the radiative (heat) balance of the atmosphere. This simplified account of a complex series of interactions provides another example of intricate connections between the activities of marine organisms and world climate. This is discussed further in Section 10.2 in the context of global warming. Recent experiments also suggest that some seabirds such as petrels may be able to smell DMS given off by phytoplankton as this is eaten by the krill on which the birds feed.

4.4   Specific gravity and pressure

The specific gravity of seawater varies with temperature and salinity and very slightly with pressure. At 20°C and atmospheric pressure, seawater of salinity 35‰ has a specific gravity of 1.026.

At salinities above 24.7‰ the temperature of maximum density lies below the freezing point. Because of this, cooling of the sea's surface to freezing point does not produce a surface layer of light water, as it does in fresh water where the temperature of maximum density is 4°C. In contrast to fresh water, surface cooling of the sea below 4°C therefore causes the density of the surface water to increase, and convectional mixing to continue, right up to the point at which ice crystals begin to separate. Consequently, there is no formation of a winter thermocline in the sea as occurs in freshwater ponds and lakes.

At sea level the atmosphere presses down with a force of approximately 1 kg for every square centimetre of the Earth's surface (14.7 pounds per square inch). In simple terms this is 1 *atmosphere* or approximately 1 *bar* (1 atm = 1.013 bar). A more accurate measure is the *pascal* and 100 00 pascals (Pa) is equal to 1 bar. Hydrostatic pressure increases by approximately 1 atm per 10 m increase in depth. In the deepest ocean trenches, pressures exceed 1100 atm.

Although water is only very slightly compressible, such enormous pressures are

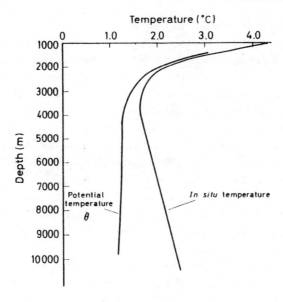

Figure 4.8 The effect of pressure on temperature. In situ and potential temperatures at deep levels.

sufficient to produce a slight adiabatic compression of the deep water, resulting in a detectable increase in temperature. In some areas, temperature readings taken from the 4000 m level down to 10 000 m show a rise of about 1°C (Figure 4.8). The water column is not unstable because this rise of temperature at the lower levels is the result of compression and there is no decrease of density. Deep water temperatures are often expressed as *potential temperature*, θ, i.e. the temperature to which the water would come if brought to atmospheric pressure without heat change. Increase of pressure also reduces the pH of seawater through its effect on the dissociation of bicarbonate (see page 115).

Although marine creatures are found at all depths, each species is restricted in the range of levels it inhabits. It is usually difficult to know to what extent this limitation is due to pressure because the associated gradients of temperature and illumination are probably in many cases the dominant factors regulating distribution in depth. However, the pressure gradient must certainly play some part. Organisms which live in the surface layers of the sea can be killed by subjecting them to the very high pressure of abyssal depths. Alternatively, organisms brought up from the deep-sea bottom require a high pressure, comparable to that of their usual environment, for normal activity. Evidently marine creatures are adapted to suit particular pressure ranges.

The physiological effects of pressure are not well understood but various responses to experimental changes in pressure have been observed (Knight-Jones and Morgan, 1966; Newell, 1976; Rice, 1964). It has a direct influence on several aspects of cell chemistry including weak acid and base dissociations, ionic interactions, protein structures and it affects the rates of enzymatic catalysis in deep-sea organisms (Somero *et al.*, 1983). Pressure changes influence metabolic rates and oxygen consumption, and are observed to produce alterations of behaviour in certain marine organisms. Experiments on small pelagic animals indicate that pressure increase commonly causes an increase of swimming activity, and movement upwards towards light. In natural conditions the pressure gradient may therefore be one factor which helps to prevent surface forms from descending too deep. Some shallow water species are remarkably sensitive to pressure variations, in some cases, e.g. the amphipod *Corophium*, responding to changes as small as 200–500 Pa. It is uncertain how organisms without gas cavities can detect pressure change. Possibly the compressibility of their tissues differs appreciably from that of seawater.

Many species are fairly limited in vertical distribution, but some are eurybathic forms found over a great range of depths. Examples from the British fauna which extend from the continental shelf to below 2000 m, listed by Ekman (1953), include the polychaetes *Lumbriconereis impatiens, Notomastus latericeus* and *Hydroides norvegica*, to at least 3000 m; the keel worm *Pomatoceros triqueter*, 5–3000 m; the polychaete *Amphicteis gunneri*, 20–5000 m; the cumaceans *Diastylis laevis* 9–2820 m; and *Eudorella truncatula*, 9–2820 m; the starfish *Henricia sanguinolenta*, 0–2450 m; the brittlestar *Ophiopholis aculeata*, 0–2040 m. An exceptional example of extensive depth distribution is reported for the pogonophore *Siboglinum caulleryi*, ranging from 22 m to over 8000 m.

Most of the organisms which are found near the surface of the sea seem to have a more restricted depth range than those which inhabit deep levels. The pressure gradient may be partly accountable because the greatest relative changes of pressure with depth occur close to the surface. Between the surface and 10 m the pressure doubles, and on descending a further 10 m the pressure increases by a further 50 per cent. An organism which changes its depth between the surface and 20 m experiences the same relative pressure change as one which moves between 2000 and 6000 m.

Earlier we mentioned the dangers of breathing air beneath the sea's surface (see page 81). Air-breathing aquatic vertebrates such as seals and whales face some risk of gas embolism when surfacing from deep dives, but the danger is far less than for human divers because the animals are not continually breathing air under pressure while diving. When they dive they take down relatively little air, the lungs being only partially inflated. Water pressure tends to collapse the lungs, driving the air into the trachea from which little absorption takes place. There is consequently not much gas dissolved under pressure in the blood. These animals contain large quantities of myoglobin from which they draw oxygen during submergence. They can accumulate a considerable oxygen debt and can tolerate a high level of CO_2 in the blood, the threshold sensitivity of the respiratory centre in the brain being higher than in terrestrial mammals. In seals the heart rate is generally reduced during diving, the peripheral circulation is shut down and the blood flow mainly restricted to supplying the brain (Hempleman and Lockwood, 1978). Little is known of the movements of seals and whales under water. It is thought that seals do not often dive very deeply, probably seldom much over 30 m, although a depth of 600 m has been recorded for a Weddell seal. Some whales go even deeper than this, with sperm whales known to descend to over 1000 m and occasionally even to over 2000 m (Clarke, 1978).

A distinctive feature of certain species living on the deepest parts of the sea floor in the ocean trenches is the exceptionally large size they attain compared with closely related forms in shallow water. This phenomenon of 'gigantism' is specially marked in some groups of small crustacea, notably amphipods and isopods, which seldom exceed lengths of 2–3 cm in shallow and middle depths but grow to 8–10 cm or more in the hadal zones. The explanation of these unusually large sizes is uncertain. The water temperature in ocean trenches is not much lower than on shallower parts of the abyssal floor, so it is possible that gigantism is an effect of pressure on metabolism, perhaps associated with a longer growing period and delayed maturation. Also, because deep-sea sediments are more radioactive than near-shore deposits, the mutation rate may be greater at hadal depths resulting in a higher rate of speciation.

4.4.1 Buoyancy problems

Most protoplasm, cell walls, skeletons and shells have a density greater than seawater, and therefore tend to sink. The specific gravity of seawater is usually within the range 1.024–1.028. The overall density of much of the zooplankton is around 1.04 and of fish tissues about 1.07. Within a floating body the distribution of weight determines its orientation in the water. Therefore one of the problems facing pelagic organisms is how to keep afloat in a suitable attitude between whatever levels are suitable for their life.

The phytoplankton must obviously remain floating quite close to the surface because only here is there sufficient illumination for photosynthesis (Smayda, 1970). Plants which sink below the euphotic zone die once their food reserves are exhausted. Although animals are not directly dependent on light, the most numerous pelagic fauna are small herbivorous planktonts feeding on phytoplankton, and these must also remain fairly close to the surface to be within easy reach of their food. However, many of these herbivores do not remain constantly in the illuminated levels but move up and down in the water, ascending during darkness for feeding and retiring to deeper levels during daylight. Many carnivorous animals, too, both planktonic and nectonic, make upward and downward movements, often of considerable extent (see pages 140 ff.), and traversing ranges of temperature and pressure. These creatures must therefore be able to control their level in the water, and make the necessary adjustments to suit the conditions at different depths.

There are broadly two ways in which pelagic organisms can keep afloat and regulate their orientation and depth: by swimming, or by buoyancy control. In many cases the two methods function together.

A wide variety of marine creatures, both small and large, swim more or less continuously and control their level chiefly by this means. For example, dinoflagellates are said to maintain themselves near the surface by repeated bursts of upward swimming, alternating with short intervals of rest during which they slowly sink. In the laboratory, copepods seem generally to swim almost continuously, mainly with a smooth, steady motion but with occasional darting movements. When there are brief resting periods they usually appear to sink slowly; but Bainbridge (1952), using aqualung equipment, observed copepods in the sea and reported *Calanus* apparently hanging motionless in the water as if suspended from its long outstretched antennae, drifting without any obvious tendency to rise or sink. Among the larger crustacean zooplankton are many forms which swim almost ceaselessly, and seem to sink fairly quickly if swimming ceases, for example mysids, euphausids and sergestids. Some of the most abundant pelagic fish also keep afloat by swimming; for example, almost all the cartilaginous fish, and certain teleosts including mackerel and some tunnies.

Apart from flotation, swimming movements also fulfil several other important functions. They are obviously important for pursuit or escape, and in many forms they serve to set up water currents for feeding and respiration.

The alternative to swimming is to float by means of some type of buoyancy device. Gas vacuoles sometimes form in planktonic algae. Some pelagic animals have a gas-filled float, which in a few cases is so large that the animal has positive buoyancy and floats at the surface, i.e. pleuston. Examples are the large pneumatophores of some species of siphonophora (*Physalia*, *Velella*), or the inflated pedal disc of the pelagic actinians (*Minyas*). The pelagic gastropod, *Ianthina*, floats at the surface attached to a raft of air bubbles enclosed in a viscid secretion. If detached from its raft, the animal sinks and apparently cannot regain the surface. In some of the attached brown algae of the seashore (*Fucus vesiculosus*, *Ascophyllum nodosum*, *Sargassum* spp.) the fronds gain buoyancy from air bladders within the thallus.

Many of the pelagic organisms which dwell below the surface have some method of controlling their overall density to match that of the water, in this way achieving neutral buoyancy so as to remain suspended. This is commonly done by means of some form of adjustable gas cavity. Some siphonophoran colonies (*Nanomia* and *Forskalia*) float below the surface suspended beneath a gas-filled pneumatophore. *Nautilus* obtains buoyancy from gas secreted into the shell, and in *Sepia* the cuttlebone is partly filled with gas. In numerous families of teleost fish there is a gas-filled swimbladder.

The volume of a gas varies inversely with pressure, so the pressure gradient presents special problems to those animals which rely for buoyancy upon gas-filled cavities. When a fish with a gas bladder moves upwards there is a tendency for the bladder to expand as the hydrostatic pressure falls. If this should occur, the fish would rapidly gain positive buoyancy and bob up to the surface, and the distension of the bladder would be likely to cause damage. Alternatively, during downward movement the bladder must tend to contract under increasing pressure, and this would cause loss of buoyancy and force the fish downward. Normally, of course, such changes of buoyancy are prevented by the ability of the animal to maintain the volume of the float virtually constant despite changes of external pressure. A fish with a closed swimbladder does this by absorbing gas during ascent, or secreting more gas during descent, thereby counteracting the tendency of the float to change volume as the pressure alters. However, these processes of absorption and secretion cannot take place very quickly, and creatures which rely for flotation on this mechanism are therefore unable to make very rapid changes of depth without danger. This effect shows dramatically in fish brought quickly to the surface in trawls, when the swimbladder often expands so greatly that it extrudes through the mouth like a balloon.

With increase of depth and pressure the density of gas in a swimbladder increases and the lift it gives decreases. For epipelagic marine fish to achieve neutral buoyancy by means of a swimbladder, the volume of the bladder must be about 5 per cent of total body volume. At deeper levels the density of gas is greater and the bladder must therefore occupy a greater proportion of body volume. Also, the greater the density of fish tissues, the larger the bladder necessary for neutral buoyancy. But the larger the bladder, the more difficult to control its size with rise or fall of pressure because of the large volume of gas that must be absorbed or secreted. Small bladders have the advantage that the fish can rapidly adjust its buoyancy as it changes level.

The lift provided by a swimbladder is equivalent to the force the fish would otherwise have to exert continually in upward swimming to maintain level if it had no swimbladder. As this force is approximately 5 per cent of body weight in air, neutral buoyancy clearly effects a great economy of energy. Also, because fish with neutral buoyancy do not need to use their lateral paired fins for dynamic lift while swimming, these fins are available for other functions. Used as paddles they give very fine control of movement and hovering. The evolution of teleosts from their heavily armoured ancestors has involved a progressive lightening of structure, giving greater ease of buoyancy control together with more efficient and precise locomotion. In some species the fins have also been modified and adapted to function as feelers, suckers, defensive spines, or for various roles in signalling and courtship.

Some creatures with gas-filled floats can release gas through an opening; for example, the siphonophoran *Nanomia* controls the buoyancy of its pneumatophore by secreting gas into it, or by allowing gas to escape through a sphincter-controlled pore. An interesting feature of the gas in this and several other siphonophora is that it contains carbon monoxide, a possible advantage being its low solubility in water. There are some fish which have a connection between the swimbladder and the exterior via the gut. This is commoner in freshwater fish than in marine species but is found in the Clupeids (herring, sprat, pilchard, etc.) and in some others. Some of these fish can release pressure during ascent by allowing gas to escape from the bladder through the mouth or anus; at the surface, air can be swallowed and forced into the bladder. However, in most marine teleosts, if a swimbladder is present, it is closed. In those that live at great depths the gas is at high pressure and gas-secreting structures are enlarged; for example in Myctophidae (lantern fish), Sternoptychidae (hatchet fish) and some Gonostomatidae. Below 1000 m few of the pelagic species have a swimbladder, although in some cases it is present during the larval stages, which are spent nearer the surface. In certain bottom-dwelling fish highly developed swimbladders are found down to about 7000 m (in some Macrurids, Brotulids and Halosaurs), probably associated with a need for neutral buoyancy to enable the fish to hover just above the sea floor, i.e. benthopelagic forms.

Apart from acting as a buoyancy device, the swimbladder has additional functions. It is a pressure-sensing organ whereby buoyancy is automatically adjusted to changes of level. Also, it probably has a role as a detector of vibrations in the water, and in some species has connections with the ear, indicating a hydrophonic function. In certain fish it is evident that the swimbladder can itself be set in vibration to function as a sound-producing mechanism, perhaps for communication, defence of territory, courtship or echo location. Examples of fish producing sounds in this way include the Drums (Sciaenidae), Gurnards (Triglidae), toad-fishes (Batrachoidiformes) and some gadoids such as haddock and cod.

Some cephalopods obtain flotation from gas but have a different method of regulating buoyancy from that of fish. In the cephalopods the problems of pressure change are avoided by enclosing the gas in a rigid-walled container which cannot appreciably change volume as depth alters. For example, *Sepia* has gas in its cuttlebone, which contains spaces filled partly with gas and partly with liquid. The overall density of the cuttlebone varies between about 0.5 and 0.7, but the mass of gas within it seems to remain constant, and the density of the bone is controlled by regulating the amount of liquid it contains. The siphuncular membrane on the underside of the cuttlebone apparently acts as a salt pump, controlling osmotically the quantity of liquid in the bone. By maintaining the concentration of the cuttlebone liquid below that of the blood, sufficient osmotic pressure is generated to balance the hydrostatic pressure of the surrounding water. The distribution of gas within the cuttlebone also assists in maintaining the normal orientation of the animal in the water. There is a similar mechanism in

Nautilus, which also has gas and liquid in its shell and adjusts its buoyancy by controlling the quantity of liquid.

Instead of gas-filled floats with their attendant difficulties of control, many pelagic organisms obtain buoyancy from liquids of lower specific gravity than seawater in a way similar to the bathyscaphe (see page 86). Liquid-filled floats have the advantage of being virtually incompressible; but because of their higher density they must comprise a much greater proportion of the organism's overall volume than is necessary with gas-filled floats if they are to give equivalent lift. We have referred earlier to the probability that the large central vacuole of diatoms contains cell sap of low specific gravity, conferring some buoyancy (see page 28). In young, fast-growing cultures, diatom cells often remain suspended, or sink only very slowly, although in older cultures they usually sink more rapidly. Studies of the distribution of diatoms in the sea suggest that some species undergo diurnal changes of depth, usually rising nearer the surface during daylight and sinking lower in darkness, possibly due to slight alterations of their overall density effected by changes in specific gravity of the cell sap, or in some cases by formation or disappearance of gas or oil vacuoles in the cytoplasm. Pelagic fish eggs during the later stages of their maturation gain buoyancy by the follicle cells secreting into them dilute fluids containing smaller quantities of salts than in seawater. In radiolaria, the vacuolated outer protoplasm is probably of lower specific gravity than the water, giving buoyancy to the whole. Changes in the number of vacuoles cause the cells to rise or sink. Some of these animals possess myonemes with which the outer protoplasm can be expanded or contracted, and this presumably alters the overall density and may provide a means of changing depth.

Many pelagic organisms contain considerable quantities of oil. This may well serve a dual purpose, acting as both a food reserve and a buoyancy device. The specific gravity of most of these oils averages about 0.91 but some, e.g. squalene (see page 134) are considerably lighter. Oil droplets are common inclusions in the cytoplasm of phytoplankton, and the thermal expansion of these oils may be of some significance in effecting diurnal depth changes. Many zooplanktonts also contain oil vacuoles, or oil-filled cavities; for example, radiolaria, some siphonophora, many copepods and many pelagic eggs. The body fluids of marine teleosts are more dilute and less dense than seawater, and therefore confer some buoyancy, and many species derive additional buoyancy from accumulations of fat. In the abundant oceanic fish, *Cyclothone*, there is fat in the swimbladder and tissues amounting to about 15 per cent of the total volume of the fish.

The thick fat layer in the skin of marine mammals provides both heat insulation and buoyancy. In sperm whales *Physeter macrocephalus*, which often weigh some 30 tonnes or more in air, additional buoyancy is obtained from the large mass of clear white spermaceti wax enclosed in a reservoir in the snout (Clarke, 1978). A 30-tonne whale probably contains about 2–3 tonnes of this oil. During deep dives these whales seem to have fine control of neutral buoyancy, enabling them to lie virtually still in the water. It is most likely that this control is effected by regulating the temperature of the oil. It seems that water is drawn into the huge nasal passages (the right one is 5 m long and up to 1 m in diameter) and that this cools the surrounding spermaceti wax just enough to increase its density such that neutral buoyancy is achieved. By controlling the amount or distribution of water of different temperature from different depths in these passages, it can change the density of the wax to suit its buoyancy needs.

The gelatinous tissues which are a feature of many pelagic organisms, for example medusae, siphonophora, ctenophora, salps, doliolids, heteropod and pteropod molluscs, have a slight positive buoyancy which affords flotation to the denser parts. The tissue fluid is isosmotic with seawater but of a lower specific gravity due to the replacement of sulphate ions with lighter chloride ions. The small deep-sea squids of the family Cranchidae, mainly 1–2 cm in length, gain buoyancy from coelomic fluid of slightly lower density than seawater contained in their capacious coelomic cavities. About two-thirds of their weight is made up of coelomic fluid, isosmotic with seawater but of low density due to its high concentration of ammonium ions. Some other families of squids also attain near neutral buoyancy in this way (Clarke *et al.*, 1979). The dinoflagellate *Noctiluca* also gains buoyancy from a high concentration of ammonium ions, exclusion of relatively heavy divalent ions, especially sulphate, and a high intracellular content of sodium ions relative to potassium.

In those mesopelagic fish which lack swimbladders there are several ways in which body weight is reduced. Compared with species having swimbladders the body fluids are more dilute, the haematocrit is lower, the blood is of lower viscosity and the tissues contain more fat and less protein. Apart from the jaw structures much of the skeleton is lightly developed and relatively poorly ossified, and the associated muscles correspondingly reduced. The heart is smaller and the red part of the myotomes which functions mainly for sustained swimming is less developed than the white musculature used for the brief, rapid movements of escape or attack. Denton and Marshall (1958) have drawn up a 'buoyancy balance sheet' comparing a coastal fish, *Ctenolabrus rupestris*, which has a gas bladder, with a bathypelagic fish, *Gonostoma elongatum*, in which the swimbladder is degenerate and filled with fat (Figure 4.9 and Table 4.4). According to this reckoning, the deep-water fish is only slightly heavier than the weight of water it displaces, and presumably does not need to make much effort to maintain its level. *Gonostoma elongatum* migrates daily through some 400–500 m, and it is possible that such fish may exert some control of buoyancy by alterations in the ionic content of their body fluids.

In elasmobranchs, none of which has swimbladders, the overall density varies considerably in different species in ways related to their different shapes and modes of life. Some species approach or attain neutral buoyancy due to the relative lightness of their skin, skeleton and muscles. Lift is provided by

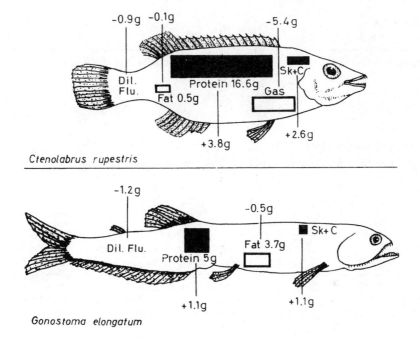

Figure 4.9   Diagram of the 'buoyancy balance sheet' for a bathypelagic fish Gonostoma elongatum *without a swimbladder (below) and a coastal fish* Ctenolabrus rupestris *with a swimbladder (above). Positive values are given for those components of the fish which are heavier than the seawater which they displace and thus tend to 'sink' the fish, whilst negative values are given for those components which displace more seawater than their own weight and thus tend to 'float' the fish. Weights given per 100 g of fish. Dil. Flu., dilute body fluids; Sk + C, skeleton and other components.*
(From Denton and Marshall (1958), published by Cambridge University Press.)

low-density oil, especially in the liver, and also from other non-fatty tissues of low density, notably the subcutaneous layer or masses of gelatinous tissue. Some elasmobranchs may be able to change buoyancy by regulating their lipid content. When elasmobranchs are swimming they gain dynamic lift from the plane surfaces of their pectoral fins, which function also for changing direction or braking.

It appears that many of the bottom-dwelling elasmobranchs of shallow water have a relatively high density, probably associated with a fairly inactive mode of life and a habit of resting on the bottom. But other bottom species are close to neutral buoyancy, especially those of deeper water; these are probably more active benthopelagic fish which cruise above the bottom in search of food, which at deep levels is more scarce.

The pelagic predatory sharks, swimming rapidly in pursuit of prey, have densities which result in nearly neutral buoyancy. This enables them to be fully manoeuvrable with fins which are quite small and therefore add little frictional

Table 4.4

(a) Balance Sheet for the goldsinny wrasse *Ctenolabrus rupestris*

Component	Percentage wet weight	Weight in seawater/100 g of fish
Fat	0.5	−0.1
Protein	16.6	+3.8
Body fluids	73.3	−0.9
Other components including bone	9.2[a]	+2.6[a]

Buoyancy. This fish without its swimbladder has a weight in seawater of +5.4 per cent of its weight in air.

[a]These values are given by difference.

(b) Balance Sheet for the deep-sea fish *Gonostoma elongatum*

Component	Percentage wet weight	Specific gravity	Weight in seawater/100 g of fish
Fat	3.7	0.91[a]	−0.5
Protein	5.0	1.33[b]	+1.1
Body fluids (water + dissolved salts)	87.6	1.013	−1.2
Other components including bone	3.2[c]		+1.1[c]

Buoyancy. These fish had no gas-filled swimbladder and their average weight in seawater was approximately +0.5 per cent of their weight in air (wet weight).

[a]*Handbook of Biological Data* (1956). Ed. W.S. Spector. Ohio, USA.: Wright Air Development Center.
[b]Hober, (1954). *Physical Chemistry of Cells and Tissues*. London; Churchill. Specific gravity taken as the reciprocal of the partial specific volume.
[c]These values are given by difference.

drag to their well-streamlined forms. There are other pelagic species which are slow-moving and feed by filtering plankton, and these are of two shapes with different densities. The shark-shaped form *Cetorhinus* (basking shark) gains little lift from its small fins but can remain afloat while swimming slowly because it has almost neutral buoyancy. This is largely due to the low-density oil *squalene* (specific gravity 0.86 at 20°C), found in the huge liver, and a mass of gelatinous tissue in the nose. On the other hand, the manta rays (*Mobula*) have a relatively high density but gain such enormous dynamic lift from their large plane surfaces that they can leap out of the water.

The buoyancy afforded by the water is of course important in reducing the skeletal needs of aquatic organisms, which can be very lightly built compared with terrestrial forms. This has made possible the evolution of some extremely large marine creatures. The enormous present-day whalebone whales, for example the blue whale *Balaenoptera musculus* and the fin whale *B. physalus*, may reach weights of the order of 100 tonnes. In comparison, the largest known terrestrial animals, the huge Mesozoic reptiles, probably did not exceed some 30 tonnes. The rapid death of large whales on stranding is often caused by the collapse of the rib cage when the massive weight is unsupported by the water.

4.5 Viscosity

The viscosity of seawater decreases considerably with rise of temperature and increases slightly with increase of salinity. The coefficient of dynamic viscosity is a measure of the drag exerted on moving objects in a fluid. In seawater of salinity 35‰ at 0°C the coefficient of dynamic viscosity is 18.9×10^{-9} N per cm^2 per unit velocity gradient, and at 30°C is only 8.7×10^{-9}. These values indicate that resistance to movement is over twice as great in the coldest parts of the sea as in the warmest. Viscosity influences both sinking and swimming speeds, and in many organisms must also have effects on feeding rates and respiration.

4.5.1 Sinking rates

The rate at which a small object sinks in water varies with the amount by which its weight exceeds that of the water it displaces, and inversely with the viscous forces between the surface of the object and the water. The viscous forces opposing the motion are approximately proportional to the surface area, and therefore, other things being equal, the greater the surface area the slower the sinking rate. Because very small objects have a large surface-to-volume ratio, they are likely to sink more slowly than larger particles of similar density. However, there is a tendency for marine micro-organisms to become aggregated with inorganic particles and tiny fragments of organic debris, into larger clumps, which may increase the sinking rate.

There are a number of structural features of planktonic organisms which increase their surface area and must certainly assist in keeping them afloat. The majority of planktonts are of small size, and therefore have a large surface-to-volume ratio. In many cases, modifications of the body surface increase its area with very little increase in weight. Comparable adaptations are found in wind-dispersed seeds and fruits of land plants, which clearly serve to keep them airborne. These modifications generally take two forms: a flattening of the body, or an expansion of the body surface into spines, bristles, knobs, wings, or fins. Many of these extensions of the surface must also have the effect of setting up water turbulence as the object moves, greatly increasing viscous drag.

Reference has already been made to the range of flattened or elaborately ornamented shapes that occur in diatoms. In dinoflagellates, also, the cell wall is in some cases prolonged into spines (*Ceratium*) or wings (*Dinophysis*, see Figure 2.5). In the zooplankton, flattened shapes are common, for example the pelagic polychaete *Tomopteris* (see Figure 2.14), the copepod *Sapphirina* and various larvae (phyllosoma of the crawfish *Palinurus*, cyphonautes of the seamat *Membranipora*). Arrays of spines and bristles also occur, for example some species of copepods *Calocalanus* and *Oithona* and many larvae (mitraria of the polychaete *Owenia*, crab zoeas). The pluteus larvae of echinoderms have long ciliated arms. The chaetognaths have flat fins, and in the polychaete *Tomopteris*, the parapodia form a series of flattened, wing-like appendages.

Warm water has a lower density and viscosity than cold, and therefore affords less buoyancy and less resistance to sinking. The plankton of warm water includes several species in which the bristles, wings and other flotation devices are exceptionally elaborate (for example *Calocalanus* spp., *Ornithocercus*). In temperate waters, there is a tendency in some species for spines to be relatively longer in the warm season than in the colder months, e.g. the diatom *Rhizosolenia hebetata* (Figure 4.6).

Reduction of the sinking rate is unlikely to be the only, or even the major, advantage gained from modifications which enlarge the surface area. In the phytoplankton, an extensive surface presumably facilitates the absorption of nutrients present only in very low concentration, and also favours light absorption. The shape and distribution of weight determine the orientation of a passively floating body, and it seems probable that diatoms float in a position which presents the maximum surface to light from above. Irregular shapes and long spines may function also as protective devices which small predators find awkward to grapple with. In certain species there are sexual differences in bristles and plumes which indicate some role in sexual display; for example, the long plume of the males of *Calocalanus plumulosus*.

4.5.2 Swimming rates

Because viscous forces are related to surface area, drag effects on swimming are relatively greater for small organisms than large. Small animals swim slowly compared with large ones. In fish, swimming speeds are roughly proportional to length. Based on experimental observations, Bainbridge (1958) formulated the following equation relating speed, length and rate of tail beat in teleosts:

$$V = 0.25L \ (3f - 4)$$

where V = swimming speed in cm s^{-1}, L = fish length in cm, and f = tail beats per second s^{-1}.

The effects of viscosity are greatly increased where movements generate turbulence; hence the advantages of streamlining. The best streamlined form for preserving laminar flow is circular in cross-section with a rounded forward end tapering aft to a point. Fast-swimming cephalopods impelled by jet propulsion adopt a shape close to this ideal streamlining. The majority of fish are elliptical rather than circular in cross-section because they swim by lateral oscillations of the trunk, lateral flattening giving extra thrust (Gray, 1968). However, the fastest fish such as large tunas are more circular in cross-section, their lateral flexures being confined mainly to the narrow posterior end of the body, and thrust is obtained chiefly from the large caudal fin. Certain tuna are reported to achieve bursts of speed exceeding 20 m s^{-1}, but normally they cruise at slower speeds of 1–2 m s^{-1}.

Swimming efficiency varies with drag, thrust and efficiency of energy conversion. Generally the speed at which a fish can swim the greatest distance for a given supply of energy appears to be about one fish length per second irrespective of size. Efficiency of energy conversion decreases at speeds above and below this optimum.

The limitations on swimming speed inherent in small size may be one factor setting a lower limit on the size of fishes. Few teleosts have adult sizes less than 2.5 cm. Other disadvantages for fish of very small size include the difficulties of osmocontrol with a relatively large surface area, and reduced capacity for egg production.

Recent research at Woods Hole Oceanographic Institute in Massachusetts, USA, has provided new data on those features of fish propulsion that result in the high swimming efficiencies seen in many fish and in dolphins (Triantafyllou, 1995). The work is primarily aimed at the development of robotic free-swimming craft which will need a very efficient means of propulsion if constraints of energy storage are to be overcome. To this end they have made considerable progress in constructing a free-swimming 'robo-tuna' of aluminium and lycra, and have added to our understanding of drag, thrust and turbulence.

4.6 Illumination

Compared to the depth of the ocean, light does not reach very far into the sea. Illumination of the surface layers varies with place, time and conditions depending upon the intensity of light penetrating the surface and upon the transparency of the water. The strength of the incident light varies diurnally, seasonally and with latitude, and is influenced by cloud conditions and atmospheric absorption. Much of the incident light is reflected from the surface, more light being reflected from a ruffled surface than a calm one, and reflection increases as the sun becomes lower in the sky. Depending on conditions some 3–50% of incident light is usually reflected. The light which penetrates the surface is quickly absorbed, partly by the water and dissolved substances but often largely by suspended matter including planktonic organisms, translucent water generally being indicative of a sparse plankton. Extinction coefficients (see Section 3.1.4) vary from about 1.0 to 0.1 between turbid inshore and clear offshore areas, but in exceptionally transparent ocean water may be as low as 0.02. Even in clear water about 80% of the total radiation entering the surface is absorbed within the uppermost 10 m, and in more turbid water the absorption is far more rapid than this. The heating effects of solar radiation are therefore confined to a very thin surface layer.

Different wavelengths of light do not penetrate equally. Infrared radiation penetrates least, being almost entirely absorbed within the top 2 m, and ultraviolet light is also rapidly absorbed. Within the visible spectrum, red light is absorbed first, much of it within the first 5 m. This is why underwater photographs taken without a flash have an overall bluish-green colour. In clear water the greatest penetration is by the blue-green region of the spectrum, while in more turbid conditions the penetration of blue rays is often reduced to a greater extent than that of the red-yellow wavelengths. This differential absorption of the solar spectrum partly accounts for the colour of the sea's surface by its effect on the spectral composition of reflected light. In bright sunlight, clear ocean water may appear very blue because the yellow and red rays are largely absorbed, and blue rays predominate in light reflected from below the surface. In more turbid coastal waters, their greener appearance may result from the relatively greater absorption of blue light. Sometimes the colour of the water is due to the pigmentation of minute organisms.

Light is of supreme biological importance as the source of energy for photosynthesis (see Section 5.3.1). Primary food production in the marine environment is virtually confined to the illuminated surface layers of the sea where there is sufficient light to support plant life. The depth of this photosynthetic or *euphotic zone* varies with conditions, extending to some 40–50 m from the surface in middle latitudes during the summer months and to 100 m or more in low latitudes if the water is fairly clear. Below the euphotic zone, down to about 200 m, is the dimly illuminated *dysphotic zone* where light is insufficient for the survival of plants. The water below 200 m is termed the *aphotic zone* because there is little or no light, but in clear tropical waters a small amount of blue radiation penetrates to at least 1000 m.

Light is certainly one of the major factors controlling the distribution of marine organisms, but in many cases its effects are not easy to understand. Plants are restricted to the euphotic zone by their dependence upon light for energy; and animals are most numerous in or near the surface layers because they derive their food, directly or indirectly, from plants. Below the productive surface zone, animals are almost entirely dependent upon food sinking to them from above, and the deeper they are, the less food is likely to reach them because much of the assimilable material is decomposed on the way down. Broadly, the deeper the level, the less the food supply and the fewer the population. But numbers do not fall off evenly with depth. Although we do not have much knowledge about the distribution of mesopelagic and bathypelagic forms, it appears that populations tend to congregate at certain levels. There is commonly a concentration of animals somewhere between 300 and 1000 m, sometimes coinciding with the oxygen-minimum zone, and lesser concentrations below this. Many organisms, however, do not remain consistently at one level but perform vertical movements, often over a considerable distance, which are related to changes of illumination (Figure 4.10).

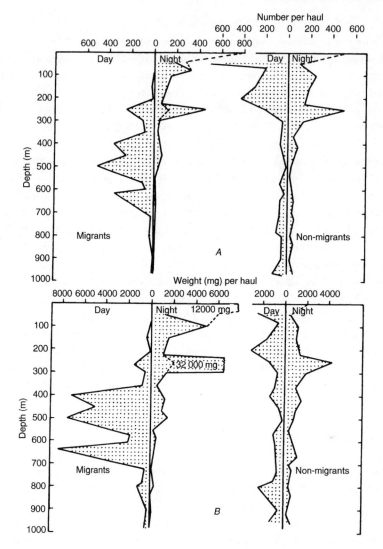

Figure 4.10 Day and night distributions of total migrant and non-migrant mesopelagic euphausids off the Canary Islands taken during a cruise by RRS Discovery, *1965.* A – *Total numbers per haul;* B – *Total weight per haul. Pecked line in upper 100 m of night distribution indicates catches adjusted by addition of species assumed to be missed at night by migrating above 50 m. Pecked line at 250 m in night distribution indicates catch when a large sample of* Nematoscelis megalops *is omitted.*
(From Baker, A. de C., *J. Mar. Biol. Ass. UK,* **50**, 301–42 (1970) published by Cambridge University Press.)

4.6.1   Diel (diurnal) changes of level

It has long been known that certain fish, for example, herring and mackerel, usually approach the surface in large numbers only during darkness, and fishermen have adapted their techniques to match this feature of fish behaviour. Since the early days of plankton research it has also been observed that tow-net catches of zooplankton from the surface are usually richer and more varied during the night than in daytime. Poor daytime catches of zooplankton at the surface may be partly due to the ability of some of the larger and more active animals to see the slow-moving net and to avoid capture by swimming; but if repeated plankton hauls are made at a series of depths throughout the day and night, it becomes clear that many planktonts alter their depth dielly, moving nearer the surface during darkness and returning to the deeper water to pass the day. At night, these organisms are not found in samples collected at the deeper levels where they can be captured during the day, but are then found in samples taken nearer the surface from which they are absent in daytime.

These vertical migrations have often been termed *diurnal*. However, they are more accurately described as *diel* – events that occur with a 24-hour rhythm. Diel changes of distribution are shown by a great variety of organisms, both plankton and nekton, including medusae, siphonophores, ctenophores, chaetognaths, pteropods, copepods, cladocerans, amphipods, mysids, euphausids, pelagic decapods, some cephalopods and fish. In shallow water there are many creatures that live on or in the bottom during the day but leave the sea-bed at night and swim in the surface water.

Deep-scattering layers

Early studies of these diel vertical movements of plankton and other creatures were mainly confined to populations of shallow water, but during World War II it became apparent that this phenomenon is of very wide occurrence throughout the deep oceans. Physicists investigating the use of underwater echoes for the location of submarines obtained records during daylight hours of a sound-reflecting layer in the deep water beyond the continental shelf (Dietz, 1962; Farquhar, 1977). On echogram tracings this layer gave the appearance of a false bottom at about 300 m where the sea floor was known to be far below this. Towards the end of the day this sound-reflecting layer was observed to rise until close to the surface. During darkness it was less distinct, but at the dawn twilight the layer formed again near the surface, and descended gradually to its usual daylight level as the sun rose. Further detailed investigations revealed that there are sometimes several of these mid-water sound-reflecting layers at depths between 250 and 1000 m and they are now usually termed sonic scattering layers (SSLs), or deep-scattering layers (DSLs).

It was first thought that the SSLs were due to discontinuities between layers of water differing in some physical property, for example temperature, but this would not account for such marked diel changes of position. It is now generally accepted that the tracings are caused mainly by echoes returned from marine creatures which change their depth between daylight and darkness. Where the SSLs comprise several sound-reflecting zones, these are caused by concentrations of different groups of animals at each depth, e.g. euphasiids and fish.

These layers have subsequently been detected in all oceans and vary in distinctness from place to place, being especially faint in the Arctic. They are generally indefinite beneath areas of low fertility, presumably because of the smaller population of these waters. They are usually strongest below warm, productive surface waters. Their detection and distinctness depend to some extent on the frequency of the sonic equipment.

The nature and number of animals existing in the layers are difficult to ascertain due to difficulties of sampling at these depths. Major groups known to contribute to the DSL include the lantern fish (Myctophidae) and other similar mesopelagic fish, various crustaceans such as euphausiids, sergestid and pasaphaeid shrimp, squid and siphonophores. Thus the traces mainly result from larger animals following their plankton prey up and down as well as from the concentrations of plankton themselves. The positions of SSLs on echogram tracings change with alterations of the sonic frequency (Greenlaw, 1979). Evidently the strength of sound reflection is influenced by structural features of organisms, different species giving strongest echoes at different frequencies. Those that contain vacuoles or gas-filled cavities are likely to give specially strong echoes by resonance at particular frequencies determined by the dimensions of the resonating spaces. Where a species includes a wide range of sizes it is likely to reflect sound over a correspondingly broad frequency band.

It is now thought that, at frequencies up to about 35 kHz, echoes come mainly from small fish possessing gas-filled swimbladders. Commonly captured within the SSLs detected at these frequencies are Myctophids (lantern fish), Sterno-ptychids (hatchet fish) and Gonostomatids, all small fish mostly measuring about 4–10 cm in length. Various Crustacea, mainly euphausids and sergestids, are often caught with these fish but probably do not give much echo at these frequencies. Strong echoes at frequencies of 10–12 kHz may also come from gas-filled nectophores of siphonophores such as *Nanomia bijuga*. Larger fish e.g. herring, cod or redfish, give echoes at 4–6 kHz, and these may account for some of the indistinct SSLs found in the Arctic where mesopelagic fish are not abundant. Sound scattering is also effected by organisms which do not contain resonant cavities. Above 40 kHz reflections may be received from euphausids or pteropods, and above 300 kHz even small copepods may give echoes.

Diel patterns

The study of SSLs together with data from deep-level net samples indicate that throughout the oceans, in both deep and shallow water, a numerous and varied assortment of animals perform vertical migrations with a diel rhythm. There are differences in the depths through which different species move, the speeds at which they ascend and descend, and the precise times at which they make their movements, and the same species may behave differently in different areas and at different times, but there is none the less a remarkable consistency in the general behaviour pattern. Most marine zooplanktonts make a single daily ascent and a single descent. During daylight, each species appears to collect near a particular level. Shortly before sunset they commence an ascent which continues throughout the twilight period. During total darkness the population tends to disperse, but shortly before dawn it again congregates near the surface and then makes a fairly rapid descent until, about an hour after sunrise, the daytime level is reached (Figure 4.11). The movements are therefore slightly asymmetrical with respect to midnight.

As well as this common pattern, two other patterns have been recognized. In *twilight migration*, there are two ascents and descents. The animals rise at sunset to their preferred level, but during the night they sink once again though not necessarily to their original depth. This is called the *midnight sink*. At sunrise the animals rise again before later decending to their normal daytime depth. A few animals exhibit *reverse migration*, ascending during the day and descending at night.

Regulating factors

There are many features of the migrations which are not understood, but it seems certain that changes in illumination play a dominant role in regulating the activity. No other factors are known to which the movements can be so exactly related. Although the relative durations of night and day vary with latitude and season,

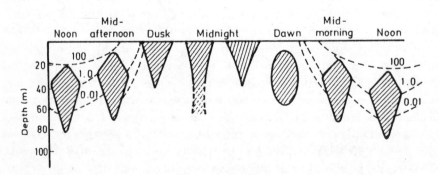

Figure 4.11   Generalized diagram illustrating diurnal changes of vertical distribution of many epipelagic zooplanktonts. The dotted lines show levels of approximately equal illumination indicated in arbitrary units. Logarithmic decrease of illumination accounts for the kite-shaped pattern of distribution in daytime.

the migrations are always closely synchronized with dawn and dusk. In high latitudes they occur during spring and autumn when there are daily alternations of light and dark, but the migrations are not observed when darkness or daylight persists throughout the 24 hours.

Many of the organisms which make these diurnal movements must be extremely sensitive to light. During the day their level rises when clouds cross the sun. At night, moonlight has a noticeable effect, some creatures remaining further from the surface in bright moonlight than on dark nights, while others appear to be attracted by moonlight and reach the surface in greatest numbers at full moon on clear nights.

The majority of migrating forms are fairly active swimmers which regulate their depth by the rate and direction of swimming. In some cases, strong light may have an inhibitory effect on swimming, preventing these animals from reaching the surface during daytime even though they make no directional response to light. However, the movements of many zooplanktonts show an orientation to the direction of light, moving away from a source of strong illumination. A creature which is both negatively phototactic (moving away from light) and negatively geotactic (moving against gravity) presumably swims towards the surface in darkness, but in daytime moves down to the level where its responses to light and gravity are in balance, moving upwards or downwards as illumination changes. Some species orientate to the plane of polarization of light, which changes with the angle of the sun. This would cause the population to rise and fall, though the migration would not be a simple ascent and descent. The movements may be partly regulated by an intrinsic 'physiological clock' operating in phase with changes of illumination. For instance, it has been shown that various copepods, if removed from the sea and kept in total darkness in a tank, continue for some days to exhibit periodic changes of activity having a 24-hour rhythm which results in diel alterations of their level (Palmer, 1974).

These mechanisms perhaps explain the movements of some organisms but do not account for the tendency of many species to disperse during total darkness. It seems as if many planktonts move towards a preferred 'optimum' intensity or spectral composition of illumination which differs according to the species. During daytime the animals tend to assemble at whatever level they find their preferred illumination, and move up or down as this level changes with rising and setting of the sun. During darkness, when there is no longer sufficient illumination to cause the population to concentrate at one level, they scatter. Experiments in which planktonic animals have been enclosed in long tubes exposed to an illumination gradient have demonstrated that some species do collect within particular ranges of light intensity, the optimum differing for different species. Blaxter (1973) observed the distribution of herring and plaice larvae in a vertical tube illuminated from above and found that, in strong illumination, changes of light intensity had little effect on the larvae; but vertical migrations took place at critical levels of illumination equivalent to intensities at late dusk or early dawn. However, the idea that organisms are following preferred light intensities is often a better explanation of their ascent than of their descent, which frequently begins before the sky appreciably lightens. There are some mesopelagic species which start their descent at midnight and have reached their daytime level well before sunrise.

To study the effects of illumination on speed and direction of swimming, Hardy and Bainbridge (1954) devised an apparatus which has been termed a 'plankton wheel'. It consisted of a circular transparent tube, 4 ft in diameter, mounted vertically by spokes on a central axle on which it could be rotated. The tube was filled with seawater, and small animals could be inserted. In effect, the animals were enclosed within an endless water column. By rotating the wheel one way or the other as the enclosed animals swam upwards or downwards, they could be kept at a fixed position relative to the observer. On the inside wall of the tube were mounted a series of small valves, automatically operated by floats and weights as the wheel was turned. These valves ensured that the water turned with the wheel, but did not interfere with the animals in the observation position. The apparatus was placed in a small glasshouse provided with movable screens to give variations in illumination. Movements of the wheel were recorded on a rotating smoked drum.

With this instrument it was possible to measure the speed and duration of upward and downward movement of small animals and to record the pattern of their activities under controlled conditions of lighting, although the apparatus did not provide the gradients of illumination, temperature or pressure that organisms encounter in the sea. Among the animals studied were copepods, euphausids, the polychaete *Tomopteris*, the arrow worm *Sagitta elegans*, and nauplii of *Balanus* and *Calanus*. The experiments confirmed that the swimming capabilities of these animals are more than adequate to account for the extent of the migrations inferred from net hauls. Results were incomplete because of the difficulty of setting up the experiments, and the time needed for recording, but many of the animals tested showed a positive movement downwards in strong light. Their responses varied at different times of day, but during the evening in dim illumination the animals moved upwards towards the light. If kept in total darkness during daytime, they did not consistently move upwards, indicating that geotaxis was not automatically operating in the absence of strong light. The results generally suggested a directional movement towards an optimum level of illumination of low intensity.

Other factors have been suggested as contributing causes of vertical migration; for example, temperature changes in the surface layers of water may have some effect. At night the surface layers undergo slight cooling, and this might permit some of the inhabitants from colder water below to approach the surface, while in daytime the warmer surface temperature would discourage their rising so far.

It has been observed in some animals that increasing temperature reverses the direction of their phototaxes. Apart from physiological and behavioural effects, the physical effects of changes of temperature upon water density and viscosity must have some influence on buoyancy. Surface cooling at night might be expected to cause passively floating forms with neutral buoyancy to float at slightly higher levels than during the day, and the effect might be more marked on small upward-swimming creatures. It has also been suggested that the weight gained by organisms during their nightly feeding in the abundant food supplies of the surface layers may cause them to sink to lower levels until weight is subsequently lost through respiration, excretion and egestion. However, these processes cannot be closely correlated with either the extent or the periodicity of the migrations. The distances covered by most species are too great to be explained simply as passive rising or sinking due to alterations of buoyancy, but must involve active swimming upwards and downwards at considerable speed. Nor does the timing of the migrations correspond closely with the change of temperature of the surface water, which falls gradually throughout the evening and night until shortly before dawn, whereas the majority of vertically-migrating animals reach the summit of their ascent soon after sunset.

In the course of their upward and downward movements, organisms traverse a pressure gradient, and often a temperature gradient, and these must play some part in regulating the extent of the migrations. Increase of pressure causes some zooplanktonts to increase their upward swimming rate, and this presumably limits their depth of descent. Where there is a sharp thermocline between the surface and deeper levels, some migrating forms do not pass through the discontinuity layer. The thermocline then acts as a boundary between two groups of organisms, those of the warmer surface layers which descend during daylight only as far as the thermocline, but not through it, and those of deeper levels to which the thermocline is the limit of ascent during darkness.

It has been suggested that 'exclusion' (see page 189) may have some connection with diurnal migration. If during photosynthesis phytoplankton liberates into the water metabolites which are distasteful or toxic to animals, the zooplankton would be expected to remain below the photosynthetic zone throughout the hours of daylight. Only during darkness, when photosynthesis ceases, would the surface water become suitable for the entry of animals. If the phytoplankton should be very abundant, the exclusion effect might persist throughout the night, upsetting the rhythm of the migration. However, it is uncertain to what extent exclusion operates in natural conditions, nor can this explanation account for the timing of the main movement downwards, which for the majority of animals begins before dawn. Some zooplanktonts become more photonegative when well fed on phytoplankton and so would be expected to move downwards out of phytoplankton patches in daytime without any external metabolites being involved.

There are many anomalies. The migratory behaviour sometimes becomes erratic or ceases completely. Animals which usually seem consistently to move up or down with changes of illumination may occasionally swarm at the surface in bright sunshine. The extent of the movements sometimes varies seasonally; for example, around the British Isles they are of more general occurrence during the summer months, and many planktonts cease this behaviour during winter when they remain at a deep level. Often the migrations are made only by particular stages of the life history. In some cases the migrations are made by only a proportion of the population, resulting in a greater spread of the population through the water column at night rather than a general change of level.

In conclusion, light appears to be the main factor initiating and regulating diel migrations. However, other factors including temperature, pressure and hunger probably also play a part. Different organisms may respond to different stimuli.

Adaptive value

Vertical migrations are performed by so great a range of species that this behaviour must presumably be an adaptation of major importance, but the benefits are not altogether clear. Many of these migrating forms are herbivores grazing on the phytoplankton, and we may wonder what advantage it is to them to spend so much time away from the surface layers where they find their food. Different species must gain different advantages, and some of the possible benefits are as follows:

(a)   Invisibility
(b)   Pursuit of prey
(c)   Lateral mobility
(d)   Genetic exchange
(e)   Control of population density
(f)   Protection from ultraviolet light
(g)   Energy conservation
(h)   Shoaling advantages

Invisibility

Probably a major advantage for many organisms from these migrations is that they gain some safety from predation by those animals that use sight for food capture. Just as there are many terrestrial animals which are nocturnal, emerging from the safety of their hiding places only during darkness, innumerable pelagic creatures may also find safety during daytime in the darkness of deep water, ascending only at night to feed. For many of the herbivorous zooplanktonts sight is not essential for food capture. They gather their food by various processes of filtration and can feed effectively in darkness, but their chances of survival are greater if they avoid well-illuminated water because sight is far more important to most of their predators for the detection of prey. Many of the migratory species which inhabit deep levels appear to make use of bioluminescence for camouflage (see page 154) and in these species it seems likely that their changes of level are adjustments of position to the appropriate intensities of dim illumination at which they are virtually invisible.

However, many aspects of these migrations are not well explained solely in terms of protection from predation. Why are there great differences in the extent of migrations in closely similar forms? Does poor illumination really confer much protection when it is clear that the majority of visual predators feed quite effectively during the night, and many epipelagic planktonts even at their lowest daytime levels are well within the visual threshold of many mesopelagic fish? Why do many transparent, virtually invisible species make migrations at least as great as some easily visible, pigmented forms? Why do some migrating species give displays of bioluminescence which must surely make them conspicuous to predators? Evidently visual protection cannot be the only advantage of vertical migrations.

Pursuit of prey
It is obviously of advantage to predators which depend for food on vertically migrating species to make migrations similar to those of their prey.

Lateral mobility
Hardy (1956) has pointed out that vertical migrations influence the horizontal distribution of planktonic organisms because the deeper layers of water usually move more slowly than the surface and often in a different direction. Planktonts which move downwards at dawn will therefore return to a different mass of surface water each night. This may be advantageous in several ways; for example, by enabling the organisms continually to sample fresh feeding grounds by simply moving up and down through the water, or, where surface and deep layers move in opposite directions, by stabilizing distribution within a particular locality, the population drifting one way at night and back again during the day. When the phytoplankton is sparse the surface water is likely to be clear and therefore strongly illuminated in daytime. This would lead to deeper descent and consequently quicker drift to beneath a different body of surface water. On reaching an area of richer phytoplankton, where more light would be absorbed near the surface, zooplanktonts would tend to rise into the better food supply. In localities where the deep levels move in opposite direction to the surface, descent is likely to carry organisms towards areas of upwelling (see Figure 5. 7), which are always regions of high fertility and usually correspondingly rich in planktonic food.

Genetic exchange
Differences between individuals in the timing and extent of their vertical movements through layers of water travelling at different speeds and directions,

resulting in individual differences in lateral transport, must effectively intermix a population. For small or feebly swimming creatures this process of population mixing must have the advantage of greatly increasing the possibilities of genetic exchange and recombination, maintaining the variability of the species with the resultant benefits of adaptability to environmental changes.

Control of population density

Wynne-Edwards (1962) has suggested that swarming, patchy distribution and the vertical migrations of zooplanktonts may involve aspects of social behaviour related to control of population density. By congregating within thin layers and moving up and down, animals which would otherwise be widely distributed in a homogeneous environment can compete for food without much risk of too seriously depleting the food stocks of the whole volume of water through which they pass. In many species the quantity of egg production varies with the food supply and therefore the number of progeny is to some extent regulated by the availability of food for the adults. In this way competition for food provides a natural mechanism which may prevent the population reaching a size which would over-exploit the food resources. Also, concentration of populations within thin layers or patches facilitates display behaviour in sexual competition, whereby reproduction may be limited at supportable numbers.

Protection from ultraviolet light

Ultraviolet light has a destructive effect on unprotected protoplasm, and animals that live on or very close to the sea surface develop a protective pigmentation, usually blue. It is necessary that organisms lacking this protection should not remain near the surface in sunlight. Ultraviolet light is quickly absorbed by seawater and even in the clearest ocean little penetrates more than a few metres. Protection from ultraviolet light can therefore explain the migrations of only the small group of organisms that make small changes of level near the surface.

Energy conservation

Although some energy must be used in vertically migrating, it is probable that the overall effect for many species is an appreciable energetic advantage. This results from the effects on metabolism of the thermal gradient through which the migrations occur. Many species feed mainly at night near the surface where food is abundant and the water relatively warm. They then move downwards to colder water during the day. It is an advantageous utilization of energy for the animal to be most active and capable of rapid movement for the short period of feeding, and then to retire for the remaining hours to rest in colder water in a relatively inactive state where conversion of food can be largely directed to growth and fecundity.

Shoaling advantage

Aggregation of a population within a narrow zone may confer some of the advantages of shoaling behaviour. Observation and experiment show that predators are often more successful in capturing prey which have become separated from a shoal than from the shoal itself. Individual prey are easily pursued and taken whereas the presence of a shoal lessens the efficiency of capture. The attacker is distracted, first one and then another target is chased; thus energy is wasted and time given for the prey to recover from pursuit. A shoal also possesses innumerable watchful eyes to detect the approach of a predator, with possibilities of imitation or communication within the shoal leading to evading action (Shaw, 1962).

4.6.2   Adaptations for life in the darkness

Vision

Certain creatures living at levels down to about 1000 m show modifications for vision in very dim illumination (Herring, 1996). The eyes of some deep-water fish are large and have exceptionally wide pupils, and, like many nocturnal land vertebrates, the light-sensitive cells of the retina are all rods. These rod cells are unusually long, and numerous rods supply each fibre of the optic nerve, an arrangement thought to form a retinal surface of extremely high sensitivity to light. The visual pigments (Bowmaker, 1976) are also different from those of shallow water fish, giving a maximum light absorption at the shorter wavelengths of the bluer light of deep ocean levels.

In coastal fish the main retinal pigments are the red-purple rhodopsins with maximum absorption at about 500 mμ; but the majority of deep-sea species of the mesopelagic zone have yellow or gold chrysopsins with a maximum absorption around 480 mμ and a much increased quantity of visual pigment in the receptor cells. It is estimated that the eyes of some deep-sea fish are 60 to 120 times as sensitive as human eyes and are able to detect daylight to depths of about 1150 m. In some deep-water species the eyes are tubular structures directed either forwards (*Gigantura*) or upwards (*Argyropelecus*, Figure 4.12a). Tubular eyes have a much increased resolving power, and probably also provide a degree of binocular vision to assist perception of distance, a particular difficulty in very dim light. Below 1000 m where darkness is virtually complete, there are fish with eyes which are very small (many ceratioids), degenerate (*Bathymicrops regis*) or even absent (*Ipnops*), although in larval stages living near the surface the eyes may be well formed.

Other senses

The ears and lateral line organs of fish are sensitive to vibrations in the water, and enable fish to detect and locate objects in their vicinity. Many species have very acute hearing, and in some deep-level fish the lateral line system is very well developed, for example, in myctophids and macrourids, and this must compensate

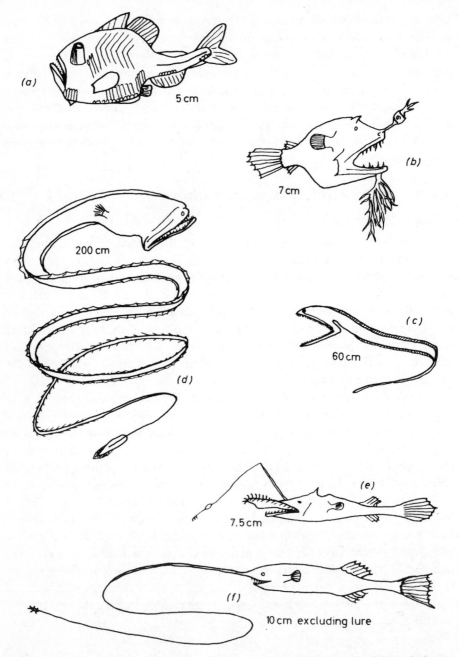

Figure 4.12   Some bathypelagic fishes showing various adaptations for life in darkness.
(a) Argyropelecus – *photophores and tubular eyes.* (b) Linophryne – *luminous lure and elaborate barbel.* (c) Eurypharynx – *large mouth and distensible stomach.* (d) Saccopharynx – *large mouth, distensible stomach and luminous tissue at end of elongate tail.* (e) Lasiognathus – *luminous lure with hooks.* (f) Gigantactis – *luminous lure. Note the small size of most of them (approximate lengths given in cm).*

to some extent for poor illumination and reduction or loss of vision. In some cases it appears that the swimbladder and acoustico-lateralis system function together for production and detection of vibration (Hawkins, 1973).

The olfactory organs of some species are exceptionally well developed and assist the detection and recognition of organisms and possibly also have a function in navigation. Very long tactile appendages are a frequent feature of animals that live in darkness, and are found in several forms among deep-sea creatures. Some fish have barbels of extraordinary length or elaboration (*Stomias boa*, *Linophryne*, *Ultimostomias*) (Figure 4.12b). Others have feeler-like fin rays (*Bathpterois*) or delicate, elongate tails (*Stylophorus*). The abundant family of deep-sea prawns, Sergestidae, have antennae of great length which are minutely hooked and possibly are used both for detecting and entangling their prey.

Colouration

The colouration of marine creatures is obviously related to the illumination of their surroundings. Many creatures of shallow water are protectively coloured, usually dark on the upper surface and whitish underneath. The upper parts are often mottled or patterned in a way which makes them very inconspicuous against their normal background, and in some cases they can rapidly change colour to match different surroundings. Some pelagic fish are difficult to see in water because their dorsal surfaces are darkly pigmented and their scales have a structure and arrangement which reflect an amount of light closely equivalent to the background illumination from almost any angle of view (Denton, 1971). In most cephalopods the skin contains both chromatophores and reflecting elements which together enable the animal to match the background. Deep-water creatures which at times come close to the surface may have a reflecting surface or be almost transparent. At middle depths, where only very faint blue light penetrates, there are many highly pigmented species, usually black, red, or brown. By reducing reflection these colours must in the dim blue illumination give virtual invisibility below 500 m. In the total darkness of great depths there are numerous non-pigmented forms, light or buff in colour.

Bioluminescence

The marine fauna includes a wide variety of bioluminescent species, with numerous examples known in almost all the major groups (Boden and Kampa, 1964; Herring, 1978; Tett and Kelly, 1973). Although the most numerous bioluminescent organisms are very small, chiefly dinoflagellates, the phenomenon is also commonly exhibited by many larger animals, especially fish, crustacea and cephalopods living in the mesopelagic zone. The common occurrence of bioluminescence in the sea, in contrast to its rarity on land, is probably the result of the much lower intensities of light in the sea. On land, even on the darkest

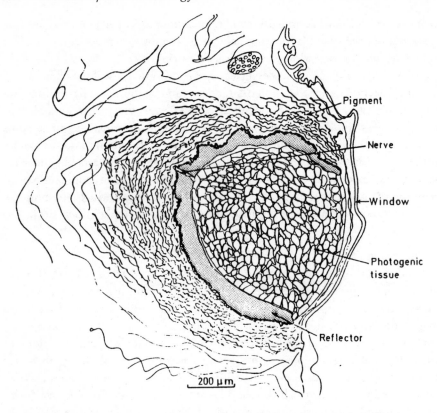

Figure 4.13   Transverse vertical section through a subocular light organ of a stomiatoid fish, Astronesthes elucens.
(From J.A.C. Nicol (1960). Studies on luminescence. On the subocular light organs of Stomiatoid fish. *J.M.B.A, UK*, 39, 529.)

nights, there is ample illumination for good vision by animals adapted for nocturnal life, but darkness in the sea is virtually total except for bioluminescence. It is especially common in warm surface waters, but evidently the deep levels are by no means uniformly dark because so large a part of the population carries light-producing organs (photophores).

The overall reaction involved in bioluminescence is the oxidation of a substrate, known generally as *luciferin*, catalysed by an enzyme, *luciferase*. The light energy released is transferred into another fluorescent compound which then emits its own light (of a different and characteristic wavelength). Almost all organisms that bioluminesce use a similar process with slight variants in the structure of the luciferin and luciferase between species. Only a few of the chemicals involved have been identified. Some animals discharge luminous secretions into the water. In others the reaction is intracellular, the mass of photogenic cells often being backed by a reflecting layer (Figure 4.13) and sometimes covered by a lens. Masses of luminous bacteria within the tissues are responsible for the light production in a few cases.

Bioluminescence is employed by marine animals in myriad ways some of which are described below (Robison, 1995). These include improving vision, attracting prey, recognition, for defence, attracting mates and territorial defence and for camouflage. However, why it should be advantageous to certain bacteria to luminesce is not obvious. It is possible that in some cases biochemical reactions that produce light are fulfilling some important function in connection with intracellular oxidations, the emission of light being merely incidental. However, although light may be of no direct use to bacteria, the value of this light to other organisms is so great that it has led to the evolution of symbiotic relationships of remarkable refinement in the elaboration of various types of photophores containing luminous bacteria.

In some species, photophores evidently emit beams of light which illuminate the field of vision. The fish *Pachystomias* has light organs close to the eyes which emit flashes of red light (Denton, 1971). In contrast to most deep-level fishes, which are sensitive mainly to blue light, the eyes of *Pachystomias* respond to red light enabling it to see without being seen. Light may also serve to attract prey, and a number of deep-water fish have remarkable luminous lures. These may take several forms, such as the luminous barbels in *Linophyrne* (Figure 4.12b) or structures resembling a fishing line with luminous bait (*Gigantactis*, Figure 4.12f) in some cases complete with hooks (*Lasiognathus*, Figure 4.12e). In *Galatheathauma* there is a luminous lure within the capacious mouth.

Photophores are often arranged in highly distinctive patterns which differ between closely related species and between the sexes, probably providing a means of recognition and facilitating shoaling. Some creatures can flash their lights on and off, and this may serve to confuse, alarm or dazzle attackers, or in some cases may even attract the attackers' own enemies. Luminous clouds squirted into the water, such as the luminous ink of the squid *Heteroteuthis*, and some worms, are presumably a means of defence, perhaps by dazzling or possibly by making attacking creatures themselves visible to their own predators. In some cases the attacker becomes coated in sticky luminous material.

The luminescence of surface water usually associated with blooms of dinoflagellates may possibly have some protective function by discouraging the upward migration of grazing copepods. They produce their light when stimulated mechanically. Such 'contact flashing' also occurs in mid-water amongst the larger gelationous animals and, once started, can spread as previously quiescent fish and other animals move away causing a 'storm' of light.

A notable feature of the photophores of many migratory species of the upper mesopelagic zone between about 200 and 700 m is that they are positioned so that their light is apparently directed mainly downwards, even in animals whose eyes are directed upwards, for example the hatchet fish *Argyropelecus*, and

Figure 4.14   Diagram to illustrate how downwardly directed photophores may reduce the conspicuousness of animals at deep levels when seen out of focus. The two objects represent the silhouettes of fish, one with and one without photophores, seen from below against a background of weak illumination from above. If the diagram is viewed in a dim light or with the eyes sufficiently closed to blur the outlines, the right-hand object becomes less visible than the left.

Opisthoproctus. These photophores can hardly play any part in illuminating the field of vision, and the most likely explanation of their function is that they camouflage the shadow of the animal when viewed from below against the faint background of light from above. If these photophores emit downwards a light equal in intensity to the background illumination, the silhouette of the creature must virtually disappear (Figure 4.14). Many mesopelagic euphausids, decapods and fish of the upper 700 m have bodies which are mainly transparent or silvery, though with some pigmentation of the dorsal surfaces, and have downwardly-directed photophores close to the most opaque parts of the body, and in many cases internally situated. Some of these animals are able to adjust both reflectivity and light emission to suit changing illumination, and their vertical migrations may well serve to keep them within the light conditions under which they are least visible. Detailed examination of the reflectors associated with downwardly directed photophores in the hatchet fish *Argyropelecus aculeatus* (Denton, 1971) reveals that these reflectors direct part of the light laterally, the light intensity varying with direction in a way similar to the reflection of light from above by the scales of silvery fishes, whereby they are made inconspicuous when seen from the side.

The mesopelagic macrofauna can be broadly divided into two main groups,

one living above about 700 m and the other below this depth. The two groups
have rather different adaptations with respect to bioluminescence and pigmentation. Above 700 m there are many species, both fish and crustacea, which are
transparent but possess a few relatively large chromatophores. Many of the fish
have silvery scales covered by dark chromatophores which can be expanded or
contracted fairly rapidly to regulate the reflectivity of the scales. The crustacea
have semi-transparent cuticles but are termed 'half-red' species because they
become scarlet if their chromatophores are fully expanded. In both the fish and
crustacea of this group, ventrally-placed, downwardly-directed photophores are
common, usually internally positioned beneath the densest parts of the body, and
the eyes are well developed. Many of the fish have swimbladders and make diel
vertical migrations.

Below 700 m the animals are more opaque; the fish are usually dark in colour
and the crustacea, the 'all-red' species, have heavily pigmented cuticles of orange
chitin with numerous small chromatophores covering the surface. In this group
bioluminescence is less common, and photophores, if present, are superficially
placed, not directed downwards, and evidently have functions other than camouflage by counter-illumination. Fewer of these species make vertical migrations, the
eyes are generally smaller and many of the fish lack swimbladders.

There can be little doubt that these differences between the two groups relate
mainly to different light conditions in the two zones. The upper group lives where
solar light fluctuates appreciably, and can adapt to these changes in several ways.
These animals can vary from transparent to fully opaque, and from reflective to
non-reflective, by expansion of their chromatophores. The fish can become darkly
pigmented and the crustacea scarlet – a colour non-reflective in the blue light which
is the only part of the solar spectrum to penetrate to these depths. Many of these
creatures use counter-illumination for camouflage, matching the background
illumination as they make changes of level. The lower group lives where the light is,
at most, very dim. These are not translucent, nor do they need to make use of
counter-illumination. They have slight control of reflectivity by changes in their
chromatophores, the crustacea changing colour between orange and scarlet.

Bioluminescence is by no means limited to the pelagic species. There are many
luminescent benthic animals, notably polychaetes, echinoderms and the common
piddock (*Pholas dactylus*). Quite why piddocks should be luminescent since the
adults are permanently trapped in the holes they bore into rock and wood,
remains a mystery. Recently the chemicals responsible for the piddock's
luminescence have been successfully extracted and used as a medicinal 'tracer' in
the human body in place of radioactive tracers.

Behaviour

Observations have been made from submersibles on the behaviour of certain
mesopelagic fish. These have revealed that some species of Myctophid at night swim

horizontally but during the daytime rest in a head-upwards or head- downwards position. This orientation must obviously reduce the size of their shadow when viewed from below. These fish have forwardly directed eyes, so the head-up attitude must assist detection of shadows above them. A variety of animals including fish such as eel pouts, have been observed to curl up and hang motionless when startled or attacked (Robison, 1995). Some scientists believe this may be a form of mimicry designed to hide them from predators. In the gloom, rounded shapes are usually associated with unpalatable animals such as medusae.

Reproduction

In the sparse population of the dark levels of the sea, finding a mate must present a problem. Possession of photophores or the ability to communicate by sound or scent must greatly increase the chances of success. Light signals can provide a means of communication in social and courtship displays and in warning and territorial behaviour. An additional adaptation for this purpose has been the evolution of dwarf parasitic males. For example, in some ceratioid angler fish the male is much smaller than the female, and relatively undeveloped except that his eyes and olfactory organs are large and his teeth sharp and specialized for gripping the female. Once she has been found, perhaps partly by scent, the male apparently bites into her skin and remains permanently attached. His eyes and olfactory organs then degenerate and a partial fusion of tissues occurs between the two individuals, the male drawing nutriment from the female by intimate, placenta-like association of the two blood systems. The testes develop and ripen so that sperms can be shed in the immediate vicinity of the eggs.

Feeding

Certain fish of deep levels are equipped with relatively huge mouths and jaws and remarkably distensible stomachs, for example *Saccopharynx*, *Eurypharynx* (Figure 4.12c and d) and *Gigantura*. The jaws are often the only well-developed parts of the skeletal and muscular systems, and their swimming movements must be sluggish. In zones of darkness and scarce food it is evidently important to be able to take very large mouthfuls on the infrequent occasions when prey is found. These deep-level fish are mostly quite small, seldom more than about 10 cm in length. There is insufficient food within the water column to support many large bathypelagic species, but on the deep sea floor aggregation of food provides sufficient for a more numerous population of larger, bottom-feeding animals. Some of the macrurid fish at 5000–6000 m grow to over 1 m in length.

4.7   Currents

The major ocean currents have been described earlier (see page 9 ff.). Currents keep the water well mixed, influence the distribution of salinity and heat, bring to the surface the nutrients necessary for plant growth and carry down supplies of oxygen to deep levels. Because the bottom layers of water are in movement, it has been possible for a diverse benthic population to evolve which includes many forms living attached to or embedded in the sea bottom, and relying upon the flow of water to carry food and oxygen to them and waste products away. The benthic population is also influenced by the speed at which the bottom water moves because of effects on the nature of the sediment and the settlement of pelagic larvae.

Currents have a direct influence on the distribution of many species by transporting them from place to place, especially the smaller holoplanktonic forms. Many benthic and nektonic species also start life with a planktonic phase, and the direction of drift of their eggs and larvae must to some extent determine their areas of colonization. Currents probably also provide a means of navigation for animals such as turtles (Carr, 1974) and fish (Harden Jones, 1968). Some fish show a tendency to swim against the stream, at least during certain phases of life, and at the approach of the spawning period this behaviour is important in determining the position of spawning areas. Afterwards, the success of the brood depends upon the drift of larval stages to suitable nursery grounds.

How and to what extent can aquatic creatures detect water currents? Close to the bottom or any other fixed object they can obviously be aware of water movements through sight or touch; but where the water is flowing in a straight line at constant velocity, a floating organism out of sight or contact with fixed reference points can have little evidence of movement. Probably all animals are sensitive to acceleration and may therefore respond to velocity gradients or rotational flow of water. Because different water layers usually move at different velocities and often in different directions, there are possibilities of detecting water movement from effects of pressure or turbulence, or by observation of the movement of other floating objects above or below. There is also a theoretical possibility, though no firm evidence supports the idea, that organisms may be sensitive to the slight electromotive forces generated in the water or within themselves by movement through the earth's magnetic field.

4.7.1 Plankton indicators

Because different communities of organisms are found in different parts of the sea, it is possible to distinguish particular bodies of water not only by their physical and chemical features, but also to some extent by their characteristic populations. Moving water carries with it an assortment of planktonts which can be regarded as natural drift bottles, and by observing the distribution and intermingling of different planktonic populations it may sometimes be possible to trace the movement and mixing of the water.

Large easily identified planktonts which are characteristic of particular bodies of water serve as convenient labels or 'indicators' of the water. Pelagic larvae of certain benthic species sometimes have special value because when they appear in an area where the adults do not occur, or if present do not breed, these larvae must have been carried there by the flow of water. The study of the distribution of plankton indicators has several advantages as a method of investigating water movements. It may be simpler to obtain plankton samples than hydrographic data; and the biological evidence may be more informative because, where mixing occurs between different bodies of water, their characteristic populations may remain recognizable longer than any distinctive hydrographic features can be detected. Where conditions are changing, some species quickly succumb while hardier organisms survive longer. The distribution of a range of species of different tolerances may therefore give some indication of alterations in the quality of water as it moves from place to place. Plankton indicators are not as important as they used to be and their use has now largely been superseded by modern detection methods including the use of anthropogenic tracers (see page 51). However, they can still signify changes which may have been missed by current meters and other hydrographic indicators and it is still interesting to look at some of the classic work done in this field. As an example, we will consider the distribution of some planktonic species around the British Isles.

During the period 1920–1927, Meek (1928) studied the distribution in the northern part of the North Sea of the two commonest chaetognaths of the area, *Sagitta setosa* and *Sagitta elegans*, and correlated his findings with hydrographic data. He observed that the two species are seldom found together. *S. setosa* occurs throughout the greater part of the North Sea, and is the only planktonic chaetognath normally found in the southern part of the North Sea. Inshore along the east Scottish coastline *S. setosa* is less common, and here *S. elegans* usually predominates. Further south along the Northumbrian coast, the abundance of the two species fluctuates. Meek concluded that when hydrographic conditions indicate a strong flow of Atlantic water into the northern part of the North Sea, *S. elegans* spreads further south, mainly in the inshore waters along the British coast. Alternatively, when the flow of Atlantic water into the North Sea is weak, the distribution of *S. elegans* retreats and *S. setosa* extends further north.

S. elegans and *S. setosa* both occur in the English Channel, and their distribution there was investigated by Russell (1935, 1939). As in the North Sea, the two species have rather different distributions which vary from time to time. Prior to 1931, *S. elegans* was dominant to the west of Plymouth, *S. setosa* to the east. In the 'elegans' water Russell found several other planktonic animals often present, including the medusae of *Cosmetira pilosella*, the hydroid stage of which does not usually occur in the Channel but is found in deeper water along the Atlantic shores of the British Isles, also the trachymedusan *Aglantha digitale*, the pteropod *Clione limacina* and the euphausids *Meganyctiphanes norvegica* and *Thysanoessa inermis*. These were absent from the 'setosa' water further east. This group of organisms found in 'elegans' water was thought to come from the west, flowing into the Channel from the Celtic Sea area to the south of Ireland and around the Scillies. Russell designated this water as 'western' water to distinguish it from the water of the main part of the English Channel, and also from 'south-western' water which occasionally flows into the Channel from the Bay of Biscay bringing a different characteristic collection of warmer-water forms, notably the copepod *Euchaeta hebes* and the medusa *Liriope tetraphylla*. It is now known that the species associated by Russell with 'western' water usually enter the Channel from the north-west rather than from the west, and consequently they are nowadays referred to as 'north-western' forms.

After 1930, the boundary between Channel water and 'north-western' water, as indicated by the distribution of the two species of chaetognath, lay further west than previously, approximately at the longitude of Land's End. This shift of the boundary was accompanied by changes in the quality of the water off Plymouth. Mean water temperature increased by about 0.4°C compared with the previous thirty years. Chemically, a lower concentration of phosphate during the winter months was observed. Biologically, it was noted that in subsequent years there were changes in the population, the plankton becoming sparser and less varied with a poorer survival of most types of fish larvae during the summer months. Herring shoals became so much reduced that the Plymouth fishery was eventually abandoned. The changed conditions, however, seem to have favoured the pilchard, which increased in numbers in the Channel as the herring declined.

Work by Russell (1973) and Southward (1974) showed that after 1965 there was a return to the pre-1930 conditions in the English Channel. *S. elegans* occurred in much greater numbers off Plymouth than during the 1930–60 period, the plankton was richer with much larger numbers of fish larvae surviving, and there were more herring and fewer pilchards. These changes were associated with a slight climatic shift towards cooler conditions following the exceptionally cold winter of 1962–3, favouring north-western species rather than southern (Reid, 1975). The mean sea temperature dropped by about 0.4°C and the seasons were slightly later.

Russell also related the distribution of the two species of chaetognath to the rate of flow of the current through the Strait of Dover, normally a net flow of water eastwards from the Channel into the North Sea. Russell showed that the stronger this current through the Strait, the further eastwards *S. elegans* spreads up the English Channel and vice versa. In both the western part of the Channel and the northern part of the North Sea, *S. elegans* and *S. setosa* are regarded as indicators of different qualities of water. Studies of the overall distribution of the two species have shown that, around the British Isles, *S. setosa* is restricted to the neritic water of lowish salinity found in the Channel, the North Sea, the Bristol Channel and Irish Sea. *S. elegans* is also a neritic form but occurs in slightly more saline water than *S. setosa*. It predominates in those regions where oceanic and coastal water become mixed, and is consequently limited around the British Isles chiefly to the western and northern areas and the northern part of the North Sea. One factor influencing its distribution is the extent to which Atlantic water flows over the continental shelf. The boundary between the two species is by no means distinct. In the western English Channel they sometimes exist together, one or the other being more numerous at different times and places, and it is often difficult to know to what extent these variations are due to either circulatory or climatic changes.

Following the early investigations by Meek and Russell, plankton surveys around the British Isles have provided further information about the distribution of organisms which can be related to particular areas of water (Lee and Ramster, 1977; Colebrook *et al.*, 1991). It is beyond the scope of this book to attempt any detailed account of this complicated subject but the following is a brief summary. First we should note that the water of the North Sea and English Channel is derived from several sources, being a mixture of ocean water entering through the western mouth of the Channel and the northern mouth of the North Sea, and diluted by appreciable quantities of low-salinity water from the Baltic and fresh water from rivers (Lee, 1970). The plankton of the area therefore includes endemic neritic forms, to which may sometimes be added oceanic species carried in from deep water, and estuarine or brackish species carried out to sea.

The plankton of the eastern end of the English Channel, the southern part of the North Sea and much of the Irish Sea consists predominantly of organisms which prefer a slightly reduced salinity and are appreciably euryhaline and eurythermal to suit the somewhat fluctuating conditions of the region. This water is indicated by the presence of *S. setosa*. The copepods *Temora longicornis*, *Centropages hamatus*, *Isias clavipes*, *Labidocera wollastoni* and *Oithona nana* are common in the area, and the phytoplankton often includes *Biddulphia sinensis*, *Asteroniella japonica*, *Nitzchia closterium*, *Eucampia zodiacus* and *Bellarochia malleus*. Because this water is shallow and close to extensive coastlines its plankton also contains many meroplanktonic forms, eggs, spores, medusae and larval stages of innumerable benthic and intertidal organisms which are not often found in more remote, deeper ocean water. These constituents vary seasonally according to the breeding habits of the different contributors and are a characteristic feature of neritic plankton.

In the northern part of the North Sea, the western part of the English Channel and off the west and north coasts of the British Isles, there is a higher proportion of water of oceanic origin than occurs in the southern North Sea and eastern Channel. The higher concentration of nutrients derived from deep water gives a rather richer and more varied plankton than that of 'setosa' water. We have already referred to this area of mixed oceanic-neritic water in connection with the distribution of *S. elegans*, and other organisms found in this water include the trachymedusa *Aglantha digitale*, the copepods *Metridia lucens*, *Candacia armata* and *Centropages typicus*, the euphausids *Meganyctiphanes nervegica*, *Nyctiphanes couchi*, *Thyanoessa inermis* and *T. longicaudata*, the polychaete *Tomopteris helgolandicus* and the pteropod *Spiratella retroversa*. All these animals are not confined to the same water as *S. elegans*. Most of them are also abundant in deep water further west and north. *T. inermis* and *N. couchi* are essentially shallow-water forms approximately limited to 'elegans' water. Another euphausid, *T. raschii*, occurs in 'elegans' water and also extends right across the northern part of the North Sea to the Skagerrak. The distribution of several species often alters seasonally in a somewhat irregular way. Along the east coast of Scotland and England there is a general tendency for the 'elegans' association to spread southwards during summer and autumn towards the southern part of the North Sea. But *T. inermis* does not usually penetrate far south in the North Sea, where it is typically found only in spring in the northern part. *N. couchi* is usually absent from the North Sea in summer but appears in the north in autumn and spreads south during winter. The population of 'elegans' water also contains many meroplanktonts from the shore and sea bottom, which change seasonally. As in the English Channel, the 'elegans' water of the North Sea has larger numbers of surviving fish larvae than the 'setosa' water.

Ocean water flowing over the continental shelf towards the British Isles comes from three main sources: (a) the North Atlantic, (b) the Arctic and (c) the Bay of Biscay and Mediterranean (Figure 4.15):

(a) The major influence is the North Atlantic Drift, driven north-eastwards from lower latitudes by the westerly winds and bringing with it a population of oceanic species. These may reach the western end of the Channel and Irish Sea, or, when the flow of North Atlantic Drift water is strong, may be carried right round the north of Scotland and enter the northern part of the North Sea. The water derives from a vast expanse of ocean, and the population it carries varies somewhat with the direction of inflow. Distinctive species from the west of Ireland include the chaetognath *Sagitta tasmanica*, the copepods *Rhincalanus nasutus*, *Pleuromamma robusta* and *Mecynocera clausi*, and the euphausids *Nematoscelis megalops* and *Euphausia krohnii*. An inflow of water from further south (south-western water) may bring with it the chaetognath *S. serratodentata (atlantica)*, the trachymedusan *Liriope tetraphylla*, the siphonophoran *Muggiaea kochi*, the copepods *Euchaeta hebes* and *Centropages bradyi* and the pteropod *Spiratella leseuri*. Depending upon movements of the water, all these Atlantic forms may become very intermingled and are sometimes carried far to the north. In the northern hemisphere, all salps, doliolids, heteropods and species of *Pyrosoma* are said by Russell to be of warm-water origin, but *Ihlea asymmetrica* and *Salpa fusiformis* are carried well to the north of the British Isles.

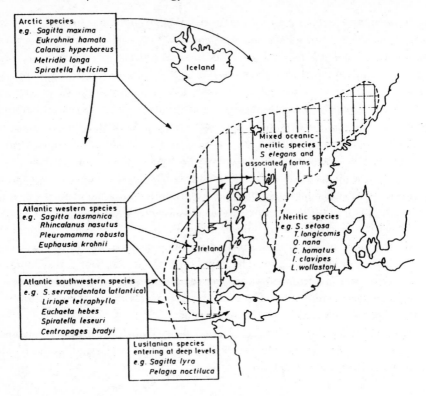

Figure 4.15   *The main areas of distribution around the British Isles of neritic species and the species of mixed oceanic-neritic water, and the chief directions from which oceanic species may enter the region.*

(b)   If water from the Arctic spreads southwards around the British Isles, it may bring with it a cold-water population including the chaetognaths *S. maxima* and *Eukrohnia hamata,* the copepods *Calanus hyperboreus* and *Metridia longa,* and the pteropod *Spiratella helicina.* Being cold, this water tends to move under the surface layers.

(c)   Water of Mediterranean origin reaching the British Isles derives from the bottom outflow through the Strait of Gibraltar. Most of this water joins the Atlantic deep current, but a stream appears to diverge northwards and sometimes upwells along the continental slope to the north-west of the British Isles. It has been named 'Gulf of Gibraltar' water and carries a Lusitanian plankton which may include the chaetognath *S. lyra,* species of *Sapphirina,* the scyphomedusan *Pelagia noctiluca* and various siphonophorans, salps and doliolids.

There are, of course, many common species which are ubiquitous throughout the north-east Atlantic, and therefore not indicative of any particular body of water in the area. Examples are *Calanus helgolandicus, Acartia clausi, Pseudocalanus elongatus* and *Paracalanus parvus*. Also, many of the organisms mentioned here as indicators have an extensive distribution beyond this area. For instance *Centropages hamatus* and *Temora longicornis*, both listed above as restricted mainly to the water of slightly lowered salinity in the North Sea, eastern English Channel and Irish Sea, occur also on the western side of the Atlantic in neritic water off Nova Scotia and Newfoundland. However, neither of them is normally to be found in the intervening deep water far from the continental shelf.

Much caution is necessary in interpreting plankton indicator studies. The distribution of any planktont changes not only with movements of water but also with variations of temperature or other parameters. Consequently the gradual extension of a particular plankton population into an adjacent area may be due less to an actual inflow of water than to some alteration in water quality associated with climatic or biological change, favouring one group of species at the expense of another. So far as attempts have been made to correlate direct current measurements with observed changes in distribution of plankton indicators, these have not been invariably successful. Changes in boundaries between different plankton populations are certainly not caused solely by water movements. Nevertheless, when planktonts sporadically appear outside their normal range, it is very likely that they have been carried there by the water.

References and further reading

Student texts

Lalli, C.M. and Parsons, T.R. (1993). *Biological oceanography: an Introduction*. Pergamon Press.

Menzies, R.J. (1965). Conditions for the Existence of Life on the Abyssal Sea-floor. *Oceanogr. Mar. Biol. Ann. Rev.*, 3, 195.

Norman, J.R. (1963). *A History of Fishes*. 2nd edition. J.H. Greenwood, ed. London, Benn.

Summerhayes, C.P. and Thorpe, S.A. (Eds) (1996). *Oceanography: An Illustrated Guide*. Southampton Oceanographic Centre, Manson Publishing.

References

Alvarino, A. (1965). Chaetognaths. *Oceanogr. Mar. Biol. Ann. Rev.*, 3, 115.

Bainbridge, R. (1952). Underwater Observations on the Swimming of Marine Zooplankton. *J. Mar. Biol. Ass. UK*, 31, 107.

Bainbridge, R. (1958). Swimming Speeds of Fish. *J. Exp. Biol.*, 35, 109–32.

Blaxter, J.H.S. (1973). Illumination and migration in fish larvae. *J. Mar. Biol. Ass. UK*, 53, 635–47.

Boden, B.P. and Kampa, E.M. (1964). Planktonic Bioluminescence. *Oceanogr. Mar. Biol. Ann. Rev.*, 2, 341.

Bowmaker, J.K. (1976). Vision in Pelagic Animals. In *Adaptation to Environment*, R.C. Newell, ed. London, Butterworths.

Carr, A. (1974). Migration of the Green Turtle. *Nature, London,* **249**, 128.

Clarke, M.R. (1978). Buoyancy Control in the Sperm Whale. *J. Mar. Biol. Ass. UK,* **58**, 27–71.

Clarke, M.R., Denton, E.J. and Gilpin-Brown, J.B. (1979). Buoyancy of Squid. *J. Mar. Biol. Ass. UK,* **59**, 259–76.

Colebrook, J.M., Warner, A.J., Proctor, C.A., Hunt, H.G., Pritchard, P., John, A.W.G., Joyce, D. and Barnard, R. (1991). *60 years of the continuous plankton recorder survey: a celebration.* The Sir Alister Hardy Foundation for Ocean Science.

Denton, E. (1971). Reflectors in Fishes. *Scient. Am.,* **224**, January, 65.

Denton, E.J. and Marshall, N.B. (1958). The Buoyancy of Bathypelagic Fish without a Gas-filled Swimbladder. *J. Mar. Biol. Ass. UK,* **37**, 753.

Dietz, R.S. (1962). The Sea's Deep Scattering Layers. *Scient. Am.,* **207**, August 1962.

Earll, R. and Farnham, W. (1983). Biogeography. In: *Sublittoral ecology; the ecology of the shallow sublittoral benthos.* R. Earll and D.G. Erwin, eds. Clarendon Press.

Ekman, S. (1953). *Zoogeography of the Sea.* London, Sidgwick and Jackson.

Erwin, D.G., Picton, B.E., Connor, D.W., Howson, C.M., Gilleece, P. and Bogues, M.J. (1990). *Inshore marine life of Northern Ireland.* Ulster Museum. HMSO Belfast.

Farquhar, G.B. (1977). Biological Sound Scattering in the Oceans. *Mar. Science,* **5**, 493–523.

Glasby, G.P. (1973). Role of Submarine Volcanism in Genesis of Marine Manganese Nodules. *Oceanogr. Mar. Biol. Ann. Rev.,* **11**, 27–44.

Grasshof, K. (1976). *Methods of Seawater Analysis.* Weinheim and New York, Verlag Chemie.

Gray, J. (1968). *Animal locomotion.* Weidenfeld and Nicholson.

Greenlaw, C.F. (1979). Acoustical Estimates of Zooplankton. *Limnol. Oceanogr.,* **24**, 226–42.

Harden Jones, F.R. (1968). *Fish Migration.* London, Edward Arnold.

Hardy, A.C. (1956). *The Open Sea. Part 1. The World of Plankton.* London, Collins.

Hardy, A.C. and Bainbridge, R. (1954). Experimental Observations on the Vertical Migrations of Plankton Animals. *J. Mar. Biol. Ass. UK,* **33**, 409.

Hawkins, A.D. (1973). Sensitivity of Fish to Sounds. *Oceanogr. Mar. Biol. Ann. Rev.,* **11**, 291–340.

Hempleman, H.V. and Lockwood, A.P.M. (1978). *The physiology of diving in man and other animals.* The Institute of Biology's Studies in Biology no. 99. Edward Arnold.

Herring, P.J. (ed.) (1978). *Bioluminescence in Action.* London, Academic.

Herring, P.J. (1996). Light, Colour, and Vision in the Ocean. In: *Oceanography: An illustrated guide.* C.P. Summerhayes and S.A. Thorpe, eds. Southampton Oceanographic Centre, Manson Publishing, 212–27.

Herring, P.J., Campbell, A. K., Whitfield, M. and Maddock, L. (eds) (1990). *Light and Life in the Sea.* Cambridge University Press.

Hill, M.N. (ed.) (1963). *The Sea.* Vol. 2. Chap. 1. The oceans as a chemical system. New York and London, Interscience.

JNCC (1995-7). *Coasts and Seas of the United Kingdom.* The JNCC Coastal Directories project. Joint Nature Conservation Committee (17 vols covering different areas).

Johnston, R. (1969). On Salinity and its Estimation. *Oceanogr. Mar. Biol. Ann. Rev.,* **7**, 31.

Jorgensen, C.B. (1976). Dissolved Organic Matter in Aquatic Environments. *Biol. Rev.,* **51**, 291–328.

Kinne, O. (1963). The Effects of Temperature and Salinity on Marine and Brackish Water Animals. 1. Temperature. *Oceanogr. Mar. Biol. Ann. Rev.,* **1**, 301.

Kinne, O. (1964). The Effects of Temperature and Salinity on Marine and Brackish Water Animals. 2. Salinity and temperature–salinity relations. *Oceanogr. Mar. Biol. Ann. Rev.*, **2**, 281.

Knight-Jones, E.W. and Morgan, E. (1966). Responses of Marine Animals to Changes in Hydrostatic Pressure. *Oceanogr. Mar. Biol. Ann. Rev.*, **4**, 267.

Lee, A. (1970). Currents and Water Masses of the North Sea. *Oceanogr. Mar. Biol. Ann. Rev.*, **8**, 33–71.

Lee, A.J. and Ramster, J.W. (1977). *Atlas of Seas around the British Isles*. Fisheries Research Technical Report No. 20. Lowestoft, MAFF (this has now been superseded and updated by a computer database called: 'United Kingdom Digital Marine Atlas, 1992 (UKDMAP)' available from the British Oceanographic data centre).

MacIntyre, F. (1970). Why the Sea is Salt. *Scient. Am.*, **223**, 104–15.

Meek, A. (1928). On *Sagitta elegans*, and *Sagitta setosa* from the Northumbrian Plankton. *Proc. Zool. Soc. Lond.*, 743.

Newell, R.C. (1976). *Adaptations to Environment: Essays on the Physiology of Marine Organisms*. London, Butterworths.

Palmer, J.D. (1974). *Biological Clocks in Marine Organisms*. London, Wiley.

Pytkowicz, R.M. (1968). The Carbon Dioxide-Carbonate System at High Pressures in the Oceans. *Oceanogr. Mar. Biol. Ann. Rev.*, **6**, 83.

Reid, P.C. (1975). Large-Scale Changes in North Sea Phytoplankton. *Nature, London*, **257**, 217–19.

Rice, A.L. (1964). Observations on the Effects of Changes of Hydrostatic Pressure on the Behaviour of some Marine Animals. *J. Mar. Biol. Ass. UK*, **44**, 163.

Robison, B.H. (1995). Light in the Ocean's midwaters. *Scient. Am.*, **273** (1), 49–56, July 1995.

Russell, F.S. (1935). On the Value of Certain Plankton Animals as Indicators of Water Movements in the English Channel and North Sea. *J. Mar. Biol. Ass. UK*, **20**, 309.

Russell, F.S. (1939). Hydrographical and Biological Conditions in the North Sea as indicated by Plankton Organisms. I. *Cons. perm. int. Explor. Mer.*, **14**, 171.

Russell, F.S. (1973). Summary of Observations on Occurrence of Planktonic Stages of Fish, 1924–1972. *J. Mar. Biol. Ass. UK*, **53**, 347–55.

Shaw, E. (1962). The Schooling of Fishes. *Scient. Am.*, **206**, June.

Smayda, T.J. (1970). Suspension and Sinking of Phytoplankton in the Sea. *Oceanogr. Mar. Biol. Ann. Rev.*, **8**, 353–414.

Somero, G.N., Siebenaller, J.F. and Hochachka, P.W. (1983). Biochemical and physiological adaptations of deep-sea animals. In: *The Sea*, Vol. 8, G.T. Rowe, ed., pp. 331–70. New York, Wiley-Interscience.

Southward, A.J. (1974). Abundance of Pilchard Eggs. *J. Mar. Biol. Ass. UK*, **54**, 641–9.

Tett, P.B. and Kelly, M.G. (1973). Marine Bioluminescence. *Oceanogr. Mar. Biol. Ann. Rev.*, **11**, 89–173.

Toggweiler, J.R. (1988). Deep-sea carbon, a burning issue. *Nature*, London, **334**, 468–9.

Triantafyllou, M.S. and G.S. (1995). An efficient swimming machine. *Scient. Am.*, **272** (3), March 1995, 40–8.

Williams, P.J. Le B. (1975). Biological and chemical aspects of dissolved organic matter in sea water. In Chemical oceanography, 2nd edn. J.P. Riley and G. Skirrow, eds., pp. 301–63. London, Academic Press.

Wynne-Edwards, V.C. (1962). *Animal Dispersion in Relation to Social Behaviour*, Chap. 16. Vertical Migration of the Plankton, p. 366. Edinburgh and London, Oliver and Boyd.

5 Organic production in the sea

5.1 Primary production

The synthesis of organic compounds from the inorganic constituents of seawater by the activity of organisms is termed *production*. It is effected almost entirely by the photosynthetic activity of marine plants. The raw materials are water (H_2O), carbon dioxide (CO_2) and various other substances known as nutrients. The latter are mainly inorganic ions, principally nitrate and phosphate. Chlorophyll-containing plants, by making use of light energy, are able to combine these simple substances to synthesize complex organic molecules. This is termed *gross primary production*. The chief products are the three major categories of food materials, namely carbohydrates, proteins and fats (Steeman Nielsen, 1975). Oxygen, derived from the water, is produced as a byproduct. The process involves a number of steps but can be summarized by the following very general equation:

$$\underset{\substack{\text{Carbon}\\\text{dioxide}}}{6CO_2} + \underset{\text{Water}}{6H_2O} \overset{\text{Photosynthesis} \rightarrow}{\underset{\text{Respiration}}{\rightleftharpoons}} \underset{\substack{\text{Carbohydrate}}}{C_6H_{12}O_6} + \underset{\text{Oxygen}}{6O_2}$$

Some of the organic material manufactured by plants is broken down again by an oxidative reaction, to an inorganic state by the plants themselves in the course of their respiration. Hence the equation is written as a reversible one. The remainder is referred to as *net primary production* and much of this becomes new plant tissue. This is of major importance as the source of food for herbivorous animals. The animal population of the sea therefore depends, directly or indirectly, upon the net primary production.

By far the greater part of primary production in the sea is performed by the phytoplankton (Raymont, 1963, 1966). Under favourable conditions this is capable of remarkably rapid growth, sometimes producing its own weight of new organic material within 24 hours, a rate greater than that achievable by land plants. The large marine algae (seaweeds) growing on the sea bottom in shallow water make only a relatively small contribution to the total production in the sea

166

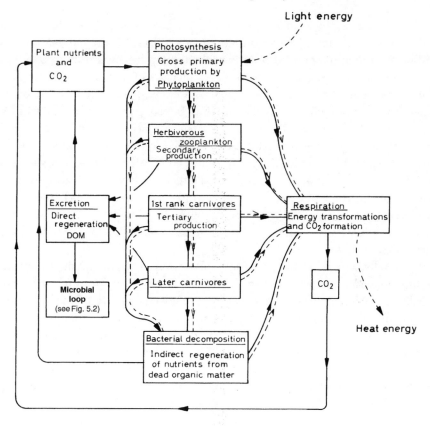

Figure 5.1 An outline of the main stages of the organic cycle in the sea. Dotted lines show flow of energy; unbroken lines show flow of materials.

because they are of very restricted distribution. There is also some primary production by bacterial chemosynthesis (see Section 5.1.4).

5.1.1   The food web

According to the simplest concept of feeding relationships, consumption of plants by herbivorous animals leads to the formation of animal tissue. This is *secondary production*. Herbivores provide food for the first rank of carnivorous animals i.e. *tertiary production*. These may in turn fall prey to other carnivores, and so on (Figure 5.1). Each of these successive stages of production of living tissue is a link in a food chain, each link being termed a *trophic level*. Because many animals take food from several trophic levels, food chains become interconnected to form intricate *food webs*.

Obviously there are large losses of organic material between each trophic level, caused in several ways. For instance, a proportion of the organisms at each trophic level are not eaten by animals but simply die and decompose by autolysis and bacterial action. Some of the food that animals consume is egested unassimilated and most of their assimilated food is broken down by respiration, leaving only a small proportion to form new tissue. The efficiency of transfer of organic matter from one trophic level to the next varies with the types of organisms, herbivores generally doing rather better than carnivores. In broad terms about 100 g of food are consumed for every 10 g of animal tissue formed, i.e. a gross conversion efficiency of 10 per cent (see page 248). Herbivorous zooplanktonts sometimes exhibit efficiencies of about 30 per cent and certain larval stages do somewhat better, but even allowing for these higher efficiencies only a small part of the original plant production becomes incorporated in animal tissue. Thus, primary production can be regarded as the broad base of a *food pyramid*, the successive smaller trophic levels being a series of steps towards the apex of final rank carnivores.

Eventually, as a result of respiration and excretion, death and decomposition, organic materials become broken down and returned to the water as simpler substances which plants can utilize in primary production. In this way, matter is continually cycled from inorganic to organic forms and back to inorganic state. The initial synthesis of organic material involves the intake of energy to the system, and this is supplied by sunlight. The transference of organic matter from one trophic level to the next is part of the energy flow of the cycle (Figure 5.1), energy being continually lost from the system and in due course becoming dissipated as heat. This is discussed further in Chapter 7.

5.1.2   The microbial loop

In the marine ecosystem the foregoing elementary account of the organic food cycle must be extended to take account of the significance of dissolved organic matter (DOM) in seawater. As mentioned earlier (see Section 4.3.3) an appreciable proportion of the products of photosynthesis become released from plant cells and soon appear in the water as DOM. Although some of this component of primary production may be reabsorbed by phytoplankton, much of it is rapidly taken up by planktonic bacteria. The importance of these bacteria in the organic food cycle of the sea has only recently been realized. The development of new nucleopore filters that can retain the smallest bacteria of 0.5 μm or less has shown that such free-living bacterioplankton are much more abundant in the water column than was previously thought (Hobbie and Williams, 1984).

These bacteria are thought to utilize the greater part of DOM as a nutrient source. By virtue of their small size and correspondingly large surface-to-volume ratio, the bacteria are well adapted to absorb nutrients at low concentrations. Some estimates suggest such direct uptake may account for up to 50 per cent of the total annual production of dissolved organic carbon (Andrews and Williams,

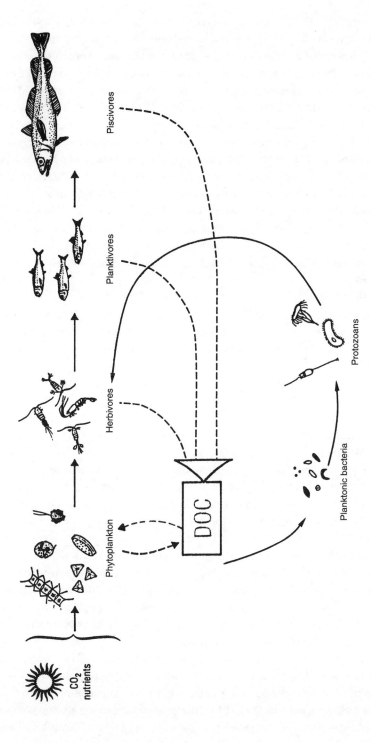

Figure 5.2 A simplified schematic illustration of the microbial loop (bacteria and protozoans) and how it fits in with the basic pelagic grazing food chain (phytoplankton to piscivorous fish). DOC is dissolved organic carbon (or dissolved organic matter). (From Lalli and Parsons (1997) by kind permission of Butterworth-Heinemann.)

1971). This uptake of DOM results in increased bacterial production which in turn is grazed by small pelagic protista, notably the smallest heterotrophic zooflagellates. These provide food for microzooplanktonts, probably chiefly ciliates and dinoflagellates, which are then preyed upon by other zooplanktonts. In this way a proportion of the energy from primary production released into the water as DOM eventually becomes returned to the main food-web several stages along a subsidiary food-chain known as the 'microbial loop' (Pomeroy, 1974).

The relationships are complex and not yet well understood. For example, some pigmented protista can ingest bacteria, thus obtaining energy both by photosynthesis and from bacterial protoplasm. Furthermore at each stage of this microbial food-chain, animal metabolism returns to the water some DOM as well as inorganic nitrogen and phosphorus compounds, thereby restoring to some extent the supply of major nutrients for photosynthetic organisms. This simplified account illustrates the intricacy of the 'microbial loop', which in recent years has become increasingly recognized as a significant part of marine food-webs.

5.1.3   Regeneration

The processes of return of plant nutrients to the water following the degradation of organic compounds are termed *regeneration*. Regeneration is 'direct' when the products set free into the water by metabolism are directly utilized by plants, which is the case with most of the excretory products of marine animals. Phosphorus is excreted mainly as phosphate with some soluble organic phosphorus (Marshall and Orr, 1961). In most marine animals, nitrogen is excreted mainly as ammonia. Teleosts excrete part of their waste nitrogen as trimethylamine oxide (see page 114). Some animals also excrete aminoacids, uric acid or urea. All these compounds are utilizable in varying degrees by plants. However, the greater part of the nutrients taken in by plants is probably regenerated by 'indirect' processes involving bacterial activity.

Bacteria are an essential part of the organic cycle, necessary for the decomposition of particulate organic matter from faecal pellets and the bodies of dead organisms. After death, the tissues of plants and animals become converted by degrees into soluble form. Dissolution may be initiated by autolysis, the tissues being broken down by the dead organisms' own enzymes, but decomposition is brought about mainly by bacterial action. Bacteria are abundant on the surface of organisms and detritus and are specially numerous in the uppermost layers of bottom deposits. Bacterial metabolism converts solid organic matter into organic solutes, and eventually into an inorganic form.

Regeneration of phosphorus is mainly to phosphate, although to some extent plants can also absorb certain dissolved organic phosphorus compounds. Following the death of marine organisms, much of the phosphorus in their tissues returns very quickly to the water as phosphate, indicating that decomposition of organic phosphorus compounds is probably largely by autolysis and hydrolysis. Particulate organic phosphorus is acted on by bacteria producing various solutes, e.g. glucosephosphate, glycerophosphate, adenosine phosphate, some of which may be utilized by plants via their phosphatase enzymes, or further degraded by bacteria to phosphate. Some links of the phosphorus cycle are illustrated in Figure 5.3.

Nitrogenous organic materials are broken down more slowly than phosphorus compounds, mainly by bacterial action. Particulate and dissolved organic nitrogen are converted by bacteria first to ammonium ions and then further oxidized to nitrite and finally to nitrate. Plants absorb nitrogen mainly as nitrate, also as nitrite, ammonium and various simple organic solutes. Certain marine plants (mainly blue-green algae) and bacteria (*Azotobacter, Clostridium, Desulfovibrians*, etc.) are capable of fixing dissolved elemental nitrogen, but this is probably not a major addition to nitrogenous compounds in the sea because of the high energy intake required for this reaction. Some bacteria in anaerobic conditions obtain energy from organic carbon compounds by oxidations involving the reduction of nitrate to elemental nitrogen i.e. nitrogen-freeing. Aspects of the very complex nitrogen cycle of the sea are shown in Figure 5.4.

Sulphur-containing compounds are regenerated mainly as sulphate. Bacterial decomposition of organic sulphur commonly yields hydrogen sulphide, which can be oxidized to elemental sulphur (by *Beggiatoa*) and thence to sulphate (by *Thiobacillus*). The nitrogen and sulphur cycles are closely interconnected by a variety of bacterial reactions involving both groups of compounds.

In the course of these various processes of regeneration and recycling, bacteria themselves grow and multiply, and constitute an important component of the food supply. Bacteria therefore perform two major functions in marine food cycles:

(*a*) the breakdown of dead organic matter into soluble forms, mainly inorganic ions, which can be utilized by plants, and

(*b*) the transformation of dead organic matter into bacterial protoplasm which is directly utilizable as food by some animals.

There are continuous losses of organic material from the euphotic zone due to sinking and to movements by animals down to deeper levels after feeding. Some of this material may reach the bottom and become lost from the cycle by being permanently incorporated in the sediment. The greater part is regenerated in deep layers of water or on the bottom, and nutrients therefore accumulate below the euphotic zone. The continuation of production depends upon the restoration of nutrients from the deep to the surface layers by vertical water mixing (see page 183).

5.1.4 Chemosynthesis

Some of the bacteria involved in recycling and regeneration are autotrophic. They function as primary producers of organic compounds by reduction of carbon

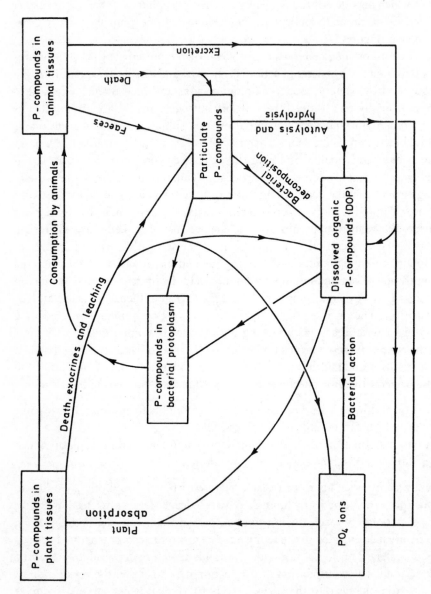

Figure 5.3 Phosphorus cycle in seawater.

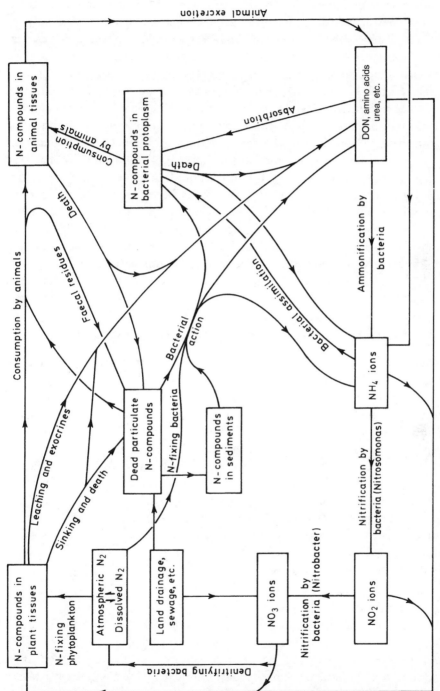

Figure 5.4 Nitrogen cycle in seawater. DON is dissolved organic nitrogen.

dioxide in chemosynthetic reactions which parallel the photosynthetic processes of plants, but derive energy from inorganic chemical sources rather than from light. For instance, the bacteria *Nitrosomonas* and *Nitrobacter*, which oxidize ammonia to nitrite and thence to nitrate, utilize the energy released by these reactions for synthesizing organic materials within their protoplasm. *Beggiatoa* and *Thiobacillus* are chemosynthetic autotrophs obtaining energy from oxidation of sulphide and sulphur. *Beggiatoa* can be seen as a white mat covering the sea-bed where conditions are anoxic, such as under floating fish farm cages. Oxidation of iron to the ferrous and thence the ferric form is another energy source for chemosynthesis. It is now known that primary production by bacterial chemosynthesis contributes a significant fraction of the food available at depths remote from the euphotic zone. Deep-sea hydrothermal vent communities (see Section 6.4.4) are entirely dependent for their energy requirements on dissolved hydrogen sulphide and particles of sulphur in the hot vent water. If the vents become inactive, the communities die. Dense clouds of chemosynthetic bacteria form the base of the food pyramid. The vent communities are thus self-contained and do not rely either directly or indirectly on sunlight. Measurements have also indicated that these communities are among the most productive in the world.

5.2   Measurements of organic production

There are several ways in which estimates of the amount of primary production in the sea have been attempted, and certain of these methods are outlined below (UNESCO, 1973; Vollenweider, 1969). All these techniques are inherently inaccurate and estimates of production will always be just that – estimates. Production rates are usually expressed as the weight of carbon fixed in organic compounds beneath unit area of sea surface in unit time; for example grams of carbon per square metre per day or year ($gC/m^2/day$; $gC/m^2/year$) or sometimes as weight of carbon fixed in unit volume of water in unit time, for example $gC/m^3/day$. Estimates of net primary production by phytoplankton mostly fall within the range 0.05–0.5 $gC/m^2/day$ with values as high as 5 $gC/m^2/day$ in the most productive sea areas. Differences in productivity between various ocean regions are discussed in Section 5.6 where comparisons are also made between production on land and in the sea.

5.2.1   Standing stock measurements

The earliest attempts to measure organic production in the sea were indirect, being based on estimates of the total amount of plant material in the water, i.e. the standing stock. This does not give a direct indication of the rate of production because account must be taken of the rate of turnover. If the plants are being very rapidly eaten, high production may maintain only a small standing stock. For example, on a well-lit coral reef, as much as 1–5 kg of seaweed can grow on every square metre in a year. However, at any one time, only a few small pieces may be seen because herbivorous fish and urchins quickly graze the growths down (Shepard, 1983). Alternatively, where the consumption rate is very low and the plants are long lived, a large standing stock is not necessarily the result of a rapid production rate. A large standing stock may itself limit production by reducing the penetration of light through the water and diminishing the supply of nutrients.

The size of the standing stock depends upon the balance between the rate of production of new plant cells and the rate at which they are lost by animal consumption and by sinking below the photosynthetic zone. To determine production rates from standing stock measurements it is therefore necessary to estimate both the rate of change in size of the population and also the rate of loss, the latter being particularly difficult to assess with any certainty. The following methods have been used for measuring the standing crop.

Direct counts

It is possible to count the number of plant cells in a measured volume of water. Because nets are not fine enough to filter the smallest plants, the phytoplankton is usually obtained by collecting water samples and removing the cells by centrifuging or sedimentation. Subsamples are made with a haemocytometer and counted under a microscope. Alternatively, a Coulter counter can be adapted for counting and recording numbers of small planktonts. This electrical instrument responds to changes in an electrical field as small particles in an electrolyte pass through an aperture. When the number of plant cells in unit volume of water has been determined, the weight of plant protoplasm must be calculated. For this it is necessary to know the size and weight of the plant cells, and to make due allowance for the inorganic content of the cells.

Chlorophyll estimations

The quantity of chlorophyll that can be extracted from unit volume of seawater depends upon the number of plant cells present, and it is possible to calibrate a scale of pigment concentration against quantities of plant tissue (Harvey, 1950). A measured volume of raw seawater is filtered and/or centrifuged to collect all cells. These are then treated with a standard volume of acetone or alcohol to extract the chlorophyll. The intensity of colour in the extract is measured colorimetrically or absorptiometrically to determine the concentration of pigment, and the results are expressed as chlorophyll concentration or arbitrary units of plant pigment (UPP).

Surveys of chlorophyll concentration in the surface layers of the sea can also be based on measurements made by photometric equipment mounted on aircraft or satellites. This apparatus analyses the spectral composition of light reflected from the sea surface. Chlorophyll content can be estimated from the relative intensities of green and blue light or from measurements of chlorophyll *a* fluorescence. Modern methods of satellite recording of sea-surface chlorophyll measurements, have been in place for about fifteen years. Computers are used to produce maps with various colours indicating the chlorphyll concentration in milligrams of chlorophyll pigment per cubic metre of surface water.

Carbohydrate estimations

It has been shown that the tissues of most zooplankton contain little carbohydrate. The carbohydrate content of material filtered from seawater derives almost entirely from the plants present. Measurements of carbohydrate therefore provide a means of estimating the amount of plant material in a sample (Marshall and Orr, 1962). Carbohydrate can be determined absorptiometrically from measurements of the intensity of brown colouration developed by the action of phenol and concentrated sulphuric acid, or with anthrone.

Zooplankton counts

To estimate the rate of loss of plant cells due to grazing by animals, quantitative zooplankton samples are needed to determine the number of herbivores present. It is also necessary to measure their feeding rates. If the size of the animal population and its food requirements are known, it is possible also to make a calculation of the primary production necessary to support this number of animals. Feeding rates of zooplankton can be calculated from measurement of clearance rate of particles or ammonium excretion rates.

5.2.2   ATP measurements

A source of inaccuracy in many methods of estimating biomass is the difficulty of distinguishing living from non-living tissue. In samples containing appreciable amounts of non-living organic matter there are possibilities of greater accuracy from biomass estimates based on measurements of the adenosine triphosphate (ATP) content, as this is a constituent of all live protoplasm but is virtually absent from dead cells. The technique measures the amount of light emitted when ATP is added to an appropriate preparation of luciferin and luciferase. Calibrated against known amounts of ATP, the method can measure very small amounts of ATP if photo-multiplier tubes are used. For plankton studies raw samples of seawater are filtered to collect particulate matter, which is then treated chemically to extract ATP. This extract can be stored deep-frozen for several months without loss of activity prior to measurement.

5.2.3   Measurement of nutrient uptake

Where production is seasonal, estimates of production can be based on measurements of the decrease of nutrients in the water during the growing period. In temperate areas, concentrations of nitrate and phosphate in the surface layers reach a maximum during the winter months when photosynthesis is minimal and convectional mixing is occurring. The concentrations fall during the spring and summer due to the absorption of these nutrients by the phytoplankton. The N- or P-content of the phytoplankton being known, measurements of the reduction of nitrate and phosphate in the water enable estimates to be made of the quantity of new plant tissue formed. Allowances are necessary for the regeneration of nutrients within the photosynthetic zone by decomposition of tissues, replenishment from deeper water, and utilization of other sources of nitrogen, for example nitrite, ammonium or organic nitrogen. Production estimates can also be based on measurements of change of oxygen, carbon dioxide or silicate content of the water.

5.2.4   Measurement of photosynthesis

In photosynthesis, carbohydrate is formed according to the equation

$$nCO_2 + nH_2O \rightarrow [H_2CO]n + nO_2$$

The rate of photosynthesis may be determined by measuring either the evolution of oxygen or the absorption of carbon dioxide. These are not exactly equal because proteins and fats are also formed, but allowance can be made for this.

Oxygen-bottle experiments

This technique was used widely during the first part of this century. Samples of seawater with their natural phytoplankton populations, are collected from several depths within the photosynthetic zone and their oxygen contents measured. Pairs of bottles are then filled from each sample and sealed. The bottles are identical except that one of each pair is transparent – the light bottle – and the other is covered with black opaque material – the dark bottle.

The pairs of bottles are next suspended in the sea at the series of depths from which their contents were obtained and left for a measured period. Alternatively, if temperature and illumination at the sampling depths are known, the bottles can be immersed in tanks at corresponding temperatures and provided with artificial illumination at the correct intensity. This is advantageous at sea because the vessel does not need to remain hove-to at each station for the duration of the experiment. With bottles in tanks, it is necessary to keep them in sufficient motion to prevent settlement of the plant cells. Whichever method is used, the amount of oxygen in each bottle is measured again after the set time interval. The reduction in oxygen in the dark bottles with respect to the original measurements is due to the respiration of the plant, animal and bacterial cells contained. On the assumption that respiration is not influenced by light, the difference in oxygen content between the light and dark bottles of each pair is regarded as being due to the production of oxygen by photosynthesis. This assumption is probably not justified but the method has been widely used and can give tolerably consistent results. One complication which applies to all experiments in which seawater is enclosed in bottles is the rapidity of bacterial growth in these conditions, and this may vary with the intensity of illumination.

Measurement of carbon dioxide uptake: the Steeman Nielsen ^{14}C method

This is a method of measuring carbon fixation by using the radioactive isotope of carbon, carbon-14 (^{14}C), as a tracer. The experimental technique is very similar to the oxygen bottle method described above. Samples of seawater are collected from a series of depths and the carbon dioxide content in each is measured. Bottles are filled from these samples and a small measured quantity of bicarbonate containing ^{14}C is added to each. The bottles are then sealed and suspended in the sea at appropriate depths for a measured period or incubated in tanks as described above.

When the bottles are hauled in or removed from the tanks, the water is filtered to collect the phytoplankton. The cells are washed and their ^{14}C-content estimated by measurement of the beta-radiation. The total carbon fixation is calculated from the known amounts of $^{14}CO_2$ and total CO_2 originally present in the water, making due allowances for the slight differences in rates of assimilation of $^{14}CO_2$ and $^{12}CO_2$.

There are several problems in interpreting the results of these measurements associated largely with the difficulties of allowing for losses of the organic products of photosynthesis in solution. This method measures the amount of $^{14}CO_2$ retained on particulate matter filtered from the water. However, there is evidence that part of the compounds formed by primary production pass into the water in soluble form (see pages 120 and 194). This method gives results of the same order as are obtained by oxygen-bottle methods, and is generally thought to give the most accurate measurements of net primary production in particulate form. It is necessary that some estimate is made of losses of soluble organic material.

Another source of inaccuracy inherent in all methods of estimating production for the complete photosynthetic zone from discrete water samples is the extremely patchy distribution of phytoplankton which occurs in some conditions (see Section 5.3.4). At times the phytoplankton is concentrated in narrow lines or layers at particular depths, which may easily be missed in taking water samples.

5.3   Some factors regulating production

5.3.1   Light and the compensation depth

In the process of photosynthesis the energy of solar radiation becomes fixed as chemical energy in organic compounds. The efficiency of the ocean surface in this energy transformation must vary with locality and conditions, but is probably on average about 0.1–0.5 per cent overall, an efficiency a little lower than that of the land surface.

The ability of plants to absorb and utilize light in photochemical reactions is due to their possession of the green pigment *chlorophyll* and certain other accessory photosynthetic pigments. These are contained in organelles known as *chloroplasts*, except in blue-green algae and photosynthetic bacteria where they are diffused in the protoplasm. Chlorophyll occurs in several forms, but only chlorophyll *a* is common to all photosynthetic organisms. Diatoms and dinoflagellates also contain chlorophyll *c* together with various xanthophyll and carotenoid pigments which give a golden to brownish appearance to their chloroplasts.

It appears that chlorophyll *a* is the essential form for conversion of light energy to chemical energy. This pigment is green in colour because it absorbs light in the blue and red parts of the spectrum. Red light is rapidly absorbed by water; consequently in aquatic plants the absorption of radiant energy by chlorophyll *a* must be mainly limited to the blue wavelengths. The possession of accessory red and yellowish pigments is important in extending the range of wavelengths which can be absorbed, presumably for energy transfer to chlorophyll *a*. The red and brown colours of many seaweeds are due to these additional pigments and seaweed depth distributions are related to the possession of these pigments.

Photosynthesis is confined to the illuminated surface zone of the sea, and a useful measure of the extent of this productive layer is the *compensation depth*, i.e. the depth at which the rate of production of organic material by photosynthesis exactly balances the rate of breakdown of organic material by plant respiration. Below the compensation depth there is no net production. The compensation depth obviously varies continually with changes of illumination, and must be defined with respect to time and place. In clear water in the tropics, the noon compensation depth may be well below 100 m throughout the year. In high latitudes in summer, the noon compensation depth commonly lies somewhere between 10 and 60 m, reducing to zero during the winter months when virtually no production occurs.

Photosynthesis varies in proportion to the light intensity up to a limit at which plants become light-saturated, and further increase of illumination produces no further increase of photosynthesis. Exposure to strong light is harmful and depresses photosynthesis, the violet and ultraviolet end of the spectrum having the most unfavourable effects. In bright daylight the illumination at the sea surface seems often to be at or above the saturation level for most of the phytoplankton, and measurements of photosynthesis in these conditions show that maximum production occurs some distance below the surface, usually somewhere between 5 and 20 m depending upon light intensity, and falls off sharply above this level (Figure 5.5).

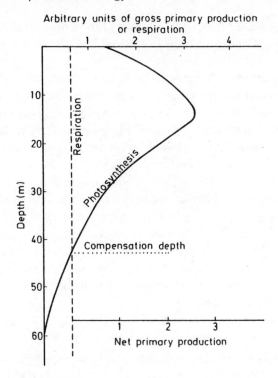

Figure 5.5 Generalized diagram relating primary production rate to depth in middle latitudes during bright sunshine. Below the compensation depth there is no net production.

Correspondingly, the maximum quantity of phytoplankton is seldom found very close to the surface, and except for a few species that seem to thrive in the uppermost few centimetres the greater part of the phytoplankton can be regarded as 'shade plants'. By absorbing light the plants themselves reduce light penetration through the water, and as the population increases the compensation depth tends to decrease.

Above the compensation depth the rate of photosynthesis exceeds the rate of respiration and there is a net gain of plant material; below it there is a net loss. At a particular level the total loss by algal respiration in the water column above may exactly equal the total gain by photosynthesis. This level is termed the *critical depth*. The distance between compensation depth and critical depth depends upon the proportions of the phytoplankton stock above and below the compensation depth. This is determined mainly by vertical water movements.

For the standing stock of phytoplankton to increase, its total photosynthesis must exceed its total respiration. This is possible in a stratified water column when the depth of surface wind-mixing is limited by a thermocline and there is very little transport of phytoplankton below the compensation depth. Around the British Isles these are the conditions of spring and summer when the water column is stabilized by thermal stratification. There is then an overall gain of organic material within the water column and the critical depth must lie below the sea-bed. But in autumn, once vertical mixing begins to distribute much of the phytoplankton to levels well below the compensation depth, a stage is soon reached where total losses by plant respiration are greater than total gains by photosynthesis, the critical depth rises and the standing stock is sharply reduced (Figure 5.6).

Survival of phytoplankton over the winter, when little light energy is available, is effected in several ways. During productive periods the plants build up food reserves, notably as oil droplets, on which they can draw when there is insufficient light for net production. Some species develop resting spores which pass unfavourable periods in a state of dormancy, germinating when conditions become propitious. The dissolved organic matter in seawater provides an energy source which some phytoplankton can utilize if light is inadequate for their needs.

5.3.2   Temperature

The rate of photosynthesis increases with rising temperature up to a maximum, but then diminishes sharply with further rise of temperature. Different species are suited to different ranges of temperature, and photosynthesis is probably performed as efficiently in cold water by the phytoplankton of high latitudes as it is in warmer water by the phytoplankton native to the tropics.

Seasonal variations of production rate in temperate latitudes are related to changes of both temperature and illumination. Apart from its direct effect on rate of photosynthesis, temperature also influences production indirectly through its effects on movement and mixing of the water, and hence on the supply of nutrients to the euphotic levels.

5.3.3   Nutrients

In addition to dissolved carbon dioxide, which is present in seawater in ample quantities to support the most prolific naturally occurring plant growth, there are other substances, the nutrients, which plants also extract from the water and which are essential for their growth. Many of these are minor constituents of seawater, present only in very low concentration, and their supply exerts a dominant control over production. Nitrate and phosphate are of special importance. Where the quantities of these ions are known, theoretical estimates of the potential productivity of the water generally accord well with observed values. Iron, manganese, zinc and copper are other essential nutrients, silicon is required by diatoms, and molybdenum and cobalt and probably other elements

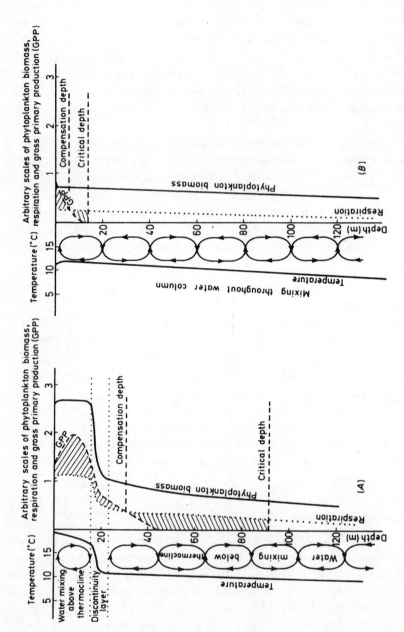

Figure 5.6 Generalized diagram to illustrate changes in depth distribution of phytoplankton, compensation depth and critical depth between (A) late summer and (B) late autumn. With onset of winter, further decline of illumination causes the compensation and critical depths to ascend further until extinguished at the surface. The biomass of phytoplankton decreases rapidly.

are necessary for some plants. Organic compounds dissolved in the water (DOM) may be important in some cases (see Section 4.3.3).

The absorption of nutrients by the phytoplankton reduces the concentration of these substances in the surface layers, and this limits the extent to which the plant population can increase. A certain amount of the nutrients absorbed by phytoplankton may be regenerated and recycled within the euphotic zone, but in deep water plants are continually being lost from the surface layers through sinking and by consumption by zooplankton which moves to deeper levels during the daytime. Many of the nutrients absorbed from the surface layers are therefore regenerated in the deeper and darker layers of water where plants cannot grow. Consequently, nutrients accumulate at deep levels due to the continuous transfer of material from the surface. This loss of nutrients from the productive layer of the sea to deep levels contrasts with the nutrient cycle of the land surface. In soil the breakdown of organic compounds releases nutrients where they are quickly available for reabsorption by plant roots, thereby maintaining the fertility of the land. In the sea the continuance of plant growth depends to a great extent upon the rate at which nutrients are restored to the euphotic zone by mixing with the nutrient-rich water from below. The lower overall productivity of the sea compared with the land is largely a consequence of the regeneration of nutrients in the sea far below the zone of plant growth, with recycling dependent upon relatively slow processes of water movement. The greater productivity of coastal areas compared with deep water is partly a consequence of more rapid recycling of nutrients where the sea bottom is closer to the productive layer.

Some of the vertical mixing processes which restore fertility to the surface layer of the sea are summarized below.

Upwelling

Offshore winds, by setting the surface water in motion, may cause water from deeper levels to be drawn up to the surface (Figure 5.7). We have mentioned earlier the upwelling which occurs in low latitudes along the western coasts of the continents to replace the westward-flowing surface water in the equatorial currents (see Section 1.3.3). Although this upwelling water probably does not rise from depths greater than some 100–200 m, this is deep enough to supply nutrients to the Canaries Current, Benguela Current, Peru Current, California Current and West Australia Current, and these are all areas of high fertility. In the Southern Ocean, continuous upwelling ensures that, even during the highly productive period of the Antarctic summer, plant growth is probably never limited by shortage of nutrients. In the Arctic, where upwelling is less, there is some depletion of surface nutrients during the summer months and production is correspondingly reduced (Smith, 1968).

Upwelling also occurs at *divergences*, i.e. areas where adjacent surface currents

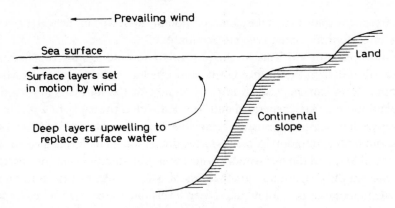

Figure 5.7   Upwelling due to wind action at the surface.

move in different directions. In low latitudes, divergences between the equatorial currents and countercurrents cause upwelling close to the Equator and along the northern boundary of the equatorial counter-currents, and these are important in maintaining the productivity of tropical waters.

Turbulence (eddy diffusion)

Turbulence is a term loosely applied to various complex and irregular movements of the water in which different layers become mixed by vertical eddies. The effects of turbulence on production depend upon the circumstances. It may promote production by bringing nutrients to the surface, or it may sometimes reduce production by carrying down a considerable part of the plant population below the compensation depth. Broadly, high production is likely to follow a limited period of turbulence once the water column becomes sufficiently stabilized for the phytoplankton to flourish in the replenished surface layer without undue loss of plants by downward water movements. In temperate latitudes these conditions occur at the end of winter.

The following are examples of how turbulent water movements may arise:

- **Convection**   When surface water cools, its density increases and convectional mixing commences once the density of the surface layers begins to exceed that of underlying water, the surface water sinking and being replaced by less dense water from below. In high latitudes, convectional mixing is virtually continuous because heat is continually being lost from the surface. In temperate latitudes, convectional mixing occurs during the winter months to depths of some 75–200 m, but ceases during the summer. The corresponding seasonal changes in production are discussed later (see Section 5.4). In low latitudes where the surface waters remain warm throughout the year there is little if any convectional mixing, and the concentration of nutrients at the surface is generally low unless vertical mixing is occurring from some other cause, such as wind action on the surface causing upwelling.

- **Currents** Vertical eddies may arise where adjacent layers of water are moving at different speeds, or where currents flow over irregularities on the sea-bed. On the continental shelf, especially where the bottom is uneven, strong tidal currents may cause severe turbulence and keep the water well mixed throughout its depth. Tidal flow in the eastern part of the English Channel and the southern part of the North Sea produces sufficient turbulence to mix almost the full depth of water, preventing the development of seasonal thermoclines and helping to maintain the fertility of the area throughout the summer months.

- **Internal waves** The depth to which surface waves appreciably move the water is rather less than their wavelength. Although swell waves in deep water occasionally exceed 100 m in wavelength, most surface waves are considerably shorter than this and do not mix the water column to any great depth. However, where the water column is not homogeneous, it is evident that very large internal waves can exist far below the surface. Over a great part of the ocean the water is stratified in layers of different density – density increasing with depth – and these layers do not remain still but oscillate about a mean level. Internal wave movements of at least 200 m in height have been detected in the deep ocean.

  Internal waves may be produced in several ways. For example, strong onshore winds driving light surface water towards the continental slope will bend the equal-density layers downwards. When the wind ceases, the low-density water which has been forced down the continental slope will return to the surface, and the displaced density layers may not return simply to the horizontal but will probably oscillate up and down, transmitting their motion over great distances as waves in the deep layers.

  Oscillations in the deep layers of the North Atlantic may also arise from irregularities in the rate of southward flow of deep water from the Arctic over the north Atlantic ridges, due perhaps to changes of Arctic climate influencing the rate of sinking of surface water. Cooper (1961) has suggested that this flow is intermittent, and that sometimes enormous 'boluses' of cold water flood down the south side of the ridge, displacing the deep density layers and setting up internal waves.

  Cooper (1952) maintains that internal waves can cause vertical mixing where they impinge upon the continental slope, their motion here becoming translated in a manner comparable with that of waves breaking on the shore, carrying deep water up the continental slope much as surface waves run up a sloping beach. In this way, oceanic deep layers rich in nutrients may sometimes spill over the contintental edge, mixing with and increasing the fertility of shelf water.

If onshore winds are sufficiently strong and continuous, the displacement of the density layers may become so severe that the water column becomes unstable. A profound disturbance may then ensue which Cooper has termed 'capsizing' or 'culbute mixing'. There is no exact English equivalent for the French word 'culbuter', meaning to upset violently resulting in a confused heap or jumble, which well describes this process.

Capsizing or 'culbution' is a cataclysmic disturbance of the water column which is believed to occur if low-density water is forced so far down the continental slope that it eventually comes to lie beneath water of greater density. This unstable, topsy-turvy condition is thought to resolve by the low-density water, which has been forced downwards, bursting up towards the surface, i.e. the water column capsizes. The ensuing upheaval must produce a homogeneous mass of water from surface to bottom. This water mass, being a mixture of surface and deep water, is denser than adjacent surface water and will subsequently subside to its appropriate density level. The process must be continuous as long as onshore winds of sufficient strength persist, the line of capsizing gradually receding seawards and involving progressively deeper water. The water blown over the continental shelf will be capsized water containing a component of deeper water, and therefore richer in plant nutrients than ordinary oceanic surface water. Its nutrient content may be expected to increase as the sequence proceeds and extends to deeper levels. This process is described in detail in Figure 5.8.

- **Frontal mixing** (Pingree *et al.*, 1975; Pingree, Holligan and Mardell, 1978, 1979). During the summer months, blooms of phytoplankton are often associated with boundary mixing zones between bodies of colder, vertically

---

Figure 5.8   From Cooper (1952), by courtesy of Cambridge University Press.
(a) A frequent pattern of isosteres south of the Celtic Sea in winter when the uppermost 75–100 m of water in the ocean is homogeneous and is overlying water with density and content of nutrient salts increasing downwards.
(b) A pattern of isosteres in winter over the continental slope, drawn to a scale of 1:4, with no forces operative. The amount of light water which will later lie to windward is considered unlimited.
(c) and (d) Cushioning of light oceanic surface water against the continental slope brought about by on-slope gales. The resistance of the solid slope to further progress of the foot of the light water is, however, absolute causing the isosteres to curl and ultimately to become vertical. Restoring Archimedian forces in (d) have then become zero.
(e) The drag of the surface wind current will draw the upper strata of stratified water with it leading to an unstable density inversion as illustrated.
(f) The unstable tongue of heavier water will capsize violently, leading to a homogeneous mass of mixed water extending from the surface to the depth of the capsizing water mass. New isosteres, bracketing the capsized water mass, are so created.

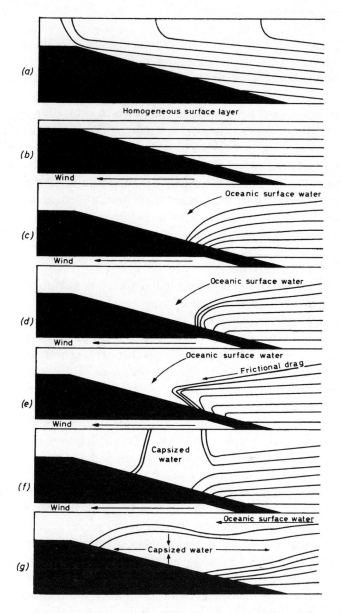

(g) *The newly formed surface water will be heavier than the surface water inshore and to seaward and will subside to form a lens of homogeneous water at its appropriate density level and will be replaced by fresh oceanic surface water blown in from seaward. Processes (d)–(g) must be considered as simultaneous parts of a continuous process, with the line of capsizing receding from the continental slope and the maximum depth of the phenomenon increasing as the gale proceeds. Consequently, the water blown on to the continental shelf of the Celtic Sea by a strong southerly gale should be entirely capsized water and richer in nutrients than it would be if it were solely oceanic surface water. Moreover, the nutrient content of the water passing on to the shelf should increase as the gale proceeds, and the depth of capsizing becomes greater.*

mixed water and warmer, strongly stratified water. Around the British Isles such zones occur mainly at the western end of the English Channel and across the central North Sea. These thermal fronts move irregularly under the influence of tides and wind, mixing the water in complex ways, sometimes characterized by extensive cyclonic eddies which persist for several days. In such transitional areas, stratified and unstratified water are closely inter-mingled in changing, unstable relationships. Over deep water these conditions appear to favour rapid phytoplankton growth by combining sufficient stratification to prevent excessive loss of plants below the compensation depth while at the same time effecting sufficient vertical mixing to supply ample nutrients.

5.3.4   The grazing rate

Although the interactions between plant and animal populations are difficult to elucidate, the grazing rate of the herbivorous zooplankton is certainly one of the factors which regulates the size of the standing stock of phytoplankton, and therefore influences the production rate. The quantity of epipelagic zooplankton generally correlates more closely with the quantity of plant nutrients in the surface layer than with the size of stock of phytoplankton, indicating how greatly grazing reduces the number of plants in fertile water. In the long term, the primary productivity of an area must determine the size of the animal population it supports, but in the short term there are often wide, and sometimes rapid, changes in both numbers and composition of populations due to a variety of causes. Interactions between species often involve a time lag, and there is consequently a tendency for numbers to fluctuate about mean levels. Although some natural populations show homoeostatic mechanisms which control reproduction within limits which do not exhaust the food resources of their environment, there is obviously a general trend for animal numbers to increase as long as there is sufficient food until consumption diminishes the food supply. Food shortage may then cause a decline in the feeding population, and eventually the reduction in food consumption may allow the food supply again to increase, and these oscillations may involve many links of the food web. If the inorganic environment were to remain uniform the system might in due course settle to a steady state, but in nature the physical conditions fluctuate and the equilibrium is forever being disturbed.

A dominant cause of short-term fluctuations in the plankton of middle and high latitudes is seasonal variation of climate which influences both the production rate and the sequence of species which predominate. These changes are discussed later (see Section 5.4) but in the present connection we should note that the sharp reduction in numbers of diatoms which follows their period of rapid multiplication in the spring occurs before nitrate and phosphate are fully exhausted, but coincides with the growth in quantity of zooplankton. There can be little doubt that the increasing rate of grazing is one cause of the decline of the standing stock of diatoms.

A striking feature of the distribution of marine plankton is its unevenness, with localized patches in nearby areas differing in both quantity and composition. One aspect of this patchiness is the inverse relationship of quantities of phytoplankton and zooplankton which has often been reported. Where phytoplankton is especially plentiful, herbivorous zooplanktonts are sometimes few in number; and where herbivores abound, the phytoplankton may be sparse.

This appearance of an inverse relationship may be due simply to the different reproductive rates of phytoplankton and zooplankton, and the effects of grazing on the size of the standing stock. In favourable conditions phytoplankton can multiply rapidly and produce a dense stock. Zooplankton populations increase more slowly, but as the numbers of herbivores rise the phytoplankton will be increasingly grazed and the stock correspondingly diminished. It may therefore be impossible for any abundance of phytoplankton and zooplankton to coexist for any length of time in natural conditions because of the rapidity with which plant cells can be removed from the water by herbivorous animals.

Measurements of filtering rates and food intake in various herbivorous zooplanktonts indicate that, in high concentrations of phytoplankton, large numbers of plant cells are rapidly ingested, sometimes apparently in excess of the animals' needs. Some of these cells pass through the gut virtually un-digested, suggesting a wasteful destruction of plant cells which has been termed 'superfluous feeding'. However, although superfluous feeding can be demonstrated for a time in laboratory conditions when food is exceptionally abundant, the filtering rate later reduces. It seems unlikely that in natural conditions much food is egested unassimilated. Normally, high rates of intake result in increased growth and egg production.

Another explanation for an inverse phytoplankton–zooplankton relationship involves the concept of 'animal exclusion'. According to this hypothesis, animals avoid water rich in phytoplankton because the plants have some effect on the quality of the water which animals find unpleasant. The nature of the excluding influence is uncertain, but it might perhaps be due to secretion of external metabolites by the plants. Small zooplanktonts could avoid this water by controlling their depth so as to remain at deeper levels until the relative movement of the different layers of water carries them to areas where the surface water is less objectionable. The exclusion effect is also observed with some pelagic fish. For instance, the occurrence from time to time of dense patches of *Rhizosolenia*, *Biddulphia* or *Phaeocystis* on the North Sea herring fishing grounds is usually associated with poor catches, the shoals seeming to avoid phytoplankton-laden water. In the mackerel fishing area in the southwest (see page 350), poor catches are obtained in areas which the fishermen recognize as 'stinking water', and this

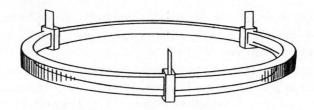

Figure 5.9   The Perspex horizontal circular apparatus used by Bainbridge to study the interrelations of zooplankton and phytoplankton, showing the three sliding doors. Openings at these three points allow for filling, and introduction of animals.
(From Bainbridge (1953), published by Cambridge University Press.)

has also been shown to contain a large amount of phytoplankton. The absence of fish from this water may be due, however, merely to the absence of the zooplankton on which they feed.

To discover if any evidence of exclusion could be demonstrated experimentally, Bainbridge (1953) devised laboratory apparatus in which the behaviour of zooplankton could be observed in the presence of high concentrations of phytoplankton. He constructed a horizontal, circular Perspex tube, 4 ft in diameter, divided into three equal compartments by sliding watertight doors (Figure 5.9). One compartment was filled with seawater enriched with cultures of phytoplankton. The other compartments were filled with filtered seawater. The doors were opened for a period to allow the phytoplankton to spread around the tube until a distinct gradient of phytoplankton concentration was seen to exist. At this stage, equal numbers of small planktonic animals were introduced into each of the three compartments, and their distribution around the tube was counted at intervals. Pure and mixed cultures of phytoplankton were tested, including the diatoms *Skeletonema, Thalassiosira, Biddulphia, Coscinodiscus, Lauderia, Eucampia* and *Nitzschia*, the flagellates *Chlamydomonas, Dicrateria, Rhodomonas, Syracosphaera, Oxyrrhis, Exuviella, Peridinium* and *Gymnodinium*, and some bacterial cultures. The animals studied included the mysids *Hemimysis lamornae, Praunus neglectus, P. flexuosus, Neomysis integer* and *Mesopodopsis slabberi*, also *Artemia salina, Calanus finmarchicus* and various other small copepods, and some decapod larvae.

The results showed definite migrations by the mysids towards concentrations of certain diatoms, notably *Skeletonema, Thalassiosira, Biddulphia, Nitzschia* and mixed cultures, and towards the flagellates *Chlamydomonas, Peridinium, Dicrateria* and *Oxyrrhis. Artemia salina* and some small copepods moved into cultures of *Nitzschia, Biddulphia* and *Thalassiosira*. Cultures of *Lauderia, Coscinodiscus, Eucampia, Syracosphaera* and *Exuviella* showed no attractive effect. The mysids showed a definite movement away from the flagellates *Rhodomonas* and *Gymnodinium II*, and the bacterial cultures. No repellent effect was demonstrated for any diatoms except *Nitzschia* at highest concentration.

In this circular tube the results with *Calanus finmarchicus* were inconsistent. Observations on this animal were also made in another apparatus consisting of a pair of straight, parallel, vertical tubes. One tube was filled with normal seawater, the other with seawater enriched with phytoplankton, and equal numbers of animals were inserted in each. The tubes were suspended in an aquarium tank under lighting of moderate intensity, and the number of animals at different levels of the tubes was counted at intervals. The experiment demonstrated significantly greater numbers of *Calanus* swimming upwards in cultures of *Coscinodiscus*, *Skeletonema*, *Ditylium*, *Chlamydomonas*, *Gymnodinium*, *Oxyrrhis* and mixed phytoplankton. *Chlorella* appeared to depress the numbers swimming up.

Bainbridge's experiments were not designed to elucidate the nature of attractive or repellent substances, but these did not appear to be associated with changes in the concentration of carbon dioxide or oxygen or pH. Positive migrations into concentrations of ammonia were observed. As this is the usual excretory product of marine invertebrates, its attractive effect might be partly accountable for the tendency of many small pelagic organisms to collect in swarms.

From his observations, Bainbridge concluded that 'exclusion' is of quite restricted occurrence in natural conditions, although it may operate during intense blooms of some toxic flagellates. It seems that, in general, natural concentrations of diatoms are likely to be attractive to grazing animals, and Bainbridge suggests that the inverse phytoplankton/zooplankton relationship may be explained in terms of a dynamic cycle of growth, grazing and migration as follows (Figure 5.10):

1   Where conditions are favourable, rapid growth produces a dense patch of phytoplankton.
2   Herbivorous animals are attracted horizontally and vertically into the phytoplankton patch. The grazing rate increases but the concentration of plants will not decline appreciably until the rate of removal by grazing exceeds the rate of increase by cell division.
3   As more animals are attracted to the area, heavy grazing rapidly diminishes the number of plants until the phytoplankton patch is virtually eliminated.
4   Meanwhile, in adjacent water, rapid growth of phytoplankton can now occur because these areas have become denuded of their grazing population by migration into the original phytoplankton patch, and conditions are set for a repetition of the cycle.

There are three species of herbivore commonly present in dense phytoplankton in the Antarctic, namely, the copepods *Calanus simillimus* and *Drepanopus pectinatus* and the mysid *Antarctomysis maxima*. These three forms are strong swimmers, and Bainbridge suggests that their presence in high concentrations of

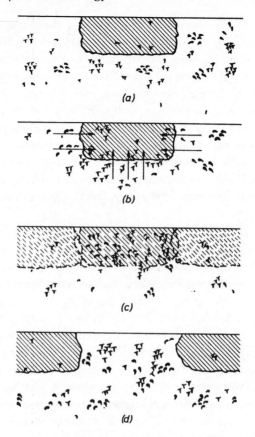

Figure 5.10 Scheme of grazing and migration cycle. (a) Initial state with inverse relationship; (b) start of migration with some grazing; (c) completion of migration and heavy grazing; (d) reversal of initial state and return to inverse relationship. Oblique hatching represents concentrations of phytoplankton.
(From Bainbridge (1953), published by Cambridge University Press.)

phytoplankton is due to the rapidity with which they move. They arrive before the other, slower-moving herbivores become numerous, but their grazing effect is insufficient to make much reduction in the number of plants.

Bainbridge's experiments were conducted using fresh cultures of phytoplankton. Others have pointed out that there is evidence that the production of antibiotic substances by algae occurs mainly in ageing cells, and this may invalidate some of Bainbridge's conclusions. His experiments at least demonstrate that certain species of phytoplankton exert an influence, either attractive or repellent, to which some zooplanktonts react.

Evidently the appearances of an inverse phytoplankton–zooplankton relationship and the patchiness of plankton distribution are both due to a variety of interconnected causes (Bainbridge, 1957). They can, for example, result from water movements. Positively buoyant organisms will tend to become aggregated along lines of convergence or in the centre of swirls, while negatively buoyant forms must be brought together in upwelling zones beneath divergences. Upward-swimming organisms may form patches above the course of cascade currents (see Section 5.6.3). Wind action sometimes sets the uppermost few metres of water in lines of spiral motion, forming so-called *Langmuir vortices*, with zones of convergence and divergence between adjacent vortices. This pattern of water movement must cause quite localized patches of organisms which differ in buoyancy or speed and direction of swimming. Any localized turbulence or vertical mixing which affects temperature, salinity or fertility must obviously influence distribution, and may sometimes result in small-scale blooms of phytoplankton.

Differences in multiplication rates of phytoplankton and zooplankton may have see-saw effects on the relative abundance of grazers and their food. Behavioural differences of species making vertical migrations must result in different organisms concentrating at various levels at various times. Attractive or repellent effects of one species on another, influencing direction of movement or the extent of vertical migrations, may cause the appearance or disappearance of certain species in particular places. Swarming behaviour, mainly associated with breeding, may account for localized patches of adults, and subsequently of eggs and larvae. The distribution of meroplankton must obviously reflect any patchiness of distribution of the benthos, which relates to differences in the nature of the sea bottom.

Whatever its various causes (Fasham, 1978; Fasham *et al.*, 1974) there is no doubt that the tendency of many planktonts to occur in patches rather than an even distribution has important implications with regard to feeding. Animals feed more economically where food is abundant than where it is scarce. By concentrating in and around patches of best food supply they have the double advantage of efficient feeding while allowing the recovery of depleted food stocks in other areas. If food were more evenly distributed, its concentration might sometimes be below the starvation threshold.

5.4 Ocean seasons

Seasonal variations in temperature, illumination and availability of nutrients in the surface layers of the sea impose a pronounced seasonal cycle in production and composition of the plankton between winter and summer, particularly in temperate latitudes (Figure 5.11). Both geographical and year-to-year patterns of fluctuation have their origins within the dynamics of the seasonal cycle. For example, continuous plankton recorder studies have shown that the copepod *Centropages hamatus* in the North Sea is most abundant in the south and in coastal areas, and the seasonal cycle in the west is about a month later than in

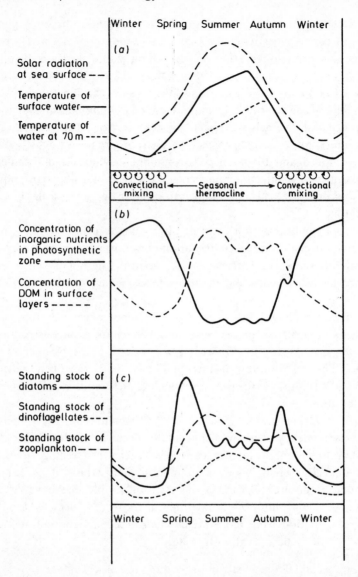

Figure 5.11   Diagram illustrating seasonal changes of (a) temperature, (b) nutrients, and (c) phytoplankton and zooplankton in the surface layers of temperate seas.

the south and east. This pattern can be related to hydrographic features (Colebrook *et al.*, 1991).

5.4.1   Seasonal features of middle latitudes
The four seasons of the sea in middle latitudes have the following general features (Bogorov, 1960).

Winter

A period when the surface of the sea is losing heat to the atmosphere, consequently the surface water is becoming progressively colder. Convectional and wind mixing extends deep into the water column, bringing nitrates, phosphates and other inorganic nutrients to the surface. By late winter the surface water reaches its annual minimum temperature and its annual maximum of inorganic nutrients. Short day-length and the low angle of the sun results in poor or absent illumination within the water. The quantity of both phytoplankton and zooplankton are minimal, except at the end of winter when many animals start their spawning, so timing their reproductive period that their larvae have the advantage of the rapidly increasing food supplies that are shortly to follow in spring.

In neritic areas from late autumn through to the end of winter the water column is often virtually fully mixed throughout its entire depth, surface and bottom temperatures being almost the same. Beyond the continental edge convectional mixing in winter usually extends to between 100 and 200 m, depending on the temperature of the deeper levels.

Spring

A period when increasing insolation is causing the surface water temperature to rise, the water column is becoming stabilized by thermal stratification, illumination is increasing and the critical depth becomes lower than the zone of wind mixing. The concentration of nutrients in the surface layer is initially high, but begins to decrease sharply due to rapid absorption by the phytoplankton which now begins to multiply very quickly in this favourable combination of temperature, light, nutrient supply and stable water column. The rate of primary production soon becomes very high and there is an enormous increase in the quantity of phytoplankton, especially diatoms, which soon reach their greatest abundance for the year (the spring diatom peak). The zooplankton undergoes a more gradual increase, but during late winter and early spring it becomes augmented by the spawning of innumerable marine animals, contributing great numbers of eggs and larvae which by late spring have developed to more advanced larval or juvenile stages. As the zooplankton increases in amount, the quantity of phytoplankton declines rapidly.

Summer

A period during which the surface water is warm and well illuminated. The concentration of inorganic nutrients at the surface is now low because they have been taken up by the phytoplankton, and there is little replenishment from deeper water because vertical mixing is restricted by a sharp thermocline. The dinoflagellates are at their greatest numbers, but the phytoplankton as a whole has declined in amount and primary production is reduced due to grazing by zooplankton and shortage of inorganic nutrients. Diatoms are often quite scarce at this time. During midsummer the numbers and production of phytoplankton are often greatest within the discontinuity layer, usually at some 15–20 m, where nutrients are to some extent available from the deeper mixed layers. The zooplankton, mainly holoplankton, now reaches its greatest amount for the year, and after that diminishes. The concentration of DOM (see Section 4.3.2) is usually highest during the summer.

Autumn
During this season the surface water is cooling and illumination is becoming less. The deeper layers are still getting slightly warmer until eventually the thermocline breaks and convectional mixing is re-established. This leads to rapid replenishment of nutrients in the surface layer, and a consequent increase in primary production, both diatoms and dinoflagellates becoming more numerous. This autumn increase of phytoplankton is always less than the spring peak. It is often followed by a slight increase in zooplankton, but these increases are short-lived. Vertical mixing disperses much of the phytoplankton below the critical depth and the size of stock falls quickly. As temperature and illumination decrease further, the quantities of both phytoplankton and zooplankton gradually reduce to their winter levels, and their over-wintering stages appear.

5.4.2   Ocean seasons in high and low latitudes
This sequence of four ocean seasons described above is a feature of middle latitudes where the temperature of the surface water undergoes the greatest seasonal change. In high latitudes the surface temperature does not vary much with season. Here, illumination is the dominant factor regulating productivity, and only two ocean seasons are apparent, a long winter period of poor or absent illumination and virtually no primary production, followed by a short period of very high production when the light becomes sufficiently good to enable the phytoplankton to grow. This productive period lasts only a few weeks, but during part of this time daylight is continuous throughout the 24-hour period. This makes possible a very rapid growth of a large quantity of phytoplankton, and this abundance of food allows a great increase in zooplankton. Because the rich food supply lasts only a short time, the developmental stages of zooplankton at high latitudes must be passed through rapidly. After that, illumination declines and primary production falls to zero. The phytoplankton virtually disappears, probably over-wintering mainly as spores locked within sea ice, and the zooplankton population decreases to its over-wintering level. This single short season of rapid growth soon followed by decline is a merging of the spring and autumn seasons with elimination of the intervening summer period, the biological winter being correspondingly prolonged.

In low latitudes, conditions are mostly those of continual summer. The surface water is consistently warm and well illuminated but there may be some limitation of production by shortage of nutrients, there being little vertical mixing across a strongly developed thermocline. However, production continues throughout the year and extends to a greater depth than in high latitudes, and the rate of turnover is probably rapid. The result in some tropical areas is a total annual production some 5–10 times greater than in temperate seas (Wickstead, 1968). Generally the rate of production in warm seas remains fairly uniform but there are parts where seasonal changes of wind, for example the monsoons, cause variations of water circulation and seasonal improvements in nutrient supply which are quickly followed by periods of increased production. In the Mediterranean the main season for the production of algae is November to April when there is vertical mixing. Except for a few areas, seasonal peaks in warm seas seldom exceed a tenfold increase of production, whereas fluctuations are sometimes as great as fiftyfold in temperate waters.

5.4.3 Seasonal changes in plankton around the British Isles

Eggs, larvae and spores of benthic plants and animals are often a conspicuous part of the plankton of neritic water, and some of the most obvious seasonal changes are related to the reproductive seasons of the benthos. We can summarize certain seasonal features of the plankton of shallow water around the British Isles as follows (Boalch *et al.*, 1978) (Figures 5.12–5.15).

Winter (November–March)

Surface cooling is continuous during the winter, reaching a minimum temperature throughout the water column, usually in early March, of about 7–8°C in the English Channel and 4–6°C in much of the North Sea. This is the season of minimum quantity of holoplankton. There is very little phytoplankton in the water, although certain diatoms are often present, mainly species of *Coscinodiscus* and *Biddulphia*. Dinoflagellates are very scarce except for *Ceratium tripos*. Copepods are few in number, overwintering mainly as the pre-adult copepodite stage V, and much of the holoplankton lies at a deeper level during winter than in summer. Many invertebrates and the majority of fishes in this area spawn during the latter part of the winter, setting free into the plankton a great number and variety of eggs and early larval stages. Usually the most abundant larvae in tow-net samples taken in late winter are nauplii of *Semibalanus balanoides*. There may also be nauplii of *Verruca stroemia* and *Balanus crenatus*, together with numerous plutei, trochophores, bivalve and gastropod veligers, zoeas and fish eggs of many species. The euphausid *Nyctiphanes couchi* often spreads southwards in the North Sea in winter, where it is not commonly found in spring and summer. In the western English Channel, water inflow in winter tends to be from the northwest, bringing in *Aglantha digitale* and associated species (see page 160).

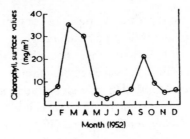

Figure 5.12   Changes in the standing stock of phytoplankton in western part of English Channel as indicated by measurements of chlorophyll concentration.
(Data from Atkins, W.R.G. and Jenkins, P.G. (1953). *J. Mar. Biol. Ass. UK*, **31**, 495–508 published by Cambridge University Press, and Jenkins, P.G. (1961). *Deep Sea Research*, suppl. to vol. 3, 58–67, published by Pergamon Press.)

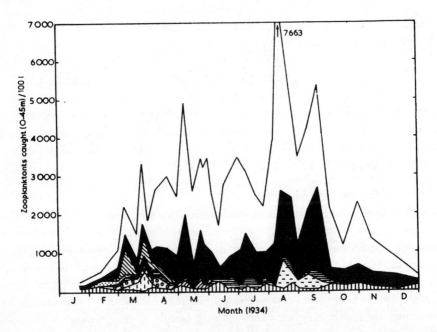

Figure 5.13   Total numbers of zooplanktonts per 100 litres caught between the surface and 45 m off Plymouth through 1934. □ *Copepod nauplii,* ■ *copepods and copepodites,* ▤ *Appendicularia,* ▧ *Cirripede nauplii,* ▨ *Cladocera,* ▨ *Polychaete larvae,* ▣ *Limacina,* ▨ *Noctiluca,* ▥ *Rotifera,* ▥ *etc.*
(From Harvey, H.W. *et al.* (1935) in *J. Mar. Biol. Ass. UK*, **20**, published by Cambridge University Press.)

Spring and early summer (March–June)

Diatoms rapidly increase in numbers to their annual maximum which usually occurs during late March or early April. The diatom population is now very mixed, *Skeletonema, Chaetoceros, Lauderia, Thalassiosira, Coscinodiscus* and *Biddulphia* often becoming very numerous. Different species tend to

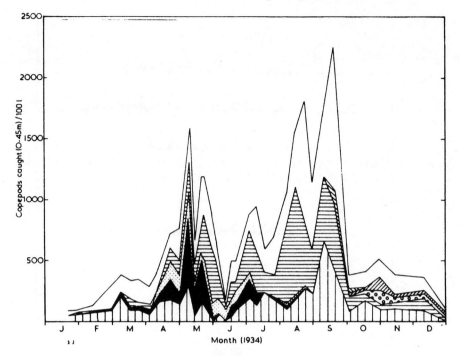

Figure 5.14 Numbers of copepods per 100 litres caught between the surface and 45 m off Plymouth through 1934. ■ Temora longicornis, ▥ Pseudocalanus elongatus, ▤ Oithona helgolandica, ▧ Acartia clausii, ▨ Euterpina acutifrons, ▩ Centropages typicus, ▦ Corycaeus anglicus, □ small copepodites, etc.

(From Harvey, H.W. *et al.* (1935) in *J. Mar. Biol. Ass. UK*, **20**, published by Cambridge University Press.)

predominate in succession, although the order is not constant from year to year. Photoperiodism may play some part in setting off the rapid growth of particular species at different times. The effects of external metabolites may be partly accountable for the successive rise and decline of each species. The earliest to become dominant must thrive mainly on inorganic nutrients. Those which achieve dominance later may be utilizing DOM produced by preceding species (Butler, 1979). Following their spring peak the diatom population declines fairly rapidly. Often the colonial flagellate *Phaeocystis pouchetii* then becomes abundant.

During early spring the plankton contains many eggs and larvae, but by early summer many of these have metamorphosed and disappeared from the plankton. The nauplii of the barnacle, *S. balanoides*, most abundant in March, develop to cypris stages in April and have mostly settled by mid-May. Over the same period, many crab zoeas advance to megalopas, and fish larvae to young fish stages. Overwintering copepodites mostly become adult in February and produce a brood of eggs in March, hatching to nauplii by April. The individuals from this brood usually grow to larger size than any others during the year. The

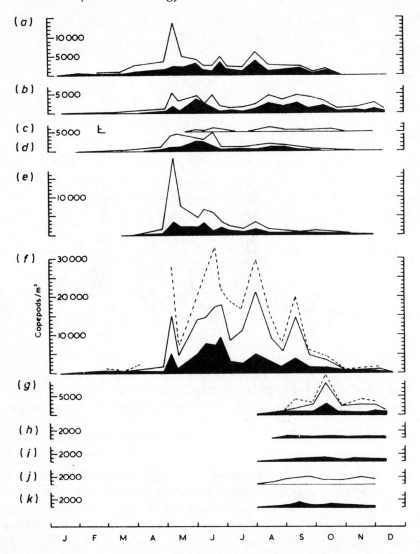

Figure 5.15 Numbers of smaller copepods per cubic metre through the year 1947 off Plymouth, all to same scale. Solid black: copepodites and adults. Continuous line: all stages including nauplii. Dotted line: total plus eggs. (a) Pseudocalanus elongatus; (b) Paracalanus parvus; (c) Centropages typicus; (d) Temora longicornis; (e) Acartia clausii; (f) Oithona helgolandica; (g) Oithona nana; (h) Oncaea venusta; (i) Corycaeus anglicus; (j) Unidentified nauplii; (k) Euterpina acutifrons.

(From Digby, P.S.B. (1950). *J. Mar. Biol. Ass. UK*, **29**, 398–438, published by Cambridge University Press.)

cladocerans *Podon* and *Evadne* often reach their greatest numbers between April and May. The euphausid *Thysanoessa inermis* sometimes enters the northern part of the North Sea.

Mid-summer (June–August)

Diatoms continue to decrease in number and are often scarce between July and August, species of *Rhizosolenia* often being the commonest at this period. Dinoflagellates are now more numerous, especially species of *Peridinium*, and usually reach their maximum abundance between June and July. The majority of larvae produced in the spring have now completed their metamorphosis and disappeared from the plankton, but other species require warm water for breeding and shed their eggs and larvae at this time. For example, the nauplii of *Chthamalus montagui*, *B. perforatus* and *Elminius modestus* appear in plankton samples taken off our south-west coastline. However, it is the holoplanktonic species, copepods, chaetognaths, ctenophores and larvaceans, which become very numerous and form the greater part of the zooplankton during the summer months. The copepods *Acartia clausi* and *Paracalanus parvus*, mainly summer to autumn forms, are often specially abundant. In the western English Channel, inflow now tends to be from the southwest, with species of *Muggiaea*, *Liriope*, etc. (see page 161) appearing. Pilchard eggs sometimes become very numerous in the English Channel in summer to autumn, though less so in recent years.

During the summer the plankton often comprises two distinct communities, one above and the other below the thermocline. Sometimes *Sagitta setosa* occurs in the upper warm layer, and *S. elegans* in the colder water below. Surface water temperature usually reaches its warmest in late August to early September, 16–18°C in the English Channel and 13–17°C in the North Sea.

Autumn (September–October)

After the thermocline breaks, diatoms show a brief but definite increase in numbers and again produce a very mixed population in which species of *Rhizosolenia*, *Coscinodiscus*, *Biddulphia* and *Chaetoceros* are often found together. Autumn plankton samples are sometimes rich in the larvae of benthic invertebrates, mainly plutei and bivalve veligers. An autumn brood of copepods appears which in some cases does not complete its development to the adult stage but survives the winter as stage V copepodites. In the English Channel the copepods *Oncaea venusta*, *Oithona nana*, *Corycaeus anglicus* and *Euterpina acutifrons*, which are not numerous earlier in the year, have their main abundance from August to the end of the year (Figure 5.14).

5.4.4 Seasonal changes: diver observations

The cold murky, winter waters around the British Isles are not an encouragement to divers. However, with the advent of effective thermal insulation in the form of modern dry suits, more divers (and biologists) are now diving all year round. They have been able to confirm, and document photographically, the dramatic seasonal changes that occur in the benthos and the plankton around the British Isles (and in other temperate seas).

Although macro-algae play only a small part in the productivity of the oceans as a whole, they form an essential part of the shallow rocky sublittoral ecosystem in temperate seas. Around the British Isles, the kelp *Laminaria hyperborea* forms forests in suitable rocky areas below the low-water mark. This kelp is, in effect, deciduous, and as the autumn storms rip through the beds, the large fronds are shed leaving behind the long-lived, tough stalks (stipes). The fronds are often cast ashore in great heaps and the nutrients in them are eventually returned to the sea via the food chain and DOM (see Section 4.3.2). As daylight starts to increase in March and April and the sea becomes less turbulent allowing greater light penetration, new fronds start to sprout from the stems. Kelp fronds make excellent surfaces for the attachment of epiphytic hydroids, brozoans, ascidians and algae and these reduce the light reaching other kelp fronds. Shedding the fronds gets rid of this epiphytic load.

Many red seaweeds are annuals and these die right back in winter leaving spores behind to start the growth cycle again in spring. Many of the smaller attached and encrusting animals, including hydroids, bryozoans, tunicates and sponges, are also annuals. They are either torn away or die back to an inconspicuous resting stage. Therefore, at very exposed sites, sublittoral rocks may appear very bare in winter with only a covering of pink encrusting, calcified algae. In early spring, attached animals such as the sea squirt *Calvelina lepadiformis*, and many others, start to re-grow as the planktonic food supply increases.

Changes in the abundance of the plankton are also evident to divers since they greatly affect the underwater visibility. In spring around March and April, visibility is often very poor. At this time increasing daylight length and intensity result in algal blooms in coastal waters, fed by an abundance of nutrients from the breakdown of tonnes of decaying marine life which died back over winter. During the summer, by about mid-June, visibility is usually good because the plankton have used up much of the nutrient excess and are being eaten more quickly than they can reproduce. The autumn often brings brief plankton blooms as storms churn up the water and release more nutrients. Poor visibility in winter is the result of storms stirring up the sea-bed and suspending fragments of plant and animal debris.

5.5   Some mathematical models

The aim of ecological studies is to achieve sufficient understanding of the interactions between organisms and their environments to be able to express these

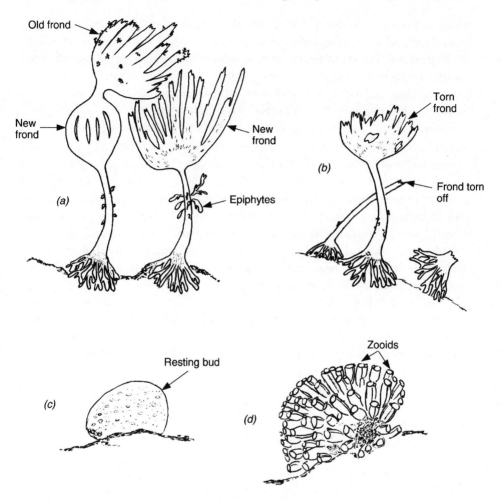

Old frond

New
frond

New
frond

Epiphytes

(a)

Torn
frond

(b)

Frond torn
off

Resting bud

(c)

Zooids

(d)

Figure 5.16   Kelp (Laminaria hyperborea) *in* (a) *spring and summer; the old frond is
discarded as the new frond grows and* (b) *in winter.* (c) *The tunicate* Diazona violacea *in
winter; only a hard bud remains and* (d) *in summer.*

relationships in precise numerical terms. If this could be done, mathematical
models could be constructed from which predictions might be made. The
processes regulating phytoplankton production can be formulated in several
ways. For example, the rate of change of a phytoplankton stock, P, obviously
depends on the rates of addition and loss of cells. Fleming (1939) expressed this
relationship as follows:

$$\frac{dP}{dt} = P[a - (b + ct)]$$

where a = rate of cell division of phytoplankton, b = initial rate of loss of cells by grazing, and c = rate of increase of grazing intensity. Taking observed values for diatom populations at the beginning and the peak of the spring bloom in the English Channel, he computed curves which fitted observed values well and indicated that, following the peak, the production rate continued to increase despite the fall in phytoplankton stock due to heavy grazing.

Cushing (1959) formulated the rate of change of phytoplankton as:

$$\frac{dP}{dt} = P(r - M - G)$$

where P = number of algae (or weight of carbon) per unit volume or beneath unit surface area, r = instantaneous reproductive rate of algae, G = instantaneous mortality of algae due to grazing, and M = instantaneous mortality rate from other causes. r was calculated for observed rates of algal cell division at various light intensities, and account taken of light penetration, compensation depth and depth of wind-mixed layer. In computing grazing rates for observed populations of herbivores, adjustments were made to allow for their reproductive and mortality rates. Applying the equation to the North Sea during the spring bloom, in an area where nutrient depletion was thought not greatly to depress production, Cushing showed fairly good agreement between observed and calculated values for standing stocks of phytoplankton and herbivores.

An alternative to these equations, based on rates of cell reproduction and loss, is a general equation advanced by Riley which takes account of energy considerations, as follows:

$$\frac{dP}{dt} = P(Ph - R - G)$$

where P = phytoplankton population, Ph = rate of photosynthesis, R = plant respiration rate, and G = grazing rate.

Riley (1946, 1947) derived expressions for the coefficients Ph, R and G which take account of illumination, and combined these into the following expanded equation:

$$\frac{dP}{dt} = P\left[\frac{pI_0}{kz_1}(1 - e^{-kz_1})(1 - N)(1 - V) - R_0 e^{rT} - gZ\right],$$

where

P = Phytoplankton population in $gC \cdot m^{-2}$.

p = Photosynthetic constant, 2.5 (gC produced per gram of phytoplankton C per day per average rate of solar radiation).

I_0 = Average intensity of surface illumination in $gcal \cdot min^{-1}$.

k = Extinction coefficient (1.7/depth of secchi disc reading in metres).

z_1 = Depth of euphotic zone (depth at which light intensity has a value of 0.0015 gcal·cm^{-2}·min^{-1}).

N = Reduction in photosynthetic rate due to nutrient depletion,

$$\left(\frac{\text{(mg At P/m}^3 \text{ at surface}}{0.55} \text{ when concentration is less than } 0.55 \right).$$

V = Reduction in photosynthetic rate due to vertical turbulence.
R_0 = Respiratory rate of phytoplankton at 0°C.
r = Rate of change of R with temperature, r being chosen so that R is doubled by a 10°C increase in temperature.
T = Temperature in °C.
g = Rate of reduction of phytoplankton by grazing of 1 gC of zooplankton.
Z = Quantity of zooplankton in gC × m^{-2}.

From this equation the likely fluctuations of the phytoplankton population over a 12-month period were predicted and showed good agreement with the observed values (Figure 5.17).

An equation was also derived for changes in the population of herbivores, H:

$$\frac{dH}{dt} = H(A - R - C - D)$$

A = Coefficient of assimilation of food
R = Coefficient of respiration
C = Coefficient of predation by carnivores
D = Coefficient of death from other causes.

Despite the many difficulties of evaluation of this equation, predicted and observed values were in substantial agreement (Figure 5.17).

For a computer simulation *see* Steele (1974).

5.6 Geographical differences of fertility

5.6.1 Global primary production

In Table 5.1, comparisons are made between primary production on the land and in the sea. The concept that open ocean areas have a productivity comparable to that of deserts on the land and much lower than that of coastal areas and upwelling zones, is well established. However, techniques for measuring organic production are constantly being refined and some recent data suggest that there may be two to three times as much organic matter per unit of surface area in the open ocean than previously reported.

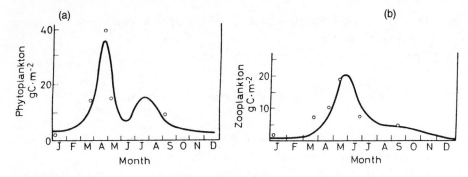

Figure 5.17   (a) Curve of calculated values of phytoplankton standing crop on Georges Bank. Circles are observed values; (b) Curve of calculated values of zooplankton population on Georges Bank. Circles are observed values.
((a) From Riley (1946), by courtesy of Sears Foundation for Marine Research; (b) from Riley (1947), by courtesy of Sears Foundation for Marine Research.)

5.6.2   Ocean primary production

Most of the primary production in the oceans originates from the phytoplankton. The large seaweeds of shallow water can achieve very high production rates. Estimates in excess of 30 gC/m²/day are reported for Californian kelps whereas 5 gC/m²/day would be high for phytoplankton. However, being limited to a very narrow belt at the sea's margin, the seaweeds contribute probably only about 0.05 per cent of the total production of the sea.

Differences in the fertility of the seas in different localities, and at different times, depend upon the availability of plant nutrients in the surface layers. Certain areas are of consistently high fertility, others are ocean deserts, and in many regions the fertility fluctuates seasonally. The general distribution of primary production in the world's oceans is shown in Figure 5.18. Such maps are the result of many years' sampling. Satellite observations (see Section 5.2.1) can now provide monthly pictures of seasonal variations on a world-wide scale.

The areas of good fertility and high productivity include most of the seas overlying wide continental shelves. There are several reasons why this shallow water is relatively rich in nutrients. Waves erode the coastline and stir up the sediments, releasing nutrients into the water. Fresh water running off the land may carry additional nutrients, including trace elements such as iron and manganese which are often scarce in deep water due to precipitation. Where there are centres of human population, sewage is usually poured into the sea and provides nutrients after decomposition. The tidal flow of the water above the shelf may cause sufficient turbulence to keep the water column well mixed, ensuring that nutrients regenerated below the euphotic zone are quickly restored to the surface, and there

Table 5.1 Comparison of Gross Primary Production on Land and in the Ocean (Source: Odum (1971) modified by Duxbury and Duxbury (1996).)

Ocean and land area	Amount (gC/m²/yr)
Open ocean	50
Deserts	50
Grassland	50
Coastal ocean	25–150
Forests	25–150
Common crops	25–150
Pastures	25–150
Upwelling zones	150–500
Deep estuaries	150–500
Rain forests	150–500
Moist crops	150–500
Intensive agriculture	150–500
Shallow estuaries	500–1250
Sugarcane and sorghum	500–1250

is unlikely to be much loss of nutrients to deep levels of the ocean unless shelf water is flowing down the continental slope, as occurs for example in cascading (see Section 5.6.3). Processes leading to enrichment of oceanic surface water near the continental slope have been mentioned earlier (see page 184), and much of this fertile water may be blown over the shelf. From time to time, some of the innumerable deep water pelagic creatures which ascend to the surface during darkness may also be carried over the shelf and augment the food supplies of neritic water. The very dense growth of benthic algae that exists on some coasts contributes greatly to the primary production of these areas.

The fertility of deep water depends largely upon the extent to which water from deep levels is brought to the surface. In temperate areas, the surface layers are well provided with nutrients by convection during the winter and early spring, but the supply of nutrients diminishes during the summer when the formation of a thermocline prevents replenishment by vertical mixing. Upwelling of deep water around Antarctica produces in the Southern Ocean the world's widest expanse of highly fertile open sea. Fertile areas are also produced by upwelling in the currents along the eastern part of the Atlantic, Indian and Pacific Oceans at low latitudes (see page 183 and Section 1.3.3 and Figure 1.6). In the north Indian Ocean, upwelling occurs in the northernmost part of the Arabian Sea during the winter north-east monsoon, and off the Arabian and Somali coasts in summer when the Monsoon Current develops.

Poor fertility occurs where vertical water mixing is minimal. Throughout the tropics, wherever a permanent thermocline is present, production rates are mainly

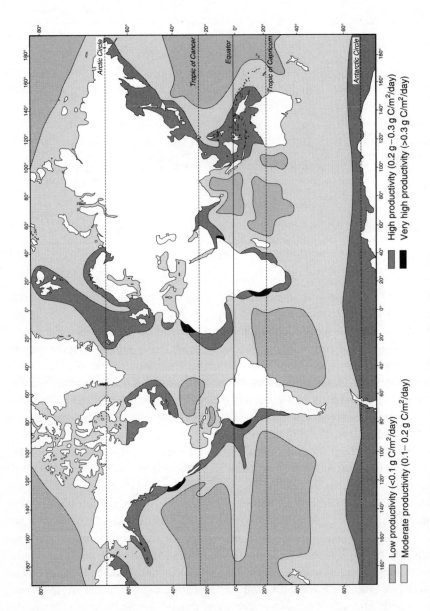

Low productivity (<0.1 g C/m²/day)

Moderate productivity (0.1– 0.2 g C/m²/day)

High productivity (0.2 g –0.3 g C/m²/day)

Very high productivity (>0.3 g C/m²/day)

Figure 5.18   The distribution of primary production in the world's oceans.

low despite rapid regeneration. However, because production continues through the year without much seasonal decline, it is probable that the total annual production in most tropical areas substantially exceeds that of temperate seas. The Sargasso is a semi-tropical marine desert area where even horizontal mixing by surface currents is slight. Here the production rate is very poor except during the brief winter period when some convectional mixing may take place.

In the Arctic there is relatively little upwelling compared with the Southern Ocean, and the Arctic is correspondingly less fertile. In the Mediterranean, fertility is low to moderate because nutrients are continually lost in the deep outflow which forms the bottom current through the Strait of Gibraltar. The inflowing surface current is derived from surface levels of the Atlantic which are relatively poor in nutrients.

5.6.3 Cascading

A process termed *cascading* is believed sometimes to cause losses of both nutrients and planktonic organisms from neritic water. During autumn and winter, loss of heat from the sea surface may cause the shallow shelf water to become appreciably colder and denser than water at similar levels beyond the continental edge. Consequently, the heavier shelf water will tend to flow down the continental slope, i.e. to cascade, to its appropriate density level. The loss of cascading water from above the shelf must be compensated by a corresponding inflow of water elsewhere.

The effect of cascading on the fertility of shelf water must depend on the quality of the compensation water. Cooper and Vaux (1949) studied cascading in the Celtic Sea (south of Ireland and west of the Bristol and English Channels), and concluded that surface layers of the Atlantic and Bay of Biscay formed the main sources of compensation water. Because the phosphate content of this water was lower than that of the Celtic Sea at the time, cascading presumably led to a net loss of nutrients from the area. Iron and other trace elements which tend to concentrate in particles in the lower layers are probably specially liable to be swept off the shelf in the cascade. Annual fluctuations in the fertility of shelf water could be caused by variations in the volume of cascading.

The distribution of organisms could be influenced by cascading in several ways. Small planktonts carried down in the cascade would be lost in deep water. Stronger-swimming forms might ascend out of the cascade as it flows down the continental slope, and this could lead to patches of large, active zooplanktonts over deep water near the continental edge. Where submarine valleys cut into the slope, cascade streams are likely to be strongest in those channels. Cascading might also occur on the sides of submarine banks where a flat top lies near the surface. The preference of certain animals for the sides of submarine valleys, troughs and mounds during winter may be due to the relative abundance of food carried by cascading to these parts of the sea floor.

5.7   Turbidity currents

When water contains a large quantity of suspended particles its overall density is increased. On occasions the excess density of highly turbid water above the continental shelf may cause it to flow as a coherent fluid down the continental slope. This downslope motion due to turbidity is termed a turbidity current. It is a major mechanism of transport of material from the shelf to the deep-sea floor. Strong onshore winds, wave action or earthquakes can cause large amounts of sediment to become suspended in the water, thus initiating a turbidity current.

The current mainly follows the course of valleys in the slope, and may be powerful and fast enough to cut its own submarine canyon down the slope, or to deepen existing canyons further. As the current emerges from its canyon, it deposits huge fan-shaped areas of sediment, sometimes reaching the abyssal plain. Following the earthquake near the Grand Banks in 1929, a turbidity current of such power was generated that material was deposited at least 500 km from its original location (see Masson *et al.*, 1996).

It is possible that turbidity currents may at times cause changes in fertility of shelf water and in distribution of organisms comparable with the effects of cascading.

References and further reading

Andrews, P. and Williams, P.J. LeB (1971). Heterotrophic utilization of dissolved organic compounds in the sea. III. Measurements of the oxidation rates and concentrations of glucose and amino acids in sea water. *J. Mar. Biol. Assoc. UK.*, **51**, 111–25.

Bainbridge, R. (1953). Studies of the Interrelationships of Zooplankton and Phytoplankton. *J. Mar. Biol. Ass. UK*, **32**, 385.

Bainbridge, R. (1957). The Size, Shape and Density of Marine Phytoplankton Concentrations. *Biol. Rev.*, **32**, 91.

Boalch, G.T., Harbour, D.S. and Butler, E.I. (1978). Seasonal Phytoplankton Production in the Western English Channel. *J. Mar. Biol. Ass. UK.*, **58**, 943–53.

Bogorov, B.G. (1960). Perspectives in the Study of Seasonal Changes of Plankton and of the Number of Generations at Different Latitudes. *Perspectives in Marine Biology.* Buzzati-Traverso, ed. University of California Press.

Butler, E.I. (1979). Nutrient Balance in the English Channel. *Estur. Coast. Mar. Science*, **8**, 195–7.

Colebrook, J.M., Warner, A.J., Proctor, C.A., Hunt, H.G., Pritchard, P., John, A.W.G., Joyce, D. and Barnard, R. (1991). *60 years of the continuous plankton recorder survey: a celebration.* The Sir Alister Hardy Foundation for Ocean Science.

Cooper, L.H.N. (1952). Processes of Enrichment of Surface Water with Nutrients due to Winds Blowing on to a Continental Slope. *J. Mar. Biol. Ass. UK*, **30**, 453.

Cooper, L.H.N. (1961). Hypotheses connecting Fluctuations in Arctic Climate with Biological Productivity of the English Channel. *Pap. Mar. Biol. and Oceanogr., Deep-Sea Research*, suppl. to Vol. 3.

Cooper, L.H.N. and Vaux, D. (1949). Cascading over the Continental Slope from the Celtic Sea. *J. Mar. Biol. Ass. UK*, **28**, 719.

Cushing, D.H. (1959). On the Nature of Production in the Sea. *Fish. Investig. Lond., Ser. II*, **22**, No. 6.

Duxbury, A.B. and Duxbury, A.C. (1996). *Fundamentals of Oceanography* (2nd edn). Wm. C. Brown Publishers.

Fasham, M.J.R. (1978). Statistical and Mathematical Analysis of Plankton Patchiness. *Oceanogr. Mar. Biol. Ann. Rev.*, **16**, 43–79.

Fasham, M.J.R., Angel, M.V. and Roe, H.S.J. (1974). Spatial Pattern of Zooplankton. *J. Exp. Mar. Biol. Ecol.*, **16**, 93–112.

Fleming, R.H. (1939). The Control of Diatom Populations by Grazing. *J. Cons. Int. Explor. Mer.*, **14**, 210.

Harvey, H.W. (1950). On the Production of Living Matter in the Sea off Plymouth. *J. Mar. Biol. Ass. UK*, **29**, 97.

Hobbie, J.E. and Williams, P.J. LeB (1984). *Heterotrophic activity in the sea*. New York, Plenum.

Lalli, C.M. and Parsons, T.R. (1997). *Biological Oceanography: an introduction* (2nd edn). Oxford, Butterworth-Heinemann.

Marshall, S.M. and Orr, A.P. (1953). The Production of Animal Plankton in the Sea. *Essays in Marine Biology*, p. 122. Edinburgh and London, Oliver and Boyd.

Marshall, S.M. and Orr, A.P. (1961). On the Biology of *Calanus finmarchicus*. XII. The Phosphorus Cycle. *J. Mar. Biol. Ass. UK*, **41**, 463.

Marshall, S.M. and Orr, A.P. (1962). Carbohydrate as a Measure of Phytoplankton. *J. Mar. Biol. Ass. UK*, **42**, 511.

Masson, D.G. Kenyon, N.H. and Weaver, P.P.E. (1996). Slides, Debris flows, and Turbidity currents. In: *Oceanography. An Illustrated Guide*. C.P. Summerhayes and S.A. Thorpe, eds. Southampton Oceanographic Centre, Manson Publishing.

Odum, E.P. (1971). *Fundamentals of Ecology*. 3rd edition. W.B. Saunders Company.

Pingree, R.D. (1978). Cyclonic Eddies and Cross-Frontal Mixing. *J. Mar. Biol. Ass. UK*, **58**, 955–63.

Pingree, R.D., Holligan, P.M. and Mardell, G.T. (1978). Effects of Vertical Stability on Phytoplankton Distribution. *Deep Sea Research*, **25**, 1011–28.

Pingree, R.D., Holligan, P.M. and Mardell, G.T. (1979). Phytoplankton Growth and Cyclonic Eddies. *Nature*, London, **278**, 245–7.

Pingree, R.D. *et al.* (1975). Summer Phytoplankton Blooms and Red Tides along Tidal Fronts in Approaches to English Channel. *Nature, London*, **258**, 672–7.

Pomeroy, L.R. (1974). The ocean's food web, a changing paridigm. *Bioscience*, **24**, 499–504.

Raymont, J.E.G. (1963). *Plankton and Productivity in the Oceans*. Oxford, Pergamon.

Raymont, J.E.G. (1966). The Production of Marine Plankton. *Adv. Ecol. Res.*, **3**, 117.

Riley, G.A. (1946). Factors Controlling Phytoplankton Populations on Georges Bank. *J. Mar. Res.*, **6**, 54.

Riley, G.A. (1947). A Theoretical Analysis of the Zooplankton Population of Georges Bank. *J. Mar. Res.*, **6**, 104.

Shepard, C.R.C. (1983). *A natural history of the coral reef*. Blandford Press.

Smith, R.L. (1968). Upwelling. *Oceanogr. Mar. Biol. Ann. Rev.*, **6**, 104.

Steele, J.H. (1974). *The structure of marine ecosystems*. Oxford, Blackwell Scientific.

Steeman Nielsen, E. (1975). *Marine Photosynthesis*. Amsterdam, Elsevier.

UNESCO (1973). *Guide to the Measurement of Marine Primary Production under Some Special Conditions*. Paris, UNESCO.

Vollenweider, R.A. (1969). *A Manual on Methods for Measuring Primary Production in Aquatic Environments*. I.B.P. Handbook No. 12. Oxford, Blackwell.

Wickstead, J. H. (1968). Temperate and Tropical Plankton; a Quantitative Comparison. *J. Zool. Lond.*, 155, 253–69.

6 The sea bottom

The sea floor is an area of great contrasts, from deep ocean 'deserts' of soft sediment to rich coral reefs. Exactly what lives there depends on a variety of conditions such as substrate type, current speeds and in particular availability of food. Even in deep water, there is generally some food available, mainly in the form of fragments of organic matter sinking from the overlying water, and in some areas this food supply may be sufficient to support a large population. Beneath shallow water there may also be primary production by benthic algae and photosynthetic bacteria to depths of about 50 m depending on water clarity. Kelps (e.g. *Laminaria, Alaria, Saccorhiza, Macrocystis*), calcareous red algae (e.g. *Lithophyllum, Lithothamnion*) and a variety of foliose algae, grow on shallow rocks where light is sufficient. A varied microflora of benthic diatoms and other photosynthetic protista occur on the surface of rock, sand or mud often visible to the naked eye as a thin brownish film.

Many bottom-dwelling creatures are able to live and grow to large size with relatively little expenditure of energy in hunting and collecting food because they can obtain adequate nourishment simply by gathering the particles that fall within their reach or are carried to them by the currents. Others simply digest the organic matter and associated bacteria contained within the sediment. Most of the sea bottom is covered with soft deposits which give concealment and protection to burrowing creatures. Where the substrate is hard it provides a secure surface for the attachment of sessile forms and affords protection for creatures which hide in crevices or burrow in rock. Compared with the pelagic division of the marine environment the sea bottom provides a far wider variety of habitats because the nature of the bottom differs greatly from place to place. The benthic population of the sea is correspondingly more diverse than the pelagic population.

Except in very shallow depths, the temperature, salinity, illumination and movements of the water at the bottom are less variable than in the surface layers. Below 500 m seasonal changes in these variables are negligible, and the deeper the water the more constant are the conditions.

6.1 The substrate

Conditions which determine the types of materials forming the sea floor include:

(a) the geology, including the proximity of land and the geographical and geological features of the coastline such as rock formations and the outflows of rivers or glaciers;

(b) the speed of the bottom current;

(c) the depth;

(d) the types of suspended matter in the overlying water, including the pelagic organisms; and

(e) the type of benthic population.

Sediments cover the sea floor except where the bottom current is strong enough to sweep away particles, or where the gradient is too steep for them to lodge. The scouring action of tidal currents may expose rocks beneath shallow water, and in deep water uncovered rock occurs on steep sides of submarine peaks or trenches.

In sediments, the type of deposit varies with the speed of the bottom current and the size and density of suspended particles. The faster the water moves, the coarser is the texture of the substrate because finely divided material is more easily held in suspension than larger particles of the same density. Stokes' equation for the settling velocity, W, applies fairly well to small particles in seawater:

$$W = \frac{2}{9} g \frac{D - d}{\mu} r^2$$

g = gravitational acceleration
D = density of particles
d = density of liquid
μ = dynamic viscosity of liquid
r = radius of spherical particles.

In many areas, parts of the sea-bed over the continental shelf are kept in motion by strong tidal currents. Once the settling velocity of particles in the sediment is exceeded, the transport of sediment increases rapidly with increase of current speed.

Marine sediments are classified into two main groups, terrigenous and pelagic deposits.

6.1.1   Terrigenous deposits

Terrigenous deposits are found near land, covering the continental shelf and upper parts of the continental slope. Much of this material is derived from weathering and erosion of exposed land surfaces, and consists largely of particles worn from the coast by wave action or carried into the sea by rivers or glaciers. Terrigenous deposits contain some organic material, often some 0.01–0.5 per cent of the dry weight, the finer-texture deposits usually having the greater proportion of organic matter. Microscopic examination sometimes reveals recognizable traces of various materials of biological origin, both terrestrial and marine. The former are mainly fragments of leaf and wood from land plants. The latter are very diverse, deriving from both benthic and pelagic sources, and often include small particles of seaweed, diatom cell walls, sponge spicules, polychaete chaetae, and fragments of the shells and skeletons of foraminifera, hydroids and corals, polyzoa, crustacea, echinoderms and molluscs. Minute animals may be seen in samples from the surface layers of the deposit; for example, flagellates, ciliates, foraminifera, nematodes and copepods. Superficial deposits from shallow water may also contain a microflora of benthic diatoms. The mud from shallow creeks is sometimes rich in fragments of marine angiosperms, for example *Spartina*.

Terrigenous deposits show considerable differences of composition from place to place, varying with the nature of the adjacent coastline, the movements of the water and the contours of the sea-bed. They range from large boulders close to rocky shores where they have been dislodged by violent wave action, through all grades and mixtures of pebbles, gravel and sands down to fine clay. These familiar terms are defined by the specific size range of their particles. The much-used Wentworth scale of particle sizes (see Table 6.1) is geometric, giving smaller intervals towards the finer end of the range. Using this scale, a pebble is a pebble only if its size lies between 4 and 64 mm diameter. A pebble larger than this is defined as a cobble or a boulder.

Geologists tend to use a different scale called the *phi* (ϕ) scale. This converts the unequal steps of the Wentworth scale into an arithmetic series of equal intervals, thereby simplifying graphical and statistical treatment. The particle diameter in millimetres is written as the equivalent negative of the power of 2. So for example to find the phi size of a pebble of 64 mm diameter, the diameter is first written as a power of 2, i.e. 2^6. ($64 = 2^6$). The power of 2 in this case is 6 so the phi size is the negative of 6, i.e. a 64 mm diameter pebble has a phi size of -6.

Likewise, a very fine sand with grains of 0.125 mm diameter has a phi size of $+3$. This is calculated as follows: convert 0.125 to a fraction = 1/8. 1/8 written as a power of 2 is 2^{-3}. The negative of -3 is $+3$.

The sizes of the sediment grains in a sample are found by drying the sample and sieving it through a series of sieves with meshes of decreasing size. The sieves are precision tools, constructed with very accurate mesh sizes relating to the phi scale. The grains that pass through one sieve but not the next, have a diameter size range between the two phi sizes of the sieves.

Marine sediments are never uniform in composition but contain particles of many grades and types. If the particles are mainly of one size, the sediment is said to be well sorted. If there are many sizes of grains then it is a poorly sorted sediment. Sediments containing more than 10 per cent dry weight of silt and clay fractions are commonly termed 'muddy sands'. If more than 30 per cent of the

Table 6.1   Wentworth Classification of Particle Grades and phi Scale.

Grade name	Particle size range (mm)	Phi units
Boulder	>256	<−8
Cobble	64–256	−6 to −8
Pebble	4–64	−2 to −6
Granule	2–4	−1 to −2
Very coarse sand	1–2	0 to −1
Coarse sand	0.5–1	+1 to 0
Medium sand	0.25–0.5	+2 to +1
Fine sand	0.125–0.25	+3 to +2
Very fine sand	0.0625–0.125	+4 to +3
Silt	0.0039–0.0625	+8 to +4
Clay	<0.0039	>8

deposit is silt and clay, the term 'sandy mud' is applied, and deposits with silt and clay fractions exceeding 80 per cent are generally described as 'mud'.

Some terrigenous deposits are exploited by man. For example, the continental shelf is nowadays an important source of sand and gravel for building operations.

6.1.2   Pelagic deposits

Pelagic deposits occur beneath deep water beyond the edge of the continental slope, carpeting the deep ocean basins. Much of this material is of fine texture, and its nature varies with the depth and with the types of organisms that abound in the overlying water. At depths of less than about 6000 m, pelagic deposits contain a considerable proportion of material of biological origin, commonly some 30 per cent or more by weight. Although these deposits are termed 'organic', they seldom contain much decomposable carbon but consist almost entirely of skeletal fragments of planktonic organisms. Organic deposits are of two main types, calcareous and siliceous. Calcareous sediments, rich in calcium carbonate and formed mainly from foraminiferan shells, are common in middle and low latitudes but only down to an average depth of around 4600 m. Below this depth, hydrostatic pressure causes some forms of calcium carbonate to dissolve. The calcite compensation depth (CCD) is the depth where the supply of clacite raining down from above is equalled by its dissolution. So below this depth, there are no sediments containing carbonate.

At high latitudes around the polar belts, great areas of the sea-bed are covered with siliceous sediments. These are areas of high productivity where the rain of material downwards consists mainly of silica-containing diatoms and radiolarians. Calcareous sediments do occur here but only down to around 3000 m. Other high-productivity areas such as the equatorial belt and coastal upwellings also have siliceous sediments especially where the sea-bed lies below the CCD. In general, the deeper organic deposits contain less calcareous material and a larger proportion of silica.

The organic deposits are named after the main organisms from whose skeletons they are made up and are classified as follows:

Calcareous oozes

- **Globigerina ooze** This is the most widespread of the deposits over the greater part of the deep Atlantic and much of the Indian and Pacific Oceans, covering nearly 50 per cent of the deep-sea bottom and extending to depths of 6000 m. It contains up to 95 per cent calcium carbonate mainly in the form of foraminiferan shells.
- **Coccolith ooze** A high proportion of coccolith material, sometimes amounting to 25 per cent or more of the total weight, is occasionally found in samples of globigerina ooze, chiefly beneath areas of warm surface water.
- **Pteropod ooze** This contains many pteropod shells and occurs below subtropical parts of the Atlantic at depths down to 3500 m.

Siliceous oozes

- **Diatom ooze** This consists mainly of siliceous material in the form of diatom fragments. It occurs as an almost continuous belt around Antarctica beneath the Southern Ocean, its northernmost limit corresponding closely with the position of the Antarctic convergence. There is also a strip of diatom ooze across the northern part of the north Pacific.
- **Radiolarian ooze** This contains many radiolarian skeletons and occurs at depths between 4000 and 8000 m beneath tropical parts of the Pacific and Indian oceans and is also recorded in the Atlantic.

Red clay

At 6000 m and below, sediments generally contain less than 10 per cent material of obvious biological origin, and at these depths the most widespread deposit is *red clay*, covering nearly 40 per cent of the deep ocean floor. It is a very finely divided sediment, usually brick-red in colour, and consisting mainly of fine-grained quartz (silica), and clay minerals such as aluminium oxide along with small amounts of various compounds of iron, calcium, magnesium and traces of many other metals. The bulk of this material is derived from 'aeolian fallout'. That is, it originates as fine mineral dust lifted from the ground by wind action and carried through the atmosphere, from which some eventually 'falls out' over the sea. These clays accumulate incredibly slowly at the rate of 1 mm or so per 1000 years. Organic oozes accumulate at around 10–30 mm per 1000 years.

The distribution of deep sediments is shown in Figure 6.1.

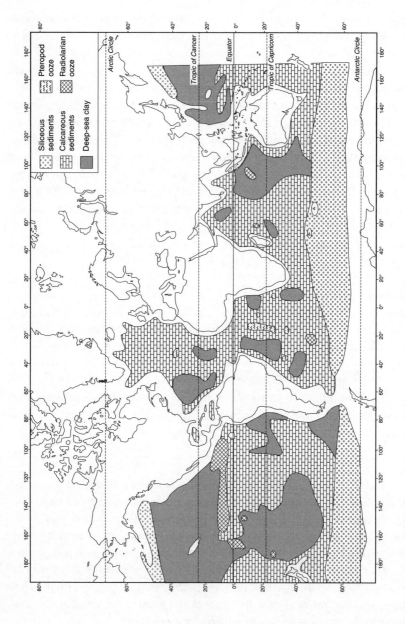

Figure 6.1. Ocean bottom deposits. Siliceous sediments = diatom ooze and radiolarian ooze. Calcareous sediments = globigerina ooze and pteropod ooze.

6.1.3 Deep-sea nodules

A feature of wide areas of the deep ocean floor below about 2000 m is the presence of sizable lumps or nodules lying on the surface of the sediment. These are known as manganese nodules or ferromanganese concretions, and occur in many irregular shapes, sizes and forms ranging in size from particles a millimetre or so in diameter to occasional huge lumps a metre or more across. They are especially numerous in the Pacific, where parts of the bottom are almost covered with nodules, but they are also found beneath other areas of deep ocean.

In structure the nodules usually have a lamellated form surrounding a central nucleus which may be a core of silty material, or sometimes a fragment of rock, a fish tooth, or even a whale ear-ossicle. Nodules from different areas are very variable in composition (Tooms and Summerhayes, 1969), but are commonly rich in manganese and iron. Mero (1960), gives the average composition by weight of 30 samples of nodules from all oceans as manganese dioxide 32 per cent, iron oxides 22 per cent, silicon dioxide 19 per cent, water 14 per cent, with smaller quantities of aluminium oxides, calcium and magnesium carbonates, and other metals including nickel, copper, cobalt, zinc and molybdenum.

Nodule formation and the way in which they become enriched with metals, is not fully understood but is basically a process whereby solids are formed by chemical reactions in sea water. The process is thought to be initiated by the precipitation of colloidal particles, for example, manganese dioxide or ferric hydroxide, which tend to attract ions of other metals from the water. These particles are electrically charged, and may be attracted to electrically conductive objects on the bottom which become the nucleus of a developing nodule. The availability of suitable nuclei may be one of the causes of their patchy distribution. The nodules form very gradually over a period of several million years. Abyssal nodules, which have the highest nickel and copper concentrations, are calculated to grow at a rate of 3 to 8 mm per million years.

Manganese nodules were first discovered by the *Challenger* expedition. Since then, specialized mining systems and ships such as the *Prospector* have been developed to lift the nodules, which can be treated to extract manganese, nickel, copper and cobalt. At present, only one area, between the Clarion–Clipperton fracture zone near Hawaii, appears to be economically attractive and so far, no-one has attempted commercial exploitation. Nevertheless, some mining claims have been filed and international consortiums formed to develop and test recovery systems.

6.1.4 Oil- and gas-bearing deposits

Some of the organic matter that settles on the sea floor becomes buried in submarine deposits. It then undergoes various processes of decomposition depending on the speed and depth of deposition. Within the uppermost few centimetres, organic matter is largely oxidized by bacteria to carbon dioxide and other substances, much of which escapes into the water. Carbon dioxide produced from organic matter deeper within the deposit mainly combines to form calcium carbonate which cements the substrate into rock. Deeper still in the substrate organic matter may undergo bacterial fermentation producing methane. Inorganic processes of decarboxylation induced by heat may also lead to the formation of methane or oil. Gas and oil in rocks tend to migrate through porous layers such as sand or limestone, but may become trapped within anticlines or unconformities beneath non-porous caps of materials such as salt or shales which prevent their escape to the surface.

6.2 Benthic communities

Using the term 'community' simply to refer to groups of species consistently occurring together in broadly similar environmental conditions, we find that different parts of the sea floor are populated by characteristic communities. The differences of environment accountable for the association of particular communities with particular parts of the sea bottom can be related to features of both the water and the substratum.

6.2.1 Physical factors affecting distribution of communities

We have already outlined in Chapter 4 the major hydrographic parameters which control the distribution of marine organisms – the temperature of the water, its composition, movements, pressure and illumination. Except on the shore and in shallow water, these vary less at the bottom than they do in the upper levels of water. They are none the less important in relation to the distribution of benthic populations, restricting certain species to particular localities and often having a major effect during the early stages of life when the majority of benthic creatures pass through pelagic phases. The planktonic eggs and larvae of many species are highly susceptible to the quality of the water in which they float, and the period prior to completion of metamorphosis is always the time of heaviest mortality. Predation and shortage of suitable food exact a heavy toll, and losses are specially severe if water temperature or salinity is unfavourable.

Substrate

The substrate material exerts a dominant influence over the distribution of organisms on the sea floor. Where the bottom is rock or large stones, the community consists chiefly of forms which live on the surface of the substrate, i.e. it is an epifauna and epiflora. The animals are mainly sessile or encrusting cnidarians (coelenterates), sponges, bryozoans, barnacles, tubicolous worms, mussels and sea squirts. Crawling among them are a variety of errant polychaetes, starfish, echinoids, gastropods and large crustaceans such as crabs, lobsters and crawfish. In shallow water where sufficient illumination reaches the bottom, seaweeds grow attached to rock or to stones heavy enough to give secure anchorage, and to each other. There is usually a wide diversity of species inhabiting a rocky bottom because the irregularities of the rock surface provide a great variety of microhabitats, with innumerable differences of living space, water movement, food supply, illumination and temperature. A rocky bottom does not support a numerous infauna, i.e. animals dwelling within the substrate, but burrowing creatures occur in accumulations of silt in rock crevices. There are also a few forms capable of boring into rock, mainly bivalve molluscs (*Hiatella*, *Pholas*, *Lithophaga*), a few annelids (*Polydora*, *Dodecaceria*), the sponge *Cliona* and certain barnacles and sea urchins. In some areas there are also species of red algae which bore superficially in calcareous rock.

Where the bottom is covered with sediment, most of the inhabitants live within the deposit. Local conditions are generally more uniform and communities less diverse than on a rocky bottom. The infauna includes burrowing sea anemones, polychaetes, bivalves, gastropods, echinoderms and crustacea. Some fish also burrow superficially in the deposit and a few, such as the red band fish (*Cepola*) make deep burrows. The particle size of a sediment is an important factor regulating the distribution of the infauna because the mode of burrowing of many creatures is specialized and suitable only for a certain grade of substrate (Trueman and Ansell, 1969). Burrowing can be done by forcing or digging through the sediment, pushing the particles aside, or eating through it, or often by a combination of methods. Large particles are more difficult to displace or ingest than small ones, and the mechanical difficulty of burrowing in coarse deposits may be one reason why these are usually less populated than finer ones. On the other hand, very fine sediment can compact into a dense, unyielding mass in which it is not easy to burrow and which requires adaptations for dealing with silt. Certain combinations of particles form thixotropic deposits which are readily reduced to a semi-fluid consistency by repeated, intermittent pressures and yield easily to burrowing. Where the deposit is exceptionally soft, such as occurs beneath some areas of very deep water, animals may simply sink into it, and here we find adaptations such as stalks or extremely long appendages to lift the main body clear of the bottom.

Although differences between communities can often be correlated with differences of particle size of sediments, other factors are also involved. For example, many sediment-dwelling animals do not actively burrow, and yet may be quite particular in their choice of sediments. The rather similar common British sea cucumbers *Neopentadactyla mixta* and *Thyone fuscus* live in quite different sediments, *Neopentadactyla* in coarse gravels and *Thyone* in mud. Here the size of sediment particles is unlikely to be the only factor controlling their distribution. The grade of a deposit depends upon the speed of bottom current, and this also controls several other features of the substrate. Slow-moving water allows organic matter to settle, giving a sediment that may be not only fine in texture but also rich in organic content. Poor or absent circulation of the contained water leads to deficient oxygenation of the subsurface layers, and high concentrations of sulphide. Beneath shallow water these conditions often support a large biomass because there is a good food supply for creatures which feed on the surface or digest organic matter from the sediment, but the infauna must be able to cope with silt and a deoxygenated medium. Where the bottom water moves more swiftly there is likely to be less settlement of food and a lower organic content in coarser sediments, but better oxygenation of the interstitial water. The poorer food supply supports a smaller biomass, but these conditions favour animals which can burrow in coarse material and capture floating food suspended in the water. Therefore, several interrelated factors must operate to limit certain species to particular substrates.

Turbidity

The turbidity of the water is another factor to which some benthic organisms are very sensitive. The quantity of suspended matter is often considerably greater in water close to the bottom than in layers nearer the surface. In shallow water this reduces illumination and may therefore restrict the distribution of benthic plants. High turbidity may also have adverse effects on animals by clogging the feeding apparatus or smothering the respiratory surfaces. Many benthic creatures are filter-feeders, notably the lamellibranchs which form a major part of the community in many sediments. These obtain food by drawing in a current of water from which they filter suspended food particles, and special adaptations are required to cope with the problem of separating food from large quantities of silt. Turbidity is also related to wave exposure and water currents. Shallow sheltered lochs or bays are likely to have poorer water clarity than similar but more open areas. Water movement carries settling silt away and prevents it from accumulating and smothering the benthos.

Water currents

Benthic organisms are also influenced by the speed of the bottom current because this controls the particle size of the substrate, its oxygenation and organic content, and also affects the dispersal of pelagic larvae and the ease with which they can settle on the bottom. The bottom current is also important in the transport of food particles, sweeping them away from some areas and concentrating them in others, especially in depressions in the sea-bed.

Pressure

At abyssal depths, where temperature and salinity are uniform over great areas, hydrostatic pressure may well be the chief factor which accounts for differences between communities within the ocean trenches and those of other parts of the deep-sea bottom.

6.2.2 Biological factors affecting distribution of communities

Although inorganic factors exert a major control, there are also biological factors which influence the distribution and composition of benthic communities. The physical and chemical features of the environment determine a range of species which compete, but success or failure in a particular habitat depends ultimately on qualities inherent in the organisms themselves. For example, they are able to some extent to choose their position. Free-living forms can move about to find areas that suit them, but the majority of adult benthic animals remain more or less stationary, confined within burrows or attached to the bottom. Most of these start life as pelagic larvae dispersed by the water (Thorson, 1950). Once they settle and metamorphose they stay in place, and die if conditions are unsuitable. Undoubtedly there are great losses, but the larvae of many species show behavioural features which influence dispersal and favour their chances of reaching situations where survival is possible. Many species of larvae have some control over the depth at which they float, often by virtue of their response to light. Larvae of shallow water species are usually photopositive for a time, collecting near the surface, while those of deeper-dwelling forms mostly prefer dim illumination or darkness, and therefore occupy deeper levels. The depth at which the larvae float must obviously influence the depth at which they settle.

Selective settlement

It has also been demonstrated that larvae of certain species can discriminate between different substrates, and have some powers of selection; for example the larvae of the small polychaete *Ophelia bicornis*. This worm has a patchy distribution in various bays and estuaries around the British coast, living in a particular type of clean, loose sand. Pelagic larvae are produced which are ready to metamorphose when about five days old. At this stage they begin to enter the deposit. Wilson (1956) discovered that the larvae are able to distinguish between sands from different areas, preferring to complete their metamorphosis in contact with certain 'attractive' sand samples, and avoiding contact with other sands which have a 'repellent' effect. When the larvae settle, they appear to explore the deposit. If the sand is of the 'repellent' type, they leave it and swim away, shortly settling again and repeating their exploration. This behaviour continues over a period of several days, with metamorphosis delayed until a suitable substrate is found. If the larvae find nothing suitable, they eventually attempt to metamorphose none the less, but then usually die. During an early series of experiments, the particle size of the sand seemed to be the main factor to which the larvae were sensitive, but further experiments with artificially constituted sands showed the importance of other factors, notably the coating of organic materials and bacteria on the surface of the sand grains. If sand samples were washed in hot concentrated sulphuric acid they became neutral, losing their attractive or repellent qualities.

Selective examination of the substrate, with metamorphosis delayed as long as possible until a suitable substrate is discovered, has now been demonstrated for a fairly wide range of organisms, including annelids, barnacles, molluscs, bryozoa and echinoderms. Larvae usually become less discriminating with age. Although the prospects of successful metamorphosis become less as they get older, some are able to continue normal development even after extended periods of pelagic life if they eventually find a satisfactory surface for attachment. Many different properties of the substrate influence choice (Bayne, 1964, 1969; Crisp, 1974; Meadows and Campbell, 1972). Particle size selection has now been demonstrated for various burrowing organisms, e.g. the polychaete worms *Ophelia*, *Protodrilus* and *Pygospio* and the horseshoe worm *Phoronis*. In most cases the attractiveness of sediments is associated with the presence of films of micro-organisms, which may be important as food. This is most evident in burrowers which actually swallow the substrate. Other qualities of surfaces between which benthic larvae discriminate include differences of chemical nature, texture, slope, contour and colour. Sensitivity to aqueous diffusing substances is most typical of species which settle on organic substrates or on sediments containing a high content of organic matter; for instance, the attraction of shipworm (*Teredo*) larvae to wood, tube worm (*Spirorbis borealis*) larvae to the seaweed *Fucus serratus*, the gastropod *Nassarius obsoletus* to mud. In other species, mainly those settling on rock, e.g. acorn barnacles, spirorbid tube worms and the honeycomb worm *Sabellaria*, it has been shown that the larvae must usually first make actual contact with the shell, cuticle or cementing substance of their own or a closely related species before settlement is attempted, the stimulus for settlement being dependent on a 'tactile chemical sense'. The majority of rock-settling larvae prefer rough surfaces to smooth, some discriminate between light or dark surfaces and some, e.g. the sponge *Ophlitaspongia seriata*, settle preferentially on overhanging surfaces. In some rocky shore animals, e.g. *Semibalanus balanoides*, water turbulence encourages settlement. Barnacle cyprids settle most readily from flowing water, different species preferring different water velocities; for instance, the velocity of maximum settlement for *Semibalanus balanoides* is higher than for *Elminius modestus*. In *Sabellaria alveolata*, a polychaete which forms tubes of cemented sand grains, the larvae settle best from swirling water in which sand grains are present, especially after contact is made with the cementing substance of adult tubes or recently metamorphosed individuals (Wilson, 1968, 1970).

Selective settlement offers several advantages. It reduces larval losses by hindering metamorphosis in unsuitable locations. By encouraging gregariousness it facilitates fertilization, especially cross-fertilization between hermaphrodite forms requiring internal fertilization, such as acorn barnacles. Settlement in areas where others of the species have already survived ensures that conditions are likely to be congenial. It is also a form of behaviour favouring close adaptation to particular habitats, with the advantages of specialization. Conversely, although this may confer great efficiency in specific conditions, it increases the risks of extermination if the environment changes and suitable sites are lost.

Competition and predation

Once the larvae have settled, many other biological factors begin to influence their chances of survival. A community is a society of organisms with many interactions between the individuals. There is often competition for living space, and the outcome of this aspect of the struggle for existence is determined by a complex of factors. After settlement many larvae exhibit some exploratory behaviour, moving about over the surface and often spacing themselves to some extent to avoid crowding. Delayed settlement is usually followed by reduced spacing movements. Some larvae tend to space themselves from their own species but not from others, on which they may actually settle and attach. Such behaviour reduces intraspecific but increases interspecific competition. The relative success of different species in competing for space is also influenced by differences of breeding periods, reproductive capacities and growth rates. One species may gain advantage by early settlement following a seasonal decline in numbers of the community, or a fast-growing species may oust a slower-growing competitor by overgrowing it, or by claiming an increasing proportion of a shared food source.

Interaction between predator and prey must regulate the numbers of both. Mortality due to predation is usually highest during the period following settlement while the individuals are still small. Certain predators, notably ophiuroids, are sometimes so numerous that they virtually carpet the bottom and it seems surprising that any small creatures suitable for food can escape. In these conditions, survival may depend upon the time of settlement. It has been observed (Thorson, 1960) that some benthic carnivores have phases when feeding diminishes or ceases, usually in association with breeding. This passive period may last several weeks, and some species which settle during this time may be able to reach a sufficient size to become relatively safe from predation before their enemies start active feeding again. The composition of a community may therefore reflect coincidences between passive periods of predators and settlement periods of other species.

Interrelationships

The individuals of a community are in various ways interdependent, and some organisms thrive only in the presence of particular associated forms. Each type of animal is dependent upon other organisms for food, and the quantity and quality of food sources obviously exert a profound control over numbers and composition of communities. Certain organisms depend upon others to provide surfaces of attachment upon which they can grow, and in some cases are seldom found elsewhere; for example, the barnacle *Pyrgoma anglicum* on the cup coral *Caryophyllia smithii*, the anemone *Adamsia palliata* on the hermit crab *Pagurus prideauxi*. Some animals share the burrows formed by others; for example the polychaete *Lepidasthenia argus* shares with another polychaete, *Amphitrite edwardsi*. The polychaete *Harmothoe lunulata* may be free-living but often shares the burrows of the polychaetes *Arenicola marina*, *Amphitrite johnstoni*, and *A. edwardsi*, or the echinoderms *Acrocnida brachiata*, *Leptosynapta inhaerens*, *Labidoplax digitata* and others. One species may even live inside another; for example, the shrimp *Typton spongicola* within sponges including *Desmacidon fruticosum*, and the barnacle *Acasta spongites* embedded in the sponge *Dysidea fragilis*.

Relationships between species certainly involve numerous associations of an epizoic, commensal, symbiotic or parasitic nature. A striking example of a group of species commonly found living in close proximity is associated with the hermit crab *Pagurus bernhardus*, widely distributed on gravelly deposits in shallow water around the British Isles. The adult crab inhabits shells of the whelk *Buccinum undatum*. Several animals are commonly found on the surface of the shell, the most conspicuous being the anemone *Calliactis parasitica*. Other inhabitants of the outer shell surface, sometimes extending into the opening, include saddle oysters *Anomia ephippium*, the hydroid *Hydractinia echinata*, the serpulid worms *Spirorbis spirillum*, *Pomatoceros triqueter* and *Hydroides norvegica*, the barnacle *Balanus crenatus* and the sponge *Suberites domuncula*. The barnacle *Trypetesa lampas* apparently lives only in *Buccinum* shells inhabited by hermit crabs, where it burrows into the shell substance just inside the shell aperture. The shell is also sometimes bored by the worm *Polydora ciliata*. Living within the shell alongside the hermit crab there is frequently a large worm, *Nereis fucata*, which feeds on fragments of food dropped by the crab. Also inside the shell the tiny crab *Porcellana platycheles* is sometimes found. The hermit crab may carry parasites; for example, the isopods *Pseudione* spp. in the branchial chamber, *Athelges paguri* attached to the abdomen or occasionally in the branchial chamber, and the parasitic barnacle *Peltogaster paguri* extruding from the abdomen.

6.3   Classification of communities

6.3.1   Petersen's soft bottom communities

During the second decade of the twentieth century, pioneer studies of marine benthos were made by Petersen, who carried out detailed investigations by grab samples of the larger animals (the macrofauna) of soft deposits in shallow water off the Danish coast. He found that different areas supported characteristic associations of animals, and he distinguished nine communities, naming each after the most conspicuous components of the population as follows:

1 *Macoma* communities, widespread in shallow muds (*M. baltica, Mya arenaria, Cardium = Cerastoderma edule*, etc.)
2 *Abra* communities in shallow, muddy sands, often in sheltered creeks. (*A. alba, A. prismatica, Macoma calcarea, Astarte* spp, etc. sometimes with *Echinocardium cordatum*.) (Note: In Petersen's original classification *Abra = Syndosmya*)
3 *Venus* communities, widespread on shallow sandy bottoms on open coasts. (*V. striatula, Tellina fabula, Montacuta ferruginosa*, often with *E. cordatum*.)
4 *Echinocardium-filiformis* communities, in sandy mud at intermediate depths. (*E. cordatum, Amphiura filiformis, Turritella communis, Nephtys* spp., etc.)
5 *Brissopsis-chiajei* communities on deeper, soft mud (*B. lyrifera, Amphiura chiajei, Abra nitida*, etc.)
6 *Brissopsis-sarsi* communities on soft mud below *Brissopsis-chiajei* depths. (*B. lyrifera, Ophiura sarsi, A. nitida*, etc.)
7 *Amphilepis-Pecten* communities on deep mud in Skagerrak. (*Amphilepis norvegica, Chlamys (Pecten) vitrea*, etc.)
8 *Haploops* communities on deep, firm mud in Kattegat. (*Haploops tubicola, Chlamys septemradiata*, etc.)
9 Deep *Venus* communities, widespread on coarse sands (*V. striatula, Spatangus purpureus, Echinocardium flavescens, Spisula* spp., etc.)

Thorson (1957) observed that in middle latitudes there are certain conspicuous genera of bivalves, echinoderms, polychaetes, etc., which occur in communities of generally similar appearance in widely separated places, but represented by different species in different parts of the range. He has therefore classified communities in terms of their most obvious genus rather than species, e.g. *Macoma* communities, *Tellina* communities, etc., the species varying with differences in local conditions, mainly temperature and salinity.

One drawback of classifying communities with respect to their most prominent constituents is that certain eye-catching organisms are so widely distributed as to occur in association with rather different assemblages of other species in different parts of their range. For example, *Venus striatula* is frequently the most obvious species in sand at shallow depths in a community that includes *Tellina fabula, Ensis ensis* and *Echinocardium cordatum* as other conspicuous components. *Venus striatula* also occurs in deeper water with a different group of associated organisms, notably *Spisula* spp., *Echinocardium flavescens* and *Spatangus purpureus*. It is therefore necessary to distinguish two different *Venus* communities, one in shallow water and one at deeper levels. Similarly *Brissopsis lyrifera* is a prominent component of different communities at different depths.

6.3.2 Jones' classification of NE Atlantic shelf communities

Because the distribution of species is determined so closely by physical and chemical parameters, these offer an alternative basis for classifying benthic communities, primarily with respect to substrate, depth, temperature and salinity. Jones (1950) divided the shelf communities of the north-east Atlantic in this way as follows. Obviously the communities defined below can be further subdivided into narrower zones of distribution with more precisely defined environmental parameters.

(1) **Shallow water and brackish communities** with upper limits of distribution extending on to the shore. Eurythermal and euryhaline within wide limits. Temperature range 3–16°C, salinity 7–34‰, but always exposed to periodical diminution below 23‰.
(A) Shallow soft bottom community
   (i) *Boreal shallow sand association* – occurring on relatively exposed coasts in north-west Europe. Important species include *Arenicola marina, Tellina tenuis, Donax vittatus, Nephtys caeca, Bathyporeia pelagica.*
   (ii) *Boreal shallow mud association* – equivalent to Petersen's *Macoma* community, occurring on more sheltered coasts of north-west Europe, and in estuaries and in the Baltic.
      Arenicola marina, Macoma baltica, Mya arenaria, Cerastoderma (Cardium) edule, Corophium volutator.
(B) Shallow hard bottom community
   (i) *Boreal shallow rock association* – shores and shallow water with rocky substrates off north-west Europe and in the Baltic.
      Balanus balanoides, Mytilus edulis, Littorina spp., *Patella vulgata, Nucella lapillus.*
   (ii) *Boreal shallow vegetation association* – algae of the shore and shallow water with associated fauna.
      Hyale prevosti, Idotea spp., *Hippolyte varians, Littorina littoralis* (= *L. obtusata* and *L. mariae*), *Rissoa* spp., *Lacuna vincta.*

(2) **Offshore communities** having upper limits of distribution below extreme low-water spring tidal level. Eurythermal and euryhaline, but between narrower limits than (A). Temperature range 5–15°C, salinity 23– 35.5‰.
(A) Offshore soft bottom community
   (i) *Boreal offshore sand association* – equivalent to Petersen's *Venus* community, occurring offshore on sandy bottom.
      Sthenelais limicola, Nephtys spp., *Ampelisca brevicornis, Bathyporeia*

guilliamsoniana, Dosinia lupinus, Venus striatula, Tellina fabula, Gari fervensis, Abra prismatica, Ensis ensis, Echinocardium cordatum.

(ii) *Boreal offshore muddy sand association* – equivalent to Petersen's *Echinocardium-filiformis* community, occurring on muddy sand offshore and in modified form in sheltered and estuarine situations. *Nephtys incisa, Goniada maculata, Lumbriconereis impatiens, Pectinaria auricoma, Diplocirrus glaucus, Eumenia crassa, Scalibregma inflatum, Notomastus latericeus, Owenia fusiformis, Ampelisca spinipes, A. tenuicornis, Nucula turgida, Dosinia lupinus, Abra alba, A. prismatica, A. nitida, Arctica (Cyprina) islandica, Acanthocardia (Cardium) echinatum, Parvicardium (Cardium) ovale, Cultellus pellucidus, Spisula subtruncata, Corbula gibba, Dentalium entalis, Turritella communis, Aporrhais pespelicani, Philine aperta, Ophiura texturata, Amphiura filiformis, Echinocardium cordatum, E. flavescens, Leptosynapta inhaerens.*

(iii) *Boreal offshore mud association* – equivalent to Petersen's *Brissopsis chiajei* community, occurring on soft mud, usually outside the muddy sand association but gradually grading into it.
Leanira tetragona, Nephtys incisa, Glycera rouxi, Lumbriconereis impatiens, Maldane sarsi, Notomastus latericeus, Eudorella emarginata, Calocaris macandreae, Nucula sulcata, N. tenuis, Abra nitida, A. prismatica, Amphiura chiajei, Brissopsis lyrifera.

(B) Offshore hard bottom community

(i) *Boreal offshore gravel association* – occurring at moderate depth wherever the deposit is very coarse, whether sand, gravel, stones or shells. Masses of *Modiolus modiolus* sometimes form a numerous epifauna.
Polygordius lacteus, Glycera lapidum, Potamilla spp., *Serpula vermicularis, Crania anomala, Balanus balanus* (=*porcatus*), *Galathea* spp., *Eupagurus* spp., *Hyas coarctatus, Nucula hanleyi, Glycymeris glycymeris, Lima loscombi, Venus casina, V. fasciata, V. ovata, Venerupis rhomboides, Gari tellinella, Spisula elliptica, Modiolus modiolus, Buccinum undatum, Asterias rubens, Ophiothrix fragilis, Ophiopholis aculeata, Echinus* spp., *Echinocyamus pusillus, Echinocardium flavescens, Spatangus purpureus.*

(3) **Deep communities** with upper limits of distribution not above 70 m depth and usually much lower. Stenothermal and stenohaline. Temperature range 3–7°C, salinity 34–35.5‰.

(A) Deep soft bottom community

(i) *Boreal deep mud association* – equivalent to Petersen's *Brissopsis sarsii* and *Amphilepis-Pecten* communities, occurring below the offshore mud association.

> *Glycera alba, Spiophanes kroyeri, Chaetozone setosa, Maldane sarsi, Clymene praetermissa, Sternaspis scutata, Notomastus latericeus, Melinna cristata, Proclea graffi, Eriopisa elongata, Nucula tenius, Nuculana minuta, Chlamys vitrea, Thyasira flexuosa, Abra nitida, Portlandia lucida, Parvicardium (Cardium) minimum, Ophiura sarsi, Amphilepis norvegica, Brissopsis lyrifera.*

(B)   Deep hard bottom community

    (i)   *Boreal deep mud association* – a little-known, deep-water epifauna of corals and associated animals.

       Lophelia pertusa, Paragorgia arborea, and *Gorgonocephalus caput-medusae.*

Although some parts of the sea floor present sharp discontinuities, for instance the change from rock to sand, alterations of substrate from one place to another are mostly gradual. Furthermore, each species has its particular distribution which is never identical with that of any other. Consequently boundaries between communities are usually indefinite with intergrading along transitional zones. On almost every part of the sea-bed the inhabitants comprise a climax community for that particular area, stable in composition within natural, short-term fluctuations. Wherever environment changes with locality there are corresponding adjustments in the make-up of the assemblage of species. The concept of a community is essentially an abstraction from studies of overlapping distributions of many species along various ecological gradients. For further discussion of the ideas underlying these classifications of benthic communities, the student may consult the paper by Glemarec (1973).

The facility with which communities may be characterized by particular conspicuous species, sometimes rather misleadingly described as 'dominants', although evident in marine benthos around the British Isles, is less apparent at low latitudes where the composition of communities generally shows a greater diversity. This diversification may be a feature of more mature communities which have evolved in stable conditions over a long period, permitting the survival of species specialized for narrow ecological niches. The communities of the north-east Atlantic may be regarded as relatively immature, having evolved in fluctuating conditions since the extremes of the Pleistocene period. This favours the evolution of polymorphic populations which survive by virtue of their wide variability, with 'dominant' species occupying broad ecological niches.

6.3.3   Modern classifications of coastal marine communities in the NE Atlantic

The classification systems described above have provided a useful working framework for a number of years. However, these systems concentrate mainly on soft bottom communities and the north-east Atlantic region still lacks a comprehensive classification of benthic marine biotopes (i.e. habitats and their associated species) that encompasses intertidal and sublittoral, rocky and sedimentary ecosystems. The littoral zone has been well studied over the years. Hoewever, it is only in the past 20 years or so that we have seen a rapid increase in our knowledge of shallow rocky sublittoral ecosystems through the use of divers. In the UK various ways of recording littoral and sublittoral habitats and species in a standard format have been developed and a large number of data has been accumulated.

Attempts are now being made by the UK government's Joint Nature Conservation Committee (JNCC, see Appendix 4), to develop such a comprehensive classification of marine biotopes in UK coastal waters using a computerized database (Hiscock, 1996; JNCC, 1996). One of the main aims is to provide a meaningful structure on which to base a conservation strategy for the marine environment. This is likely to be adopted as a standard for northern European waters. The system is intended to be practical and to be of use to anyone involved in descriptive surveys. For further information, students should contact JNCC directly (see Appendix 4). Increased use is now being made of underwater photography to help in descriptive surveys and a number of photographic guides to seabed habitats and biotopes are now available (Earll, 1992).

6.3.4 Meiobenthic communities

Studies and classifications of benthic communities have mostly been confined to examination of the larger animals, or *macrobenthos*. These communities also contain many smaller forms, the *meiobenthos* and *microbenthos*, about which an increasing amount is being learnt. Numerous small organisms just large enough to be seen by the unaided eye comprise the meiobenthos, including foraminifera, turbellarians, nematodes and various small polychaetes, bivalves and crustacea such as harpacticoid copepods. These may sometimes comprise an appreciable fraction of the total biomass, even as much as 25 per cent. Technically meiofauna are those whose size ranges between 0.1 mm and 1 mm. The smallest organisms, visible only with a microscope, comprise the microbenthos, which includes bacteria, a great variety of protozoa, mainly flagellates, ciliates and amoebae, and often other small organisms such as rotifers, crustacean larvae and the smallest nematodes. In shallow water there is often a microflora of diatoms and coloured flagellates. Organisms small enough to live in the interstices between the grains of sediment are described as an *interstitial* fauna or flora (Swedmark, 1964).

6.3.5 Stability of communities

Beneath shallow water, benthic communities show seasonal and annual fluctuations (see Section 5.4.4), but over the long term they usually remain fairly constant in numbers and composition, indicating that the major factors moulding the community are stable. Because every community includes a range of species each having slightly different tolerances, there is consequently an inherent capacity to adapt to minor changes of environmental conditions by corresponding adjustments of community structure. Changes in any parameter, for example temperature, are likely to influence recruitment and mortality of one species differently from another, with the result that the proportion of one species increases while another declines but does not necessarily become eliminated. If conditions return to their former state, the balance of species responds accordingly and the overall pattern of the community remains (Buchanan *et al.*, 1978).

Rapid, permanent changes of population are usually associated with major alterations of the environment, often the result of human activity; for example, a change of water circulation or temperature connected with industrial installations, or a change of the sea floor due to dredging or the dumping of waste at sea (e.g. Howell and Shelton, 1970). Changes may also follow the introduction of a new species to an area. Otherwise the continual minor fluctuations are largely self-cancelling. Any tendency for one species to increase is eventually counteracted by increasing competition and predation so that a dynamic equilibrium is maintained and the natural balance of the community is preserved.

Over very long periods, climatic or geological changes may slowly alter the environment, and there are also gradual and permanent modifications brought about by the activities of the organisms themselves; for example, the substrate becomes changed by accumulations of shells or skeletons or through the erosion of rocks and stones by the boring of various plants and animals, and the composition of the water is influenced by biological processes of extraction, precipitation and secretion. The continual interactions between habitat and community lead very gradually to changes in both, and the ecosystem, which comprises both environment and population, is thus a composite evolving unit.

6.4   The food supply

The food supply of the benthic macrofauna derives, directly or indirectly, almost entirely from living and dead particulate matter sinking from the overlying water. There is very little primary production of food on the sea-bed because plants can grow only where there is sufficient light for photosynthesis. Vegetation on the sea bottom is therefore limited to shallow water. Large algae produce a lush growth on and near the shore, especially in middle latitudes, and form a primary food source which supports many omnivorous and vegetarian animals, and contributes quantities of organic debris to the local sediments. This vegetation seldom extends deeper than some 40–60 m and is confined to areas of rocky bottom, or stones large enough to provide secure attachment for the plants. Rock and sediment in shallow water may also be covered with a thin microfloral film, mainly diatoms and other unicellular algae. Surface layers of sediments contain large numbers of bacteria, and traces of food are also produced by chemosynthetic species capable of metabolizing inorganic compounds.

Particulate food sinking through the water reaches the bottom in great variety. There is sometimes an appreciable amount of vegetable matter derived from the land, and even in very deep water dredging has disclosed a surprising quantity of terrestrial material in the form of fragments of wood and leaf. Much of this may be carried into the deep ocean basins by turbidity currents (see Section 5.7) flowing down the continental slope. However, in most areas the greater part of the food supply consists of the remains of pelagic organisms. In shallow water, the major component is usually planktonic diatoms and other microscopic plants, and the abundance of this food depends upon the rate of surface production. Seasonal changes in the quantity of phytoplankton produce fluctuations in the supply of food to the benthos (see Section 5.4.3). In temperate latitudes the numbers of planktonic diatoms reaching the bottom may be over a hundred times greater during the summer months than in winter, and there are associated changes in the weight of benthic populations.

The movement of food particles from the surface to deeper levels is not solely a matter of passive sinking. Active transport downwards is effected by vertically migrating organisms which ascend to feed at the surface and then move downwards, where they may be devoured by deeper-living predators or where their faecal pellets may be used as food by other organisms. Faecal pellets are often rich in organic matter, some of which may be material that was not digested during passage through the gut, and some is bacterial protoplasm rapidly multiplying on this organic substrate (Harding, 1974).

There is a tendency for planktonic micro-organisms, fragments of organic debris (e.g. discarded 'houses' of larvaceans) and inorganic particles to become aggregated by both physical and biological processes into larger clumps. These provide micro-habitats which harbour numerous and diverse communities of bacteria, algae and protozoa (Fowler and Knauer, 1986). Such material sinking through the water as particles greater than 0.5 mm in size, is referred to as 'marine snow' (Lampitt, 1996) and is sometimes observed to carpet the sea bottom as a layer of 'marine fluff'. This is probably one of the main ways in which biogenic material reaches the sea-bed. The protoplasmic content of this snow undoubtedly contributes significantly to the food web of deeper levels, especially by enabling animals to obtain food from micro-organisms which, though too small to be directly consumed individually, can be readily ingested in this aggregated state. This is also an important route whereby atmospheric CO_2, after solution in the sea surface and fixation by photosynthesis, is rapidly transferred to deep levels and eventually released, probably mainly as bicarbonate and dissolved CO_2. This has obvious implications for our understanding of the interactions between atmosphere and ocean with respect to the 'greenhouse effect' (see Section 10.2.1).

6.4.1   Feeding strategies

There are broadly four ways in which benthic animals gather food; they filter suspended particles from the water, they collect food particles which settle on the surface of the sediment, they obtain nutriment from the organic material which has become incorporated in the deposit, or they prey upon other animals. Many of course take food from several sources. For example, the shrimp *Crangon* feeds largely on surface debris but also preys on small animals; the edible crab *Cancer*, is omnivorous, taking a wide variety of plant and animal foods; the squat lobster *Galathea*, takes large pieces of animal and vegetable matter and also uses the setae on its maxillipeds to filter micro-organisms and debris from the bottom deposit. However, in many other cases the feeding mechanism is adapted to deal with a particular source of food. Suspension feeders include sponges, ascidians, bryozoans and certain polychaetes (*Chaetopterus*), crustacea (*Balanus*), gastropods (*Crepidula*), many lamellibranchs (*Cardium, Venus, Dosinia, Gari*), and some holothurians (*Cucumaria*). Surface deposit feeders include polychaetes (*Terebella, Amphitrite*), and lamellibranchs (*Angulus, Abra*). Feeding on organic matter within the sediment are polychaetes (*Arenicola*), heart urchins (*Spatangus, Echinocardium*) and holothurians (*Leptosynapta*). Benthic predators include errant polychaetes (*Nephtys, Glycera*), the majority of actinians, many crabs, some gastropods (*Natica, Scaphander, Buccinum*), starfish and ophiuroids (*Asterias, Amphiura*) and many fish, including most of the commercially fished species.

6.4.2   Biomass

Depending on the availability of food, the biomass (weight of living material per unit area) of the sea-bed varies from place to place. Biomass may be expressed as 'rough weight' of fresh material, with or without shells, or more accurately as a 'dry weight' of organic material obtained after the removal of shells and drying of the remaining tissue to constant weight, allowance being made for any inorganic material present in the residue, i.e. ash-free dry weight. Dry weights of between 1 g and 35 g m^{-2} have been found beneath shallow water off the Danish coast. Holme (1953, 1961) found a mean dry weight of about 11 g m^{-2} for the biomass of the macrobenthos at 20 stations in the English Channel off Plymouth, the percentages of different animal groups being as shown in Table 6.2. In areas of exceptional productivity, biomass dry weights as high as 100–200 g m^{-2} occur.

6.4.3   Deep-sea food supply

Whereas in shallow water much of the food supply of the sea bottom comes directly from the surface layers, in very deep water it is unlikely that much surface plankton reaches the bottom intact because most of it is consumed by pelagic

Table 6.2 Percentages of Dry Weight Biomass Contributed by Different Animal Groups in Macrobenthos of Western English Channel.

Group	Percentage
Lamellibranchs	35.46
Polychaetes and Nemertines	25.79
Echinoderms	17.8
Crustacea	10.51
Coelenterates	7.00
Polyzoa	2.46
Protochordates	0.88
Gastropods	0.02

organisms on the way down. Between the productive surface layers and the deeper parts of the ocean there is a food pyramid with many links in the food chain, and the quantity of food available to support the population at deep levels can be only a very small fraction of the surface production. Knowledge of the distribution of deep-sea benthos is scanty although increasing, but population studies in deep water generally indicate that the benthos is sparse and that the biomass decreases with depth. Broadly, the deeper the water and the further from land, the smaller the weight of animals on the sea bottom. There are certain exceptions to this, namely hydrothermal vent communities (see Section 6.4.4).

Occasionally, even at great depths, large pieces of meat, the remnants of bodies of whales or very large fish, must reach the sea bottom. This supports a population of relatively large scavengers (Issacs and Schwartzlose, 1975), including fish, amphipod and decapod crustacea and ophiuroids. Beneath areas of low production these scavengers are often more numerous than might be expected from the size of the surface population. The quantity of readily digestible food reaching the sea bottom must depend partly on the numbers of midwater organisms devouring it on the way down.

Apart from the bodies of large animals, most of the organic material of surface origin that eventually settles on the bottom in very deep water must consist of structures which cannot be digested by pelagic animals during descent; for example, cell walls, shells and skeletons. Pieces of wood and other materials from the land also find their way to the deep sea floor, sometimes carried off the continental shelf by turbidity currents (see Section 5.7). Further decomposition of these materials depends upon the activities of bacteria, which occur in the superficial layers of deposit even in the deepest water and constitute an essential link in the organic cycle. Their food is drawn from many sources. Most of them are heterotrophic, obtaining energy from oxidation of organic compounds, but they can act upon a far wider range of materials than animals are able to digest, including chitin, keratin, cellulose and lignin. Even at the deepest levels these substances are likely to reach the sea floor, and here they are digested and assimilated by bacteria. In this way, organic materials which animals cannot make use of directly are transformed into bacterial protoplasm, and in this form become assimilable by numerous animals which feed on bacteria.

DOM

A further source of food available to benthic animals is dissolved organic matter (DOM) (see Section 4.3.3). The ways in which DOM can be utilized by planktobacteria, and their role in the 'microbial loop' and the food web, are described in Section 5.1.2. However, it is believed that some benthic animals can directly utilize DOM. Some shallow-water invertebrates can utilize dissolved amino acids but it is in deep-sea benthic animals that uptake of DOM is best developed. Pogonophoran worms do not have an internal alimentary system and part of their energy requirements are thought to be met by uptake of DOM. They also utilize symbiotic chemosynthetic bacteria (see Section 6.4.5). There is experimental evidence that many other marine invertebrates, though possessing feeding mechanisms which ingest solid food, are also capable of taking in aminoacids, glucose and fatty acids from dilute solution by direct absorption through the epidermis (Sorokin and Wyshkwarzev, 1973; Southward and Southward, 1970 and 1972; Southward *et al.*, 1979). Southward and Southward (1982) have estimated that DOC (dissolved organic carbon) provides 30 per cent of the energy required by the deep water seastar *Plutonaster* and the polychaete *Tharyx*. It is not known to what extent this facility provides a useful supply of food to these organisms in normal conditions, but it is at least reasonable to suppose that when particulate food is scarce the intake of dissolved organic compounds is of some value.

Colourless flagellates as well as certain groups of pigmented plants, chiefly coccoliths and blue-green algae, sometimes occur far below the illuminated layers – even to 4000 m – in such large numbers as to suggest that they are living and multiplying there, presumably feeding as saprophytes. Many bacteria also draw nutriment from the organic constituents of the water, and the bacterial population includes chemosynthetic autotrophs which do not require organic material but are able to obtain energy by oxidizing inorganic compounds such as ammonia, nitrites and sulphides. These forms can function as primary producers of organic matter below the photosynthetic zone. It is likely that bacterial and protistan populations contribute a significant proportion of the food for animals at great depths. There is also some addition to the biomass of deep water from larval and juvenile stages which develop nearer the surface, drawing on the more abundant food available there before descending to deeper levels.

These sources of food originating far below the surface may go some way to explaining the fact that although populations of deep sediments are small compared with those of most shallow areas, they appear none the less to have a somewhat greater biomass than would be expected from food chains originating only at the surface. An overall estimate of benthic biomass at depths below 3000 m is about 0.2 g dry wt m^{-2}; but in certain areas dry weights of some 1–2 g m^{-2} of animal material occur on the bottom at depths of 5000–8000 m. These values are not so much less than mean values in shallow areas, and indicate that supplies of food must be available in deep water in addition to particulate matter sinking from above.

The food relationships of benthic communities are complex and not well known, but enough has been said to emphasize that bacteria and saprophytes are an extremely important source of food for the animal population at all depths. The microbenthos includes many flagellates, amoebae and ciliates which feed mainly on bacteria; some also ingest minute particles of organic debris, and the larger flagellates and amoebae can probably take whole diatoms. For the meiobenthos and macrobenthos, bacteria comprise an appreciable part of the food of many suspension and surface deposit feeders. Animals which obtain food from the sediment presumably derive a substantial amount of their nutriment from the bacteria and protozoa it contains. Some interconnections of the marine food web are illustrated in Figure 6.2.

6.4.4 Deep-sea hydrothermal vent communities

In some areas of the eastern Pacific and Atlantic, hot springs or hydrothermal vents escape from volcanic fissures along the global system of mid-oceanic ridges (see Section 1.2.3) where sea-floor spreading occurs. Here water emerges from cracks and crevices at temperatures of up to 250°C. Even hotter vents exist where the water flows at up to 380°C from 'black smoker chimneys', formed from precipitated minerals.

In 1977, the manned submersible 'Alvin' explored one of these newly discovered vents, on the Galapagos rift near the Galapagos Islands. Scientists were dumbfounded by the discovery of a rich community of large animals living around the vent (Corliss and Ballard, 1977). The community is characterized by the red-plumed, tube-dwelling vestimentiferan worm, *Riftia pachyptila*, which can reach a metre or more in length. Giant-sized bivalve clams, crabs, shrimps, fish and even anemones have been described from these vents (Gage and Tyler, 1991).

Such communities have subsequently been discovered associated with vents at nearly all areas of tectonic activity so far explored in the deep Pacific and Atlantic. These 'oases' of life support up to 8.5 kg wet weight per square metre, in stark contrast to the almost barren rocky areas nearby. This raises the question of what energy source can support such rich communities in the deep sea where food is in short supply. It is now known that the basis of the food chain for these vent

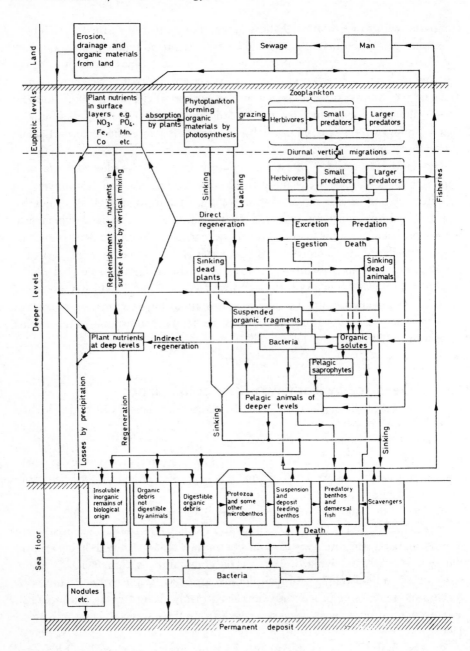

Figure 6.2   Some connections of the marine food web.

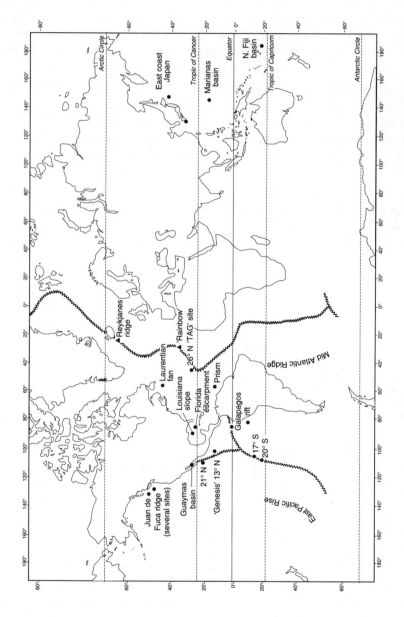

Figure 6.3 Location of major known active hydrothermal vent sites. New vents are continually being discovered particularly along active ridge areas. Based on information from several sources. ● = sites known to have vent biocommunities; ▲ = unexplored sites.

animals is chemoautotrophic bacteria. These bacteria 'feed' on sulphur-containing inorganic compounds, and chemosynthesize organic carbon from elemental carbon dioxide and methane. The most successful vent animals, including *Riftia*, have developed symbiotic relations with these bacteria and so do not need particulate feeding mechanisms. Immense numbers of free-living bacteria provide a food source for other animals that are suspension and deposit-feeders. They in turn provide food for the larger scavengers and predators. These extraordinary deep-sea communities exist in permanent darkness and are entirely independent of light and photosynthetic products originating from the surface layers.

The lifetime of hydrothermal vents and their communities may be very short, a few decades, or very long depending on the rate of sea-floor spreading in the active region. Local communities of tubeworms, destroyed by fresh lava eruptions, can regrow within 2 to 3 years (Lutz and Haymon, 1994). Vents and vent fields are often separated by hundreds of kilometres and the question of how new vents are colonized is still being studied. Biologists at the Plymouth Marine Laboratory (UK) are studying larval DNA of vent animals to help trace their dispersal and relationships with populations elsewhere. American scientists have discovered species of animals normally only associated with vents, on whale carcasses on the deep-sea bed. They speculate that these carcasses provide ephemeral stepping-stones which allow animals to cross more easily between vents. However, recently it has been discovered that in some areas, such vents are very common and can be found at distances of only 20 miles or so apart. In these cases, spread from one vent to another would be relatively easy.

An excellent account of hydrothermal vents and their associated communities is given in Gage and Tyler (1991) and German *et al.* (1996) who also site further useful references.

6.4.5 Food-fish production

Suspension, surface deposit and sediment feeding animals of the sea bottom are preyed upon by a variety of carnivores, including many of the food fish highly prized by man. Studies of biomass and composition of benthic communities therefore give some indication of the extent to which areas can support stocks of fish. The production of animal material on the sea floor can be regarded as 'useful' or 'wasteful' according to whether it contributes to the formation of commercially valuable species of fish or other creatures of no value to man. For example, starfish and brittle-stars are not used as human food and are relatively unimportant as fish food; but they are carnivores and compete with fish for the same sort of prey, especially molluscs. From a human viewpoint, the growth of these predatory echinoderms can be regarded as a wasteful type of production.

The invertebrate predators of the sea bottom are far more numerous, comprise a far greater weight of living material, and have a much higher rate of food consumption than bottom-feeding fish. Thorson (1960) reviewed what is known of feeding habits and food requirements of predatory benthos in the north-east Atlantic area, and concluded that:

> it seems reasonable to assume that most bottom-dwelling fishes in temperate waters will consume an average portion of food per day corresponding to from 5 to 6% of their own living weight during the summer half-year, and will reduce this rate significantly when the temperature decreases. During the coldest months of the year they may almost completely cease to feed ... Summarizing the data for invertebrate predators, we find that at very young stages they are extremely voracious, taking an amount of food corresponding to about 25% of their own living weight per day. Those species which remain active as adults and continue to increase in size do require somewhat less food, although their average consumption is about 15% of their living weight per day. Finally, those predators which nearly or totally stop growing when mature and are sluggish, will in their adult phase, reduce their food claims to such an extent as to be cheap or fairly cheap to run.
>
> On an average, growing invertebrate predators seem to consume four times as much food per day and unit weight as bottom-dwelling fishes, which seems reasonable when we realize that the life cycle, or at least the time from birth to maturity, is much shorter in the invertebrate predators than in most of the fishes. A species of invertebrate predator will often produce three or more generations (i.e. build up three or more units of meat) while a flounder is producing only one generation (i.e. build up one unit).
>
> On the basis of our information that invertebrate predators consume food about four times as fast as the flounders, we must recognize the amazing fact that only 1–2% of the 'fish food' on the sea bottom is actually eaten by fish; the rest is taken by invertebrates. If the standing crop of 'fish food' is increased for some reason, the invertebrate predators with their shorter life cycles and quicker growth will furthermore be ready to take the advantage of this circumstance long before the fishes are able to do so.

Hardy (1959) in his classic work *The Open Sea* discusses Thorson's ideas in the context of 'weeding' the sea-bed of some of these predators, thereby leaving more food for 'useful' fish species (see Section 9.6.2).

References and further reading

Student texts

Gage, J.D. and Tyler, P.A. (1991). *Deep-sea Biology. A natural history of organisms at the deep-sea floor*. Cambridge University Press. Chapter 15: Deep-sea hydrothermal vents and cold seeps.

Warner, G.F. (1984). *Diving and marine ecology. The ecology of the sublittoral*. Cambridge Studies in Modern Biology 3. Cambridge University Press.

Wood, E.M. (1987). *Subtidal ecology*. New Studies in Biology. Edward Arnold.

References

Bayne, B.L. (1964). Settlement of *Mytilus. J. Anim. Ecol.*, **33**, 513–23.

Bayne, B.L. (1969). The Gregarious behaviour of the larvae of *Ostrea edulis* at settlement. *J. Mar. Biol. Ass. UK*, **49**, 327.

Buchanan, J.B. *et al.* (1978). Variability in benthic macrofauna. *J. Mar. Biol. Ass. UK*, **58**, 191–209.

Corliss, J.B. and Ballard, R.D. (1977). Oases of life in the cold abyss. *National Geographic Magazine*, **152**, 441–53.

Crisp, D.J. (1974). Factors influencing the settlement of marine invertebrate larvae. In *Chemoreception in Marine Organisms*. P.T. Grant and A.M. Mackie, eds. London, Academic.

Earll, R.C. (1992). *The SEASEARCH habitat guide*. An identification guide to the main habitats found in the shallow seas around the British Isles.

Fowler, S.W. and Knauer, G.A. (1986). Role of large particles in the transport of elements and organic compounds through the oceanic water column. *Progress in Oceanography*, **16**, 147–94.

Gage, J.D. and Tyler, P.A. (1991). *Deep-sea Biology. A natural history of organisms at the deep-sea floor*. Cambridge University Press.

German, C.R., Parson, L.M. and Mills, R.A. (1996). Mid-ocean ridges and hydrothermal activity. In: *Oceanography. An Illustrated Guide*. C.P. Summerhayes and S.A. Thorpe, eds. Southampton Oceanographic Centre, Manson Publishing.

Glemarec, M. (1973). European North Atlantic benthos. *Oceanogr. Mar. Biol. Ann. Rev.*, **11**, 263–89.

Harding, G.C.H. (1974). Food of deep sea copepods. *J. Mar. Biol. Ass. UK*, **54**, 141–55.

Hardy, A. (1959). *The Open Sea II. Fish and Fisheries*. No. 37, New Naturalists series. Fontana.

Hiscock, K. (ed.) (1996). *Marine Nature Conservation Review: rationale and methods*. Peterborough, Joint Nature Conservation Committee. (Coasts and Seas of the United Kingdom. MNCR series.)

Holme, N.A. (1953). The Biomass of the bottom fauna of the English Channel. *J. Mar. Biol. Ass. UK*, **32**, 1–49.

Holme, N.A. (1961). The bottom fauna of the English Channel. *J. Mar. Biol. Ass. UK*, **41**, 397–461.

Holme, N.A. (1966). The bottom fauna of the English Channel. Part II. *J. Mar. Biol. Ass. UK*, **46**, 401–93.

Howell, B.R. and Shelton, R.G.J. (1970). The effect of China clay on the bottom fauna of St. Austell and Mevagissey bays. *J. Mar. Biol. Ass. UK*, **50**, 593–607.

Isaacs, J.D. and Schwartzlose, R.A. (1975). Active animals of the deep sea floor. *Scient. Am.*, **233**, 84–91.

JNCC (1996). *Marine Nature Conservation Review: Marine Biotopes. A working classification for the British Isles*. Version 96.7. D.W. Connor, ed. Joint Nature Conservation Committee, Peterborough.

Jones, N.S. (1950). Marine bottom communities. *Biol. Rev.*, **25**, 283.

Lampitt, R.S. (1996). Snow falls in the open ocean. In: *Oceanography. An Illustrated Guide*. C.P. Summerhayes and S.A. Thorpe, eds. Southampton Oceanographic Centre, Manson Publishing.

Lutz, R.A. and Haymon, R.M. (1994). Rebirth of a deep-sea vent. *National Geographic*, **186** (5), 114–26.

McIntyre, A.D. (1964). Meiobenthos of sublittoral muds. *J. Mar. Bol. Ass. UK*, **44**, 665.

McIntyre, A.D. and Eleftheriou, A. (1968). The bottom fauna of a flatfish nursery ground. *J. Mar. Biol. Ass. UK*, **48**, 113–42.

McIntyre, A.D. and Murison, D.J. (1973). Meiofauna of a flatfish nursery ground. *J. Mar. Biol. Ass. UK*, **53**, 93–118.

Meadows, P.S. and Campbell, J.I. (1972). Habitat selection by aquatic invertebrates. *Adv. Mar. Biol.*, **10**, 271–382.

Mero, J.L. (1960). Minerals on the ocean floor. *Scient. Am.*, **203**, October.

Sorokin, Y.I. and Wyshkwarzev, D.I. (1973). Feeding on dissolved organic matter by some marine animals. *Aquaculture*, **2**, 141–8.

Southward, A.J. and Southward, E.C. (1970 and 1972). Observations on the role of dissolved organic compounds in the nutrition of benthic invertebrates. *Sarsia*, **45**, 69–95; **48**, 61–81; **50**, 29–46.

Southward, A.J. and Southward, E.C. (1982). The role of dissolved organic matter in the nutrition of deep-sea benthos. *American Zoologist*, **22**, 647–59.

Southward, A.J. *et al.* (1979). Feeding of *Siboglinum*. *J. Mar. Biol. Ass. UK*, **59**, 133–48.

Swedmark, B. (1964). The interstitial fauna of marine sand. *Biol. Rev.*, **39**, 1.

Thorson, G. (1950). Reproductive and larval ecology of marine bottom invertebrates. *Biol. Rev.*, **25**, 1.

Thorson, G. (1957). Bottom communities. In *Treatise on Marine Ecology and Paleoecology*. J.W. Hedgpeth, ed. Vol. 1, 461–534. New York, Geol. Soc. America.

Thorson, G. (1960). Parallel level-bottom communities, their temperature adaptation, and their 'balance' between predators and food animals. p. 67, *Perspectives in Marine Biology*. A. Buzzati-Traverso, ed. University of California Press.

Tooms, J.S. and Summerhayes, C.P. (1969). Geochemistry of marine phosphate and manganese deposits. *Oceanogr. Mar. Biol. Ann. Rev.*, **7**, 49.

Trueman, E.R. and Ansell, A.D. (1969). The mechanisms of burrowing into soft substrate by marine animals. *Oceanogr. Mar. Biol. Ann. Rev.*, **7**, 315.

Wilson, D.P. (1956). Some problems in larval ecology related to the localized distribution of bottom animal, p. 87, *Perspectives in Marine Biology*. A. Buzzati, ed.

Wilson D.P. (1968). Settlement behaviour of *Sabellaria* larvae. *J. Mar. Biol. Ass. UK*, **48**, 387–435.

Wilson, D.P. (1970). Additional observations on *Sabellaria*. *J. Mar. Biol. Ass. UK*, **50**, 1–31.

7 Energetics of a marine ecosystem

Living organisms, unlike machines, cease to exist once they stop working. All biological activity depends upon continual transfers and transformations of energy, without which any natural living system almost immediately disintegrates irreversibly. We will now draw together some of the information from preceding pages in an elementary consideration of certain energy relationships of marine life, with particular reference to shelf seas around the British Isles.

Directly or indirectly the source of all energy for life is the sun, which continually emits radiant energy into space. A tiny fraction of this radiation reaches the earth, where a considerable part is lost by reflection from the earth's atmosphere, clouds and surface. Probably a global average of about 40 per cent of the incoming radiation is reflected. The remainder is absorbed by the atmosphere and the land and ocean surfaces, where its main effect is to cause the heating which generates the movements of atmosphere and ocean. However, despite the continuous absorption of solar radiation, the climate does not appear in the long term to become hotter. This indicates that there is overall an output of radiant energy from the earth equal to that received, and the total heat content of atmosphere, surface and oceans remains virtually constant except for minor fluctuations due to the elliptical form of the earth's orbit round the sun and to changes in solar activity (for example solar flares or sunspots). The incoming energy is received largely at wavelengths within the visible spectrum. The balancing emission from the earth is low-frequency heat radiation which passes out in all directions of space.

The routes through which light energy can flow between penetrating the earth's atmosphere and re-radiation into space as heat are numerous and complex. A small amount, probably only about 1–2 per cent of the light energy reaching the earth's surface, enters pathways beginning with the absorption of sunlight by plants in photosynthesis. In this process radiant energy is transformed to chemical energy by an energy-fixing reduction of carbon dioxide. For instance, the synthesis of 1 mole of glucose from carbon dioxide and water involves the intake of 673 kcal (2826.6 kJ) of light energy.

$$6 \; CO_2 + 6 \; H_2O + 673 \; \text{kcal} \; (2826.6 \; \text{kJ}) \xrightarrow{\text{chlorophyll}} C_6H_{12}O_6 + 6 \; O_2$$
$$\text{(Light energy)}$$

This energy is then available in biological processes, for when 1 mole of glucose is oxidized in respiration, 673 kcal (2826.6 kJ) of energy is released. It is by means of transfers and transformations of the energy of chemical compounds formed initially by photosynthesis that power is provided for the activity of living organisms. The movements of materials involved in nutrition occur almost entirely as means of effecting energy transfers. The global total of energy fixation by photosynthesis determines the total amount of biological activity which the earth can support. The intake of radiant energy into the living system by photosynthesis is balanced by a corresponding outflow of energy as heat through pathways of respiration and movement.

We have insufficient knowledge of the energy relationships of marine organisms to be able to trace with much certainty the passage of energy through marine ecosystems, but until we can do this our understanding of the food webs of the sea must remain at an elementary stage. As an indication of some of the processes involved we will attempt in this chapter a simple analysis of the energetics of production and feeding in the shelf waters of the north-east Atlantic, exemplified by the English Channel. **No reliance must be placed on the figures given,** which should only be regarded as reasoned guesswork based on a modicum of firm information. The exercise is illustrative rather than factual. Obviously, for simplicity, innumerable interactions within the food web have been ignored, and by averaging out all values over a period of a year no account has been taken of the highly fluctuating nature of the system. Nevertheless, something may be learned from critical examination of the figures and from comparison with various sources of data relating to these considerations. The student is recommended also to read the examination of food webs in the North Sea given by Steele (1974) in his book *The Structure of Marine Ecosystems*. Excellent explanations of the concepts contained in this chapter can also be found in Odum (1971, 1983).

The analysis is summarized diagrammatically in Figure 7.1. Energy has been quantified in calories (kcal) because this system has hitherto been most widely used in marine biological literature. The relationship between the various units of energy measurement used in the literature is as follows:

1 Calorie = 1000 calories (or 1 kcal)
1 Calorie = 4186 joules (1 calorie = 4.2 joules)

A calorie is the amount of energy needed to raise the temperature of 1 g of water by 1°C. Note that the thousand-unit Calorie is spelt with a capital C. Here the notation kcal is used to avoid possible confusion.

7.1   Primary production

Measurements of primary production in the western English Channel were made around two decades ago using the carbon-14 method (see Section 5.2.4) and indicated an average value of about $120\,\mathrm{gC\cdot m^{-2}\cdot yr^{-1}}$ for carbon fixation in particulate organic matter. We will take this to be a measure of net primary production (NPP), i.e. the formation of new plant tissue, and will ignore losses of dissolved organic matter. As conversion factors we will take 1 gC as equivalent to 10 kcal or to 2.3 g dry wt of organic matter. NPP can therefore be expressed as $1200\,\mathrm{kcal\,m^{-2}\cdot yr^{-1}}$ or $2.3 \times 120 = 276\,\mathrm{g\ dry\ wt\cdot m^{-2}\cdot yr^{-1}}$.

Estimates of energy loss in plant respiration fall between 10 per cent and 50 per cent of gross primary production (GPP). Taking the respiratory loss for phytoplankton as 20 per cent of GPP, then NPP is 80 per cent of GPP, and $\mathrm{GPP} = 1500\,\mathrm{kcal\cdot m^{-2}\cdot yr^{-1}}$. The energy loss from the system by plant respiration is $300\,\mathrm{kcal\,m^{-2}\cdot yr^{-1}}$.

A mean figure for solar energy entering the earth's atmosphere has been given as $15.3 \times 10^5\,\mathrm{kcal\cdot m^{-2}\cdot yr^{-1}}$.

Much of this is reflected or absorbed on passage through the atmosphere, and the amount reaching the surface varies greatly with locality. We will take a value of $3 \times 10^5\,\mathrm{kcal\cdot m^{-2}\cdot yr^{-1}}$ as the energy at the surface of the English Channel. Here there are losses by reflection, and on penetrating the surface the light is rapidly absorbed by the water and causes heating. Estimates of the fraction of the incident radiation which is fixed by photosynthesis in aquatic environments usually fall within the range 0.1–0.5 per cent of the energy at the surface. If the GPP for the western English Channel is $1500\,\mathrm{kcal\cdot m^{-2}\cdot yr^{-1}}$ and the energy of sunlight reaching the sea surface is $3 \times 10^5\,\mathrm{kcal\cdot m^{-2}\cdot yr^{-1}}$, then the proportion of this radiation fixed in particulate organic matter is $1500/3 \times 10^5 = 0.005$ or 0.5 per cent. Evidently the English Channel is a relatively highly productive part of the sea.

Table 7.1 presents some estimates by Harvey of the mean annual biomass of several trophic levels in the English Channel, and we will use these figures in our analysis. The value for mean annual biomass of standing stock of phytoplankton for the English Channel is $4.0\,\mathrm{g\ dry\ wt\cdot m^{-2}}$. Comparing this figure with our value for annual production, it can be seen that there is a high rate of turnover. The weight of new plant tissue produced in the year is nearly 70 times (276/4) the mean weight of standing stock.

The production of new phytoplankton does not increase the standing stock

Table 7.1 Estimates of the mean annual biomass of several trophic levels in the English channel. (From Harvey (1950) published by Cambridge University Press)

Trophic level	Dry wt of organic matter $(\mathrm{g\cdot m^{-2}})$
Phytoplankton	Ca. 4.0
Zooplankton	1.5
Pelagic fish	1.8
Demersal fish	1–1.25

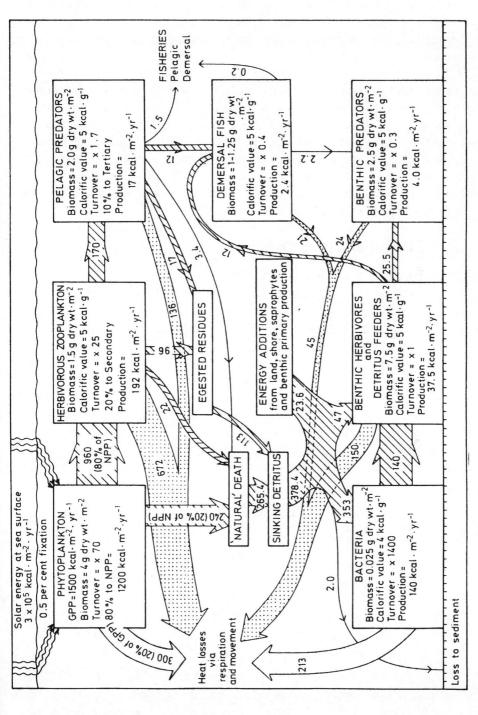

Figure 7.1 Diagrammatic representation of hypothetical energy relationships in a marine ecosystem in coastal waters of the British Isles. Figures on arrows have units of $kcal \cdot m^{-2} \cdot yr^{-1}$.

from one year to the next because it is balanced by a corresponding loss of plants from the water. Death of plant cells occurs in two main ways, by sinking and by consumption by herbivorous zooplankton. Sinking cells die through lack of light and constitute a large part of the organic detritus reaching the sea bottom. This provides a major energy source to support the benthos. However, in contrast to most terrestrial ecosystems, it appears certain that in the sea a greater proportion of the vegetation is consumed by animals than is lost by death and decomposition, the main energy transfer from NPP going to grazing zooplanktonts. We will assume that 20 per cent of NPP contributes directly to sinking detritus, with 80 per cent consumed by pelagic herbivores. The energy content of detritus from this source is therefore $0.2 \times 1200 = 240 \, \text{kcal} \cdot \text{m}^{-2} \cdot \text{yr}^{-1}$. This will be given further consideration below (see Section 7.3). The pelagic grazing population receives an energy inflow of $0.8 \times 1200 = 960 \, \text{kcal} \cdot \text{m}^{-2} \cdot \text{yr}^{-1}$.

7.2   The grazing chain

7.2.1   Secondary pelagic production

Some of the food ingested by the planktonic herbivores is not fully digested and absorbed but passes through the gut and is egested, contributing to the fall of organic debris to the sea floor. We will assume that 10 per cent of the energy content of the food consumed goes in this way to detritus. Of the assimilated food, most is used for respiration and the remainder forms new animal tissue. Herbivorous planktonts are relatively efficient converters of food to new tissue, with some early larval stages apparently using about 50 per cent of food intake for growth. The general level of efficiency is certainly lower than this (Paffenhofer, 1976; Reeve, 1969) and we will assume that 70 per cent of food intake is used for respiration and 20 per cent for secondary production. We will call this a gross conversion efficiency (GCE) of 0.2, where

$$\text{GCE} = \frac{\text{calorific value of new tissue formed}}{\text{calorific value of ingested food}}$$

We can summarize the energetics as follows:

Energy inflow by grazing	$= 960 \, \text{kcal} \cdot \text{m}^{-2} \cdot \text{yr}^{-1}$
10% egested unassimilated	$= 96 \, \text{kcal} \cdot \text{m}^{-2} \cdot \text{yr}^{-1}$ to detritus
70% utilized for respiration and movement	$= 672 \, \text{kcal} \cdot \text{m}^{-2} \cdot \text{yr}^{-1}$ lost from the system
20% utilized for secondary production	$= 192 \, \text{kcal} \cdot \text{m}^{-2} \cdot \text{yr}^{-1}$ available for predators

Calorific values for copepod and euphausid species occurring around the British Isles are generally about $5.0 \, \text{kcal} \, \text{g}^{-1}$ dry wt. Using this figure, secondary production amounts to 192/5 = 38.4 g dry wt·yr⁻¹, or (if 1 g dry wt = 0.44 gC) 38.4 × 0.44 = 16.9 gC·m⁻²·yr⁻¹.

General observation of plankton samples taken around the British Isles suggests that a mean value for biomass of standing stock of herbivores must often be appreciably greater than the standing stock of phytoplankton, and some quantitative investigations indicate this. However, taking our value from Harvey's figures (Table 7.1), the mean annual biomass of zooplankton throughout the year in the English Channel is given as 1.5 g dry wt·m⁻². At this apparently low value, secondary production amounts to about 25 times the weight of standing crop. This much lower rate of turnover compared with phytoplankton corresponds with the slower rates of growth and reproduction of zooplankton.

Some herbivorous planktonts die and reach the bottom uneaten by pelagic predators, but probably the great majority are consumed by carnivorous zooplanktonts or pelagic fish such as herring and mackerel. Assuming that dead pelagic herbivores sinking to the bottom amount to a little over 10 per cent of secondary pelagic production, this adds approximately 22 kcal·m⁻²·yr⁻¹ to sinking detritus, leaving 170 kcal·m⁻²·yr⁻¹ for consumption by pelagic carnivores.

7.2.2 Tertiary pelagic production

Compared with herbivores, predators generally use a larger proportion of their food for respiration. A figure of 10 per cent (GCE = 0.1) for conversion of food to new tissue by carnivores is generally quoted. This permits us to divide the energy transfers at this trophic level as follows:

Energy inflow by predation	= 170 kcal·m⁻²·yr⁻¹
10% egested unassimilated	= 17 kcal·m⁻²·yr⁻¹ to detritus
80% utilized in respiration	= 136 kcal·m⁻²·yr⁻¹ lost from the system
10% to tertiary production	= 17 kcal·m⁻²·yr⁻¹

Taking 5 kcal g⁻¹ dry wt as a mean calorific value for first-rank carnivores, tertiary production may be expressed as 17/5 = 3.4 g dry wt·m⁻²·yr⁻¹ or (if 1 g dry wt = 0.44 gC) approximately 1.5 gC·m⁻²·yr⁻¹.

The mean annual biomass of pelagic fish in the English Channel has been estimated at 1.8 g dry wt·m⁻² (Table 7.1). If we assume that this is 90 per cent of the weight of first-rank carnivores, the total biomass of pelagic predators is 2.0 g dry wt·m⁻². We note that annual production at this level is only 3.4/2 = 1.7 times the mean weight of standing stock. Compared to plankton, fish are long lived but slow growing.

7.2.3 Pelagic fisheries

Fishery statistics for the year 1976 give the total landings of pelagic species from the English Channel as approximately 146 000 tonnes wet weight, i.e. 146 × 10⁹ g wet wt. Taking the area of the English Channel as about $82 \times 10^9 \, \mathrm{m}^2$, this catch of pelagic fish amounts to $146/82 = 1.8 \, \mathrm{g}$ wet $\mathrm{wt \cdot m^{-2} \cdot yr^{-1}}$. Assuming the wet weight of fish to be six times the dry weight, pelagic fishing therefore took $0.3 \, \mathrm{g}$ dry $\mathrm{wt \cdot m^{-2} \cdot yr^{-1}}$. If the calorific value of the fish is $5.0 \, \mathrm{kcal \cdot g^{-1}}$ dry wt, the energy content of the catch was

$$5 \times 0.3 = 1.5 \, \mathrm{kcal \cdot m^{-2} \cdot yr^{-1}}.$$

or,

$$82 \times 1.5 \times 10^9 = 123 \times 10^9 \, \mathrm{kcal \cdot yr^{-1}}$$

for the whole of the English Channel.

According to our previous calculation, tertiary production amounts to about $3.4 \, \mathrm{g}$ dry wt, so in 1976 pelagic fishing cropped about $0.3 \times 100/3.4 = 8.8$ per cent of production at this trophic level.

7.3   The detritus chain

From the preceding sections we have derived several figures for energy content of various contributions to the organic detritus reaching the sea bottom. To these must be added a small amount to represent the bodies of dead pelagic predators sinking to the sea floor, which we will take to be equivalent to 20 per cent of tertiary pelagic production, giving a further $0.2 \times 17 = 3.4 \, \mathrm{kcal \cdot m^{-2} \cdot yr^{-1}}$. These energy sources in detrital form can be summarized as follows:

		$\mathrm{kcal \cdot m^{-2} \cdot yr^{-1}}$
From primary production	sinking, uneaten phytoplankton	240
	phytoplankton eaten but egested	96
From secondary production	dead, uneaten zooplankton	22
	zooplankton eaten but egested	17
From tertiary production	dead pelagic predators	3.4
		Total = 378.4

This gives us a figure of $378.4 \, \mathrm{kcal \cdot m^{-2} \cdot yr^{-1}}$ for food energy reaching the shallow sea bottom from several trophic levels in the water above.

Examination of the composition of shallow-water sediments often reveals a significant additional amount of organic matter recognizably derived from the seashore or the land, mainly fragments of large algae, wood or leaf. Also, where sufficient light reaches the bottom there will be primary production by plants, the compensation depth for benthic diatoms and other algae being generally lower than for phytoplankton. There will also be a slight contribution from chemosynthetic autotrophs, and also from saprophytes utilizing dissolved organic matter which may have been produced by the phytoplankton but does not register in the ^{14}C estimate of primary production. We must therefore make some addition to the energy available for biological processes on the sea bottom to allow for these sources. On the other hand we must make some deduction for organic matter lost from the living system by permanent inclusion within the sediment or by oxidations not effected by living organisms. There is as yet insufficient evidence to permit any strict quantification of these energy transfers. We will assume that $2 \, \text{kcal} \cdot \text{m}^{-2} \cdot \text{yr}^{-1}$ becomes permanently incorporated in the sediment, and that a round figure of $400 \, \text{kcal} \cdot \text{m}^{-2} \cdot \text{yr}^{-1}$ is the food energy utilized by benthic organisms.

Some of this energy is directly available to the benthic fauna by digestion and assimilation of detritus, but many detritic materials reaching the bottom cannot be digested by animals. These materials are acted upon by bacteria, which utilize these energy sources to multiply rapidly. This production of bacterial protoplasm contributes a significant proportion of the food of benthic animals. We have insufficient knowledge of the feeding metabolism of the benthos to make a reasoned generalization about the proportions of energy derived from detritus directly by digestion and assimilation of organic debris or indirectly via consumption of bacterial protoplasm, but some experimental work suggests that bacteria contribute much the greater part. We shall assume that 25 per cent of the energy intake of the benthic herbivores comes directly from detritus and 75 per cent from the consumption of bacteria.

7.3.1 Benthic fauna

For the mean annual biomass of the benthic fauna in shallow water around the British Isles we will take a value of $10 \, \text{g}$ dry wt$\cdot \text{m}^{-2}$ (see Section 6.4.2). Part of this is benthic predators, some estimates indicating about 25 per cent, leaving a standing stock of $7.5 \, \text{g}$ dry wt of benthic herbivores feeding on detritus and bacteria. With a calorific value of $5 \, \text{kcal} \cdot \text{g}^{-1}$ dry wt we have an energy content of mean standing stock of benthic herbivores of $37.5 \, \text{kcal} \cdot \text{m}^{-2}$.

The conversion efficiency of these animals is likely to be as high as in herbivorous zooplanktonts, so we will take a value of 0.2 for GCE. Compared with zooplanktonts many of the benthic animals are much bigger and live longer – years rather than weeks – so the rate of turnover is low relative to the weight of standing stock. On the assumption that the stock produces its own weight of new tissue in a year, an assumption supported by experimental evidence (Buchanan and Warwick, 1974; Hibbert, 1976; Hughes, 1970; Kay and Brafield 1973; Warwick and Price, 1975; Warwick *et al.*, 1978), the energy intake necessary to produce this is $37.5/0.2 = 187.5 \, \text{kcal} \cdot \text{m}^{-2} \cdot \text{yr}^{-1}$. Because we have also assumed that 25 per cent of this comes from direct assimilation of digested detritus and the remainder from ingestion of bacteria, we have detrital and bacterial contributions of $47 \, \text{kcal} \cdot \text{m}^{-2} \cdot \text{yr}^{-1}$ and $140 \, \text{kcal} \cdot \text{m}^{-2} \cdot \text{yr}^{-1}$ respectively.

7.3.2   Bacteria

From our assessment of 400 kcal·m^{-2}·yr^{-1} of food energy available on the sea bottom, we must now deduct 47 kcal·m^{-2}·yr^{-1} directly utilized by benthic fauna, leaving 353 kcal·m^{-2}·yr^{-1} for bacteria. We have estimated the consumption of bacteria by benthic animals to be 140 kcal·m^{-2}·yr^{-1} and for the stock of bacteria to remain constant this must be the annual production. This implies a high conversion efficiency of 140/353 = 0.4 approximately, and a respiratory loss of 213 kcal·m^{-2}·yr^{-1}.

The biomass of benthic bacteria is obviously very small compared with larger organisms. It varies greatly with the grade of deposit, being much greater in muds than in coarse substrates. Because the English Channel bottom is mainly of a gravelly texture, we will take a low figure of 0.025 g dry wt·m^{-2} for benthic bacteria. Assuming a calorific value of 4 kcal·g^{-1} dry wt for bacteria, this gives an energy content for the stock of $4 \times 0.025 = 0.1$ kcal·m^{-2}. For an annual production of 140 kcal·m^{-2}·y^{-1} this stock must reproduce itself over 1400 times, which is certainly not a very high rate of multiplication for bacteria (Zhukova, 1963).

7.3.3   Demersal fish

Harvey estimated the mean annual biomass of demersal fish in the English Channel at 1.0–1.25 g dry wt·m^{-2} (Table 7.1). With a calorific value of 5 kcal·g^{-1} dry wt the energy content of this biomass is 6 kcal·m^{-2} approximately. Harvey put the annual yield at 30–50 per cent of stock (say 40 per cent) so annual production can be calculated as $6 \times 0.4 = 2.4$ kcal·m^{-2}·yr^{-1}. If the GCE is 0.1, this level of production requires an energy intake of 24 kcal·m^{-2}·yr^{-1}.

Demersal fish feed partly on benthic and partly on pelagic prey. We have calculated above that from the annual production of 17 kcal·m^{-2}·yr^{-1} of pelagic predators 1.5 kcal. m^{-2}. yr^{-1} is removed by pelagic fisheries, and we have assumed that 3.4 kcal·m^{-2}·yr^{-1} is lost by natural death, i.e. not eaten by other predators. This leaves approximately 12 kcal·m^{-2}·yr^{-1} available as food for demersal fish. If this is all consumed by demersal fish, there is a balance of $24 - 12 = 12$ kcal·m^{-2}·yr^{-1} to be made up by devouring benthic fauna. As a fraction of the annual production of benthic herbivores this is $12 \times 100/37.5$, or about 30 per cent. Comparing this with Thorson's estimate quoted earlier (see page 241) that demersal fish take only 1–2 per cent of available fish food, with the rest consumed by invertebrate predators, it is clear that some assumptions are considerably astray. More information is necessary.

7.3.4   Demersal fisheries

The total annual landings of demersal fish from the English Channel during 1976 were approximately 20 000 tonnes. This is equivalent to approximately 0.25 g wet wt·m^{-2}·yr^{-1}, or 0.04 g dry wt·m^{-2}·yr^{-1}. If the calorific value of the fish is taken as 5 kcal·g^{-1} dry wt, the energy content of the catch was 0.2 kcal·m^{-2}·yr^{-1} or $82 \times 10^9 \times 0.2 = 16.4 \times 10^9$ kcal·yr^{-1} for the whole area. As a proportion of the annual production of demersal fish, the fishing yield amounted to $0.2 \times 100/2.4 = 8$ per cent.

7.3.5   Benthic predators

From our figure of 10 g dry wt·m^{-2} for benthic fauna we attributed 7.5 g dry wt·m^{-2} to the herbivores, leaving a biomass of 2.5 g dry wt·m^{-2} for the predators. With a calorific value of 5 kcal·g^{-1} dry wt the energy content of this biomass is 12.5 kcal·m^{-2}.

These animals feed on the benthic herbivores and may also eat dead bodies of fish. From our assumed annual production of 37.5 kcal·m^{-2}·yr^{-1} of benthic herbivores, 12 kcal·m^{-2}·yr^{-1} has been eaten by demersal fish, leaving 25.5 kcal·m^{-2}·yr^{-1} available for benthic predators. There is also the balance of production of demersal fish after the removal of 0.2 kcal·m^{-2}·yr^{-1} by fisheries, i.e. $2.4 - 0.2 = 2.2$ kcal·m^{-2}·yr^{-1}, making a total of about 27.7 kcal·m^{-2}·yr^{-1} for benthic predators. If they consume all this food energy and have a GCE of 0.15, the annual production would amount to about 4 kcal·m^{-2}·yr^{-1}, the respiratory loss about 24 kcal·m^{-2}·yr^{-1} and the annual yield a little under 32 per cent of stock weight.

7.4   The energy balance sheet

From the foregoing figures we can now draw up a statement of the energy input and output of the living system as outlined above (see Table 7.2).

7.5   Conclusion

In the preceding computations we may note that from a total energy fixation of approximately 1500 kcal·m^{-2}·yr^{-1}, the energy content of the annual crop taken by fisheries in the English Channel has been only about 1.7 kcal·m^{-2}·yr^{-1}, i.e. only a little over 0.1 per cent. This figure is a reasonably conservative estimate for sustainable yields of shelf fisheries. From the North Sea and Icelandic shelf a slightly higher proportion of GPP is harvested. But for fisheries over the deep ocean it seems unlikely that even 0.1 per cent of GPP could be gathered as fish because of greater difficulties of capture, even allowing for greater production at lower latitudes.

Despite many uncertainties there have been several attempts to compute a global quantity for total GPP over the whole ocean surface. A figure of the order of 1017 kcal·yr^{-1} seems reasonable. Taking the optimistic view that world fisheries extended over the deep ocean could take 0.1 per cent of this, we

Table 7.2 The energy balance sheet

Energy input	Value $(\text{kcal} \cdot \text{m}^{-2} \cdot \text{yr}^{-1})$	Energy output	Value $(\text{kcal} \cdot \text{m}^{-2} \cdot \text{yr}^{-1})$
GPP by phytoplankton	1500	Respiratory losses at each trophic level, i.e. heat loss:	
Additions of organic materials from shore and land, plus some chemosynthesis, utilization of DOM by saprophytes and NPP by benthic plants.	20 approx.	Phytoplankton	300
		Pelagic herbivores	672
		Pelagic predators	136
		Bacteria	213
		Benthic herbivores	150
		Demersal fish	21
		Benthic predators	24
		Loss to permanent sediment and inorganic oxidations	2
		Balance to fisheries and benthic predators approx.	2
Total approx.	1520	Total approx.	1520

arrive at a maximum sustainable annual yield for world fisheries of about $1 \times 1014 \, \text{kcal} \cdot \text{yr}^{-1}$. Taking a calorific value of $5 \, \text{kcal} \, \text{g}^{-1}$ dry wt for fish, and a wet weight of six times dry weight, this converts to about 12×1013 g wet wt or 12×107 tonnes per annum of fresh fish. For an economic return on the effort of catching fish, it is more realistic to suppose that substantially less than this could be captured. Considering that world sea fisheries are already taking over 6×10^7 tonnes of sea fish per annum, it is clear that our calculations certainly do not support ideas that the future food needs of the rapidly rising world population can be greatly alleviated by a large extension of ocean fisheries. During recent years, the world catch of fish has not increased appreciably despite greater efforts at capture (see Figure 9.25).

References and further reading

Student texts
Phillipson, J. (1966). *Ecological Energetics*. London, Arnold.

References
Beukemer, J.J. and de Bruin, W. (1979). Calorific values of *Macoma*. *J. Exp. Mar. Biol. Ecol.*, **37**, 19–30.
Buchanan, J.B. and Warwick, R.M. (1974). Bio-economy of benthic macrofauna. *J. Mar. Biol. Ass. UK*, **54**, 197–222.
Corner, E.D.S. and Cowey, C.B. (1968). Biochemical studies on the production of marine zooplankton. *Biol. Rev.*, **43**, 393.

Harvey, H.W. (1950). On the production of living matter in the sea off Plymouth. *J. Mar. Biol. Ass. UK*, **29**, 97.

Hibbert, C.J. (1976). Bivalve biomass and production. *J. Exp. Mar. Biol. Ecol.*, **25**, 249–61.

Hughes, R.N. (1970). Energy budget of *Scrobicularia plana*. *J. Anim. Ecol.*, **39**, 357–81.

Kay, D.G. and Brafield, A.E. (1973). Energy relations of *Neanthes virens*. *J. Anim. Ecol.*, **42**, 673–92.

Odum, E.P. (1971). *Fundamentals of Ecology*. 3rd ed. Philadelphia, W.B. Saunders.

Odum, H.T. (1983). *Basic Ecology*. Philadelphia, Saunders College Publishing.

Paffenhofer, G. (1976). Feeding, growth and food conversion of *Calanus helgolandicus*. *Limnol. Oceanogr.*, **21**, 39–50.

Reeve, M.R. (1969). Growth, metamorphosis and energy conversion in the larvae of the prawn, *Palaemon serratus*. *J. Mar. Biol. Ass. UK*, **49**, 77.

Rodhouse, P.G. (1978). Energy transformations by oysters. *J. Exp. Mar. Biol. Ecol.*, **34**, 1–22.

Rodhouse, P.G. (1979). Energy budget for an oyster population. *J. Exp. Mar. Biol. Ecol.*, **37**, 205–12.

Steele, J.H. (1974). *The Structure of Marine Ecosystems*. Oxford, Blackwell.

Warwick, R.M. and Price, R. (1975). Macrofauna production in an estuarine mudflat. *J. Mar. Biol. Ass. UK*, **55**, 1–18.

Warwick, R.M. *et al.* (1978). Annual macrofauna production in a *Venus* community. *Est. Coastal Mar. Sc.*, **7**, 215–41.

Zhukova, A.I. (1963). On the quantitative significance of microorganisms in nutrition of aquatic invertebrates. *Symposium on Marine Microbiology*. C.H. Oppenheimer, ed., p. 699. Springfield, Illinois, C.C. Thomas.

8   The seashore

Populations of the seashore face a number of special difficulties. The most obvious problem is how to cope with continuous alternate submergence by water and exposure to air. There are usually additional difficulties caused by the breaking power of waves. In the shallow water along the coast the physical and chemical conditions are less stable than in deep water; in particular, there are wider and more rapid changes of temperature, and fluctuations of salinity associated with evaporation, or fresh water dilution. Often the inshore water is also very turbid because quantities of suspended matter are churned up from the bottom by waves or carried into the sea by rivers.

On most shores the ecological conditions are dominated by the tides and the waves.

8.1   Tides

Although the behaviour of tides is very complex, the underlying cause of their motion has been understood since Newton (1647–1727) accounted for tides as due to the gravitational attraction of the moon and the sun upon the oceans. A brief description is given here but fuller explanations can be found in the *Admiralty Manual of Tides* (Doodson and Warburg, 1941, reprinted 1973). Other useful and practical material on tides and tidal streams can be found in the *RYA Manual of Navigation* (RYA, 1981). Most general texts on oceanography will also carry explanations and references concerning tides.

Considering the lunar effect alone, the tide-generating force is the resultant of the centrifugal force due to the revolution of the earth–moon system acting in one direction, and the moon's gravitational pull on the water acting in another. The centrifugal force on any small mass of water is the same at any point on the earth, and is equal to the moon's pull on an equal small mass at the earth's centre of gravity. Gravitational force varies inversely with the square of the distance at which it acts, and is therefore greater than the centrifugal force on the side of the earth facing the moon, and less than the centrifugal force on the side away from the moon. The horizontal components of the tide-generating forces therefore tend to move the water towards two points, one immediately below the moon and one in the same line on the opposite side of the earth (Figure 8.1). If the earth were

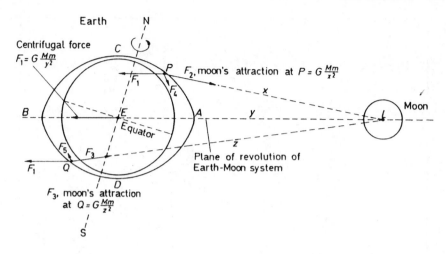

Figure 8.1   Tide-generating forces at P and Q are F_4 *and* F_5, *the resultants of* F_1, F_2 *and* F_1, F_3; M = *mass of moon*, m = *point mass of water at P or Q*; G = *gravitational constant*; E = *earth's centre of gravity*; L = *moon's centre of gravity*; x = *distance PL*; y = *distance* EL, z = *distance QL*.

covered with water, this would distort the water layer to produce two tidal bulges at A and B, where the tide would be high. Low tide would be at positions half-way between A and B, such as C and D, where the moon would be on the horizon. During the earth's rotation the tidal bulges would remain stationary relative to the moon, and at any point on the earth's surface there would be a diurnal cycle of alternate high tide, low tide, high tide and low tide within the period of the lunar day, i.e. 24 h 50 min.

Tidal bulges of such simple form cannot occur because the land surface divides the oceans into a number of more or less separate bodies of water. Nevertheless, in most parts of the sea there are tidal movements of this semi-diurnal pattern corresponding closely with the lunar period. The lunar effect on the tide varies slightly from day to day with changes in the declination of the moon and its position in its elliptical orbit.

The sun's effect on the tides is less than the moon's because of its greater distance from the earth, but is sufficient to exert an appreciable modifying influence. When the moon is new or full, the pull of the sun on the water is in nearly the same line as that of the moon. The combined pull of sun and moon then causes the specially high and low tides known as *spring tides*. At the moon's first and last quarters the sun pulls at right-angles to the moon, reducing the lunar effect. There is then less difference in the levels of high and low water, and these tides of reduced range are termed *neap tides*. The height of the tides therefore varies daily with the phases of the moon, spring and neap tides each recurring

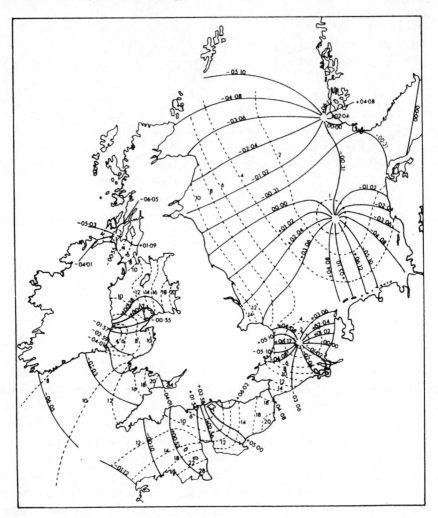

Figure 8.2   Tidal movements around the British Isles. Continuous lines are co-tidal lines joining points where high water occurs at approximately the same time. The figures against these lines give the difference of time of high water from the standard port. The standard port for the North Sea is Tynemouth, Devonport for the English Channel and Holyhead for the Bristol Channel and Irish Sea. Co-tidal lines converge on amphidromic points (three in the North Sea). Degenerate amphidromic points lie inland of Bournemouth and Wexford. Pecked lines are co-range lines joining points of the same tidal range, and the figures against them give mean range in feet.
(Reproduced from BA chart no. 5058 with the sanction of the Controller, HM Stationery Office and of the Hydrographer of the Navy.)

twice in every 28-day lunar cycle. The solar effect on the tides also varies with the sun's declination, being greatest at the equinoxes. The spring tides then have their maximum range and the neaps their minimum.

Many of the complexities of tidal behaviour arise because the oceans do not completely cover the earth, but are broken up by land. The oceans comprise many interconnected bodies of water, each with a natural period of oscillation. The tide-generating forces apply a continually varying to-and-fro pull on the water, making two complete alternations of direction within each period of complete rotation of the earth relative to the moon. The effect is to set up various complex oscillations in different parts of the sea. The number, extent and motion of these oscillating tidal systems are not completely known because of difficulties of measurement in the open ocean. In some areas the oscillation resembles a standing wave, but within each system the oscillations are to some degree deflected by the earth's rotation. This generally causes the oscillation to take the form of a progressive wave swinging around a centre, anticlockwise in the northern hemisphere, clockwise in the southern. Where the wave motion makes a complete rotation within a tidal cycle, the oscillation is termed an *amphidromic system* centred on an *amphidromic point* (see Figure 8.2). The change of water level is minimal at the amphidromic point and increases towards the periphery of the system.

Throughout the oceans the tidal movements comprise a number of interacting oscillations, the position and extent of each system being determined by the dimensions and location of particular sea areas. The range of tidal movement depends on the natural oscillation period of each basin, being strongest where this is in rhythm with the period of the tide-generating force. Areas of little tidal movement, for example the Mediterranean, are out of phase with the tidal period.

The coastal waters of the British Isles are set in oscillation by the rise and fall of the Atlantic tides. In the North Sea the tidal movements comprise three amphidromic systems as shown in Figure 8.2. The displacement of the amphidromic points to the east is the consequence of frictional loss of energy by the tides as they move anticlockwise over a shallow bottom. In the English Channel and the Irish Sea the shallow water distortion of the amphidromic systems is so great that the amphidromic points appear to lie inland, north of Bournemouth and north-west of Wexford. These are termed *degenerate amphidromic systems* (Doodson and Warburg, 1941).

Along the shore the extent of tidal movement is determined partly by the shape of the coastline. In tapering channels, where the tide enters a wide mouth and moves forwards between converging coastlines, the height of the tide is increased by the constriction of the water between opposite shores. An example is the Bristol Channel, where the tidal range sometimes exceeds 12 m at Chepstow. The average range of tides around the British Isles is about 4 m.

Tidal movements throughout each complete tidal cycle involve approximately equal ebbing and flowing of water as the tidal currents reverse their direction. Nevertheless there is usually some net transport of water in a particular direction. This resultant flow is termed a *residual current.*

Wind and atmospheric pressure have some influence on sea level, and can produce unexpected tidal anomalies. Of special importance are storm surges which occur when strong on-shore winds pile up the water along the coast and cause the tide to rise to abnormal heights. A striking instance occurred in 1953 along the east coast of England, when a great storm surge in the North Sea during the night of 31 January resulted in remarkably high tides, up to 3 m above predicted levels, overwhelming the coastal defences in many areas and causing widespread devastation and flooding.

8.1.1 Tidal levels on the shore

The intertidal zone or littoral zone is that part of the seabed that lies between the highest high water mark (extreme high water spring level or EHWS) and the lowest low water mark (extreme low water spring level or ELWS). It is sometimes convenient to refer to various levels of the shore in terms of the tidal cycle, as indicated in Figure 8.3. The strip of shore between mean high water level and mean low water level has been termed the middle shore, above MHW the upper shore and below MLW the lower shore.

The distribution of plants and animals on the shore is sometimes described in terms of the standard tidal levels but this can be misleading because, as discussed later (see Section 8.6), many other factors besides the tidal cycle influence the levels occupied by shore organisms, and their distribution varies from place to place. There is also some difficulty in determining the tidal levels with accuracy. Nevertheless, this terminology does provide a useful means of giving a general indication of the vertical zonation of different species in particular localities.

8.2 Waves

The commonest cause of surface waves is the action of wind on the surface transmitting energy to the water and setting it in orbital motion (Figure 8.4). The size of waves depends upon the speed of the wind, the length of time during which it blows and the uninterrupted distance over which the wind acts on the water (the fetch). Surface waves do not mix the water to any great depth. Their motion falls off sharply with depth, and at a depth equal to half the wavelength of the waves the water is virtually still.

When waves move into shallow water, their advance becomes slowed and the waveform changes when the depth becomes less than half the wavelength. Friction with the bottom causes the shallower parts of the wave to be slowed more rapidly than deeper parts. This is called *wave refraction.* It causes wave fronts which

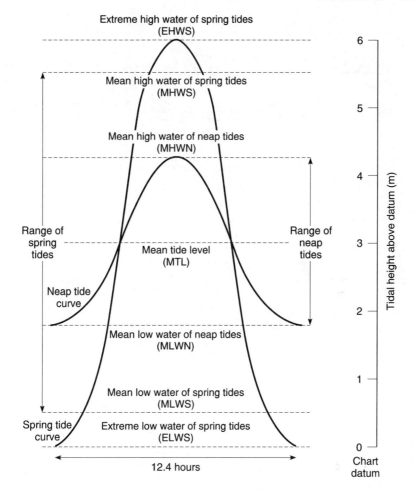

Figure 8.3 Some of the standard terms for the tidal levels of the shore, often used to describe the zonation of littoral organisms.

approach the coast obliquely to slew round and change direction in shallow water so that they reach the shore nearly parallel to the beach (Figure 8.5).

Consequently, waves converge upon headlands, but diverge and spread out within bays. The energy transmitted by the waves is therefore concentrated on promontories, and is correspondingly reduced along equivalent lengths of coastline between the headlands (Figure 8.6).

The slowing of wave advance in shallow water reduces the wavelength. As the wave crests become closer together the wave height increases and the fronts of the waves become steeper. Where the water gets progressively shallower near the shore, the waveform becomes increasingly distorted until eventually the waves

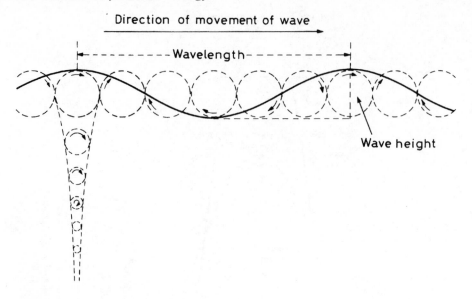

Figure 8.4   Profile of an ocean wave. The circles and arrows show the direction of movement of the water in different parts of the wave, and how this orbital movement decreases with depth.

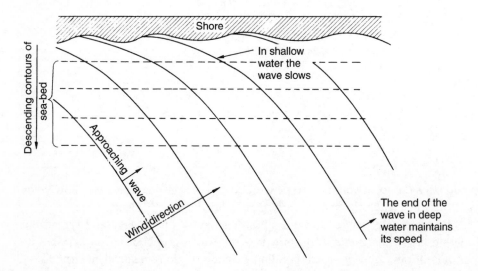

Figure 8.5   Diagram of wave refraction as waves approach shallow water obliquely.

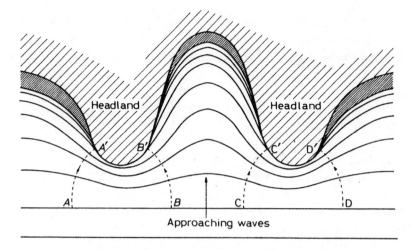

Figure 8.6 Diagram illustrating the concentration of wave energy on promontories. The energy of the wave fronts between A–B and C–D becomes concentrated by wave refraction onto the short stretches of shore A'–B' and C'–D'. Within the bay the energy of wave front B–C becomes spread out around the shoreline B'–C'.

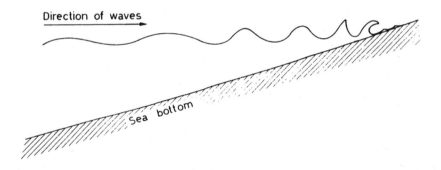

Figure 8.7 Changes of waveform on entering shallow water. At a depth of about half the wavelength, the waves become closer and higher. With decreasing water depth, the fronts of the waves become steeper until they are unstable and their crests topple forward.

become unstable, their crests overtake the troughs and topple forwards as they break. The energy of the wave motion is then translated into the energy of a forward-moving mass of water (Figure 8.7).

8.2.1 The effects of waves on beaches

The up-rush of water formed by a wave breaking on the beach is known as the 'swash' or 'send' of the wave. Part of this water percolates down through the beach and the remainder flows back over the surface as the 'backwash'. The effect of breakers on the shore may be destructive or constructive. Destructive breakers are usually formed by high waves of short wavelength; for example, those arising from gales close to the coast. As these waves break, they tend to plunge vertically or even curl seawards slightly, imparting little power to the swash, pounding the beach, loosening beach material and carrying some of it seawards in the backwash. Constructive waves are more likely to be low waves of long wavelength, sometimes the swell from distant storms. These waves move rapidly shorewards and plunge forwards as they break, transmitting much power to the swash and tending to carry material up the beach, leaving it stranded.

The continual transmission of energy from waves to shore gradually modifies the coastline, either eroding the beach by carrying away the beach material, or adding to the beach by deposition. In any sequence of breakers, there may be waves derived from many different sources which combine to form many different heights and wavelengths, some destructive and some constructive. The condition of the shore is, therefore, a somewhat unstable equilibrium between the two processes of erosion and deposition. The balance varies from time to time, and differs greatly in different regions. Where the shore is exposed to very violent wave action, erosion usually predominates. The waves break up the shore, fragmenting the softer materials and carrying them away. Harder rocks are left exposed and are gradually fractured into boulders, making the coastline rocky and irregular. Strong currents may assist the process of erosion by carrying away finely divided material. But materials carried away from the shore in one place may be deposited as beach material in another, and where the major effect of the waves is deposition, the beach is made up largely of pebbles, sand or mud.

Beach construction

Beaches consist of a veneer of beach material covering a beach platform of underlying rock. In very sheltered situations the beach material may rest on a gentle slope of rock virtually unmodified by wave action, but in wave-washed localities the beach platform has usually been formed by wave erosion. Where the land is gradually cut back, both cliff and beach platform are formed concurrently (Figure 8.8). As the beach platform becomes wider, waves crossing it lose power, and erosion of the cliff base is reduced. The seaward margin of the beach platform is itself sometimes subject to erosion by large waves breaking further out, so the cutting back of the beach platform and the cliff base may proceed together. Cliff erosion is caused largely by the abrasive action of stones, sand and silt churned up by the water and hurled against the base of the cliff, undercutting it until the overhanging rock collapses. Where the rocks are very hard, they may not be appreciably worn away, but can be cracked along lines of weakness by sudden air compression in holes and crevices when waves strike the cliff. This leads to

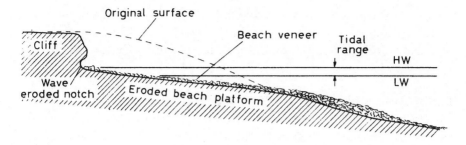

Figure 8.8 Beach section to illustrate erosion of the cliff to form the beach platform and veneer.

falls of large pieces of rock, and produces a boulder-strewn coastline on which the beach platform is often quite narrow. The range of tidal movement has a considerable bearing on the rate of erosion because the greater the depth of water covering the beach platform at high tide, the more powerful the waves that can cross it to erode the cliff base.

Various sources may contribute to the materials covering beach platforms; for example, fragments derived from erosion of adjacent cliffs, or churned up from the sea-bed, or eroded from the edge of the beach platform, or carried along the coast from other places by currents or beach drifting (see below), including material carried into the sea by rivers. In sheltered regions, finely divided material carried in suspension in the water may be deposited on the beach as sand, mud or silt.

Beach drifting and grading

When a wave breaks, the swash may carry stones or smaller particles up the beach. Waves which break obliquely on the shore carry materials up the beach at an angle, while the backwash and any particles it contains run directly down the slope of the beach. Consequently, each time a wave breaks obliquely, some of the beach material may be carried a short distance sideways (Figure 8.9). This process, known as beach drifting, can move huge quantities of material over great distances. Drift of sediments along beaches is also assisted by longshore currents generated by oblique waves.

Beach drifting has a sorting effect on the distribution of beach materials. Where cliffs are exposed to strong wave action, the shore is usually strewn with boulders too large to be moved by the waves. Boulders gradually become broken up, and the fragments may then be carried along the shore in a series of sideways hops. A short distance from the original site of erosion, the beach is likely to consist mainly of large stones which only the more powerful waves can move. Further along the shore the particle size of the deposit becomes progressively smaller, partly because the pebbles gradually wear away as they rub one against another, and partly because smaller particles are more easily transported.

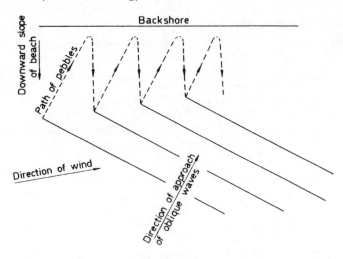

Figure 8.9   The movement of pebbles along a beach under the influence of oblique waves.

Waves also exert a sorting action on the grade of material deposited at different levels of the beach, due to the difference in energy of swash and backwash. The swash has the full force of the wave behind it, but the backwash merely flows back down the beach under the influence of gravity, and contains less water than the swash because some is lost by percolation through the beach. The swash may move large stones up the beach but the energy of the backwash may be insufficient to carry them down again. This often leads to an accumulation of large pebbles at the back of the beach, causing the slope of the beach to become steeper towards the land until a gradient is reached at which stones begin to roll down under their own weight. The smaller particles of sand and gravel are more easily carried down by the backwash, and this has a grading effect, depositing coarse material at the back of the shore and progressively smaller particles lower down. Around the British Isles there are many beaches with steeply sloping shingle at the higher levels, and flatter sand or mud on the middle and lower shore.

Drifted beach material accumulates alongside obstructions crossing the shore; for example, headlands, large rocks or groins. The purposes of erecting groins is to limit beach drifting by causing pebbles and sand to become trapped between them. Wave action then builds the beach into high banks between the groins, preventing the beach material from being carried away, thus protecting the land against erosion.

8.2.2   Tsunamis ('tidal waves')

Earthquakes, volcanoes, movements of the sea floor and submarine escapes of gas may sometimes generate immense disturbances within the oceans which cause huge waves of unusual form. The description 'tidal waves', previously widely used in this connection, is a misnomer, because these exceptional waves have no association with tides. The Japanese name *tsunami* is now generally given to this phenomenon. Such waves can have devastating effects if they strike the coastline.

The term 'tidal wave' was probably applied because the first sign of an approaching tsunami is usually a rapid change of sea level, often appearing to be a small rise of the tide quickly succeeded by a sharp fall to an abnormally low level for about 15–30 minutes. This is followed by an enormous wave suddenly rising out of the sea, often with the appearance of a near-vertical wall of water several metres high, rushing forward at great speed to overwhelm the land along low-lying coasts. The water level then falls for a period up to about an hour, when another wave of even greater size may occur. A succession of these huge waves sometimes extends over several hours, and those in the middle of the sequence usually have the greatest size, which may be as much as 30–40 m in height.

Tsunamis occur mainly around the shores of the Pacific, where they have caused terrible devastation and loss of many thousands of lives. They have also been experienced in the Indian Ocean, the Mediterranean and even the North Atlantic. Many people died in Portugal in 1755 when a tsunami swept up the river Tagus. Around 7000 years ago, a tsunami struck the east coast of Scotland and penetrated 4 kms inland. It was caused by a massive landslip off the coast of Norway (MacGarvin, 1990). One of the worst ever occurred in 1960. A very strong earthquake in southern Chile initiated tsunamis that affected not only the Chilean coast but also the Hawaiian Islands and especially Japan, over 10 000 miles away from the origin.

Such extraordinary waves at the coastline are the result of ocean disturbances quite different from wind-induced surface waves. The submarine upheavals which cause tsunamis generate deep-water waves of low height but great wavelength and velocity. Wind waves seldom have wavelengths greater than 300 m or speeds of more than 100 km/h. The wavelengths of tsunamis range between 150 and 1000 km with speeds of 400–800 km/h. The wave height of tsunamis in deep water is only a few metres and, because the wavelength is so long and the period between wavecrests often an hour or more, they are virtually undetectable in the ocean. But when they move into shallow water the velocity of the waves becomes sharply reduced, the wave height rises steeply as the crests become crowded together, resulting in a series of catastrophic waves breaking at intervals of about 15–60 minutes.

8.3 The evolution of coastlines

Erosion and deposition, continued over long periods, gradually change the configuration of a coastline, tending eventually to straighten it by wearing away the headlands. Over many millions of years these processes would have reduced long stretches of shore to virtual uniformity were it not for the changes in relative levels of land and sea which have occurred from time to time throughout the earth's history (Steers, 1969). The causes of these changes are incompletely understood, but variation of world climate has certainly been one of the major factors during the last million years by altering the volume of water in the oceans. During this period there have been a series of 'ice ages' when the world climate has become colder than at present, polar ice caps have extended to much lower latitudes, and more snow has remained on the mountains instead of melting and flowing into the sea. Because a greater proportion of the earth's water has been locked up in frozen form, sea level has fallen. During the warmer interglacial periods, the melting of ice and snow has increased the volume of water in the oceans, and raised sea level.

Changes in ocean volume do not produce equal relative changes of land and sea level in all parts of the world. The enormous weight of an ice cap can depress the level of the underlying land. When the ice cap melts, although the sea becomes deeper, that part of the land which has been relieved from the huge load of ice may rise considerably more than the sea around it so that the sea level falls relative to the rising land. Depending upon the change of relative levels of land and sea, the changes of the coastline may be submergent or emergent.

8.3.1   Submergent coasts

Where sea level rises relative to the land, deep inlets are formed by the sea flooding the lower part of river valleys. The headlands between valleys are exposed to wave action, and are gradually eroded to form cliffs. Within the valleys, sheltered conditions permit the deposition of eroded material, so the inlets gradually fill with sand or silt (Figure 8.10). This submergent type of coastline is evident around the south-west peninsulas of the British Isles and the flooded river valleys are called *rias*.

If land and sea level become stable for any long period, the headlands are progressively cut back reducing the depth of inlets until they become bays containing beaches between extensive stretches of high cliff. Eventually, when the protecting headlands have been completely obliterated, the bays themselves become exposed to erosion and the coastline then tends to become fairly uniform, consisting mainly of cliffs with little if any beach, i.e. a mature coastline.

8.3.2   Emergent coasts

Where the sea is receding from the land, the margin of the water meets a gentle slope that was originally the sea bottom. Waves tend to break a long way out because the water is shallow for a considerable distance from the shore. In the line of wavebreak the sea-bed becomes churned up, and loosened material may be thrown ahead of the breaking waves. This sometimes leads to the formation

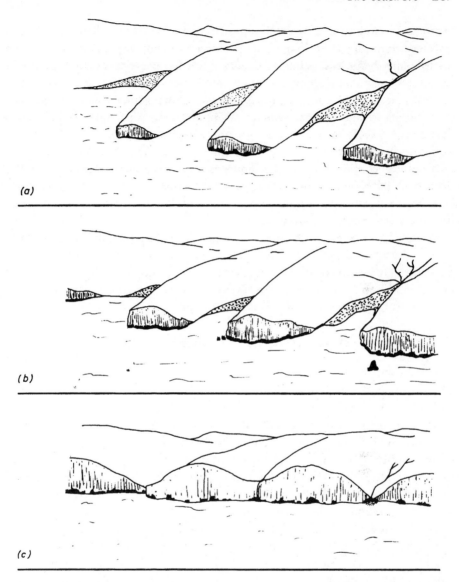

Figure 8.10   Stages of evolution of a submerged coast. (a) Early phase – flooded valleys, erosion of headlands and deposition of sand and silt within the inlets. (b) Intermediate phase headlands cut back further, lengthening the cliff line and exposing the bays to stronger wave action. Beach material becomes coarser and less stable. (c) Mature phase – promontories eliminated and cliffs virtually continuous.

of ridges or bars of sand or gravel, known as offshore bars. These are often of transient duration, their shape and position changing from tide to tide, but occasionally the process is cumulative so that a bar is eventually built up above sea level. It may then become stabilized and consolidated by plant growth along its crest. The seaward side of the bar is now the new coastline. The lagoon between the earlier shore line and the newly formed bar becomes silted up, forming an area of salt marsh which may later become converted into sand dunes, and finally into ordinary soil.

If sea level continues to fall, this sequence may be repeated several times. Once the sea level becomes constant over a long period, wave erosion is likely to encroach gradually upon the land, leading at length to the formation of cliffs and the development of a mature coastline.

Due probably to a rise of sea level since the last glaciation, much of the coast of the British Isles is the submergent type in various stages of evolution. However, there are signs of old cliffs and beaches, some now many feet above sea level and far behind the present coastline, in several places, for example, parts of Sussex and Devon, Norfolk and some areas of Scotland.

8.4   Some problems of shore life

8.4.1   Water loss

The majority of inhabitants of the seashore are essentially aquatic organisms which have evolved directly from fully marine forms. If left uncovered to the air, they lose water by evaporation and will eventually die from dehydration. To make the transition from sea to shore successfully, forms which live in exposed positions on the surface of the shore must have some means of retarding the rate of water loss sufficiently to permit survival during the periods when they are left uncovered by the receding tide. The danger of drying is most severe where exposure to air also involves exposure to sun or wind.

Water loss can be fatal in several ways. Death may be due to disturbances of metabolism resulting from the increasing concentration of internal fluids. In many cases the immediate hazard is asphyxia. Some organisms require a continuous current of water over the gills for adequate gaseous exchange. Others can survive for a time in air, but all must preserve at least a film of water over their respiratory surfaces. In a fish out of water the weight of the gill lamellae unsupported by water causes them to collapse against each other and adhere by surface tension. So little area of gill is then left exposed for respiratory exchange that the fish asphyxiates despite the high oxygen content of the air.

8.4.2   Wave action

Enormous forces are transmitted to the shore by the breaking of waves. The destructive impact of a great weight of water together with stones and other suspended matter hurled by the waves presents a major hazard to the inhabitants of the shore surface. Beneath the surface, burrowing forms are in danger of being crushed when waves churn the beach deposits. There is also the danger of dislodgement by waves, which may carry creatures up or down the shore to levels unsuitable for their survival. When masses of stones, sand, or organic debris are washed up and stranded by the waves, the shore population is in danger of becoming smothered. Continuous rapid movement of the water presents difficulties for the settlement and attachment of spores and larvae, and may prevent colonization of the shore except in crevices and sheltered parts.

8.4.3 Temperature fluctuations

During low tide, wide and rapid changes of temperature are sometimes encountered on the shore. Strong sunshine can produce high temperatures on exposed shore surfaces, and the temperature of water standing on the shore when the tide recedes may be raised well above the normal limits of sea temperature. Shallow pools on English coasts sometimes reach a temperature of 25–30°C on hot summer days. In the tropics, temperatures of 50°C or higher have been recorded in shore water.

Intertidal organisms may also be exposed to severe frosts. When shore water freezes they face the additional dangers of moving ice which may scrape them off the rocks or crush them within their hiding places. In high latitudes the shore may be kept virtually barren by the effects of moving ice.

8.4.4 Salinity fluctuations

Shore organisms often encounter water of much reduced salinity due to dilution of the shore water by rain, or fresh water flowing off the land. On the other hand, evaporation may raise the salinity of shore water above that of normal seawater. Increased salinity and raised temperature usually occur together.

8.4.5 Fluctuations of oxygen, carbon dioxide and pH

These occur in shore water in association with photosynthesis and bacterial activity. In bright light, photosynthesis by dense algal vegetation in small pools sometimes raises the oxygen content appreciably, and the withdrawal of carbon dioxide from the water raises the pH. Diminished oxygen, increased carbon dioxide and reduced pH may result from rapid bacterial decomposition in stranded detritus.

8.4.6 The range of illumination

The illumination of the shore varies widely with rise and fall of the tide. When the tide recedes, the shore is directly exposed to light. When the shore is covered, the illumination is much reduced, especially where the water is very turbid.

Inadequate light is probably one of the factors limiting the downward spread of some of the algae on the shore. Direct exposure to light may favour the growth of some plants but high intensities of light are probably lethal to many red algae. Strong sunlight is detrimental to many organisms due to the combined effects of radiation, heating and drying, but it is difficult to dissociate their separate influences.

8.4.7   Predation

The inhabitants of the seashore are exposed to a double set of predators. During submergence they are preyed upon by other marine creatures. When uncovered they encounter enemies from the land and air. On some parts of the British coast, seabirds exact a very heavy toll from shore populations during low tide. For example, oystercatchers (*Haematopus ostralegus*) eat huge numbers of cockles from the large, intertidal beds in South Wales. Studies in the Burry estuary indicate that as many as 500 cockles per bird per day can be taken. Oystercatchers prefer two-year-old cockles and in some circumstances, have been blamed for consuming up to 70 per cent of the stock of this age group. In other areas more bivalves are taken by flatfish when the tide is in than by waders when it is out.

8.4.8   Immersion

The majority of shore creatures utilize the oxygen dissolved in seawater for their respiration, but a few are air-breathing. Some of these have evolved from land forms which have spread to the seashore, for example the insects *Petrobius* and *Lipura*, and the gastropods *Otina* and *Leucophytia*. Others have evolved directly from marine forms, their respiratory organs having become adapted to absorb atmospheric oxygen, for example the sea slater *Ligia*, the amphipods *Orchestia* and *Talitrus*, and the periwinkles *Littorina neritoides* and *L. saxatilis*. In either case immersion cuts off the essential supply of air, and may be fatal if too prolonged.

During immersion the shore population experiences varying water pressure, with very rapid fluctuations where there are waves. Sensitivity to water pressure must be a factor having some influence on distribution within the intertidal zone.

8.4.9   Pollution

An additional hazard of shore life which has appeared in recent years is pollution from accidents at sea, especially by stranding of oil. Even more toxic than oil itself are some of the chemicals used to clean beaches (see Section 10.1.11).

8.5   Food sources

Despite its dangers, the shore is often densely populated by a variety of organisms excellently adapted to the difficult conditions. So great are the numbers on some shores that every available surface is colonized, and there is severe competition for living space. A numerous population indicates an abundant supply of food, and this is derived from several sources. On rocky coasts there is often a thick cover of seaweed. The rapid growth of these plants is favoured by the excellent lighting conditions and by a good supply of nutrients. The nutrients are continually released and replenished by wave disturbance of sediments, weathering of the coastline and input from fresh water flowing off the land, and are well distributed by water movements. Some animals browse directly on these seaweeds. Many more obtain food from the masses of organic debris formed by the break-up and stranding of pieces of seaweed. Even where there are no large plants, the surface of the beach may be covered by a film of microscopic algae.

In addition to primary production on the shore, large quantities of food are brought in by the sea. Inshore water often contains a rich plankton on which innumerable shore creatures feed, replenished at each rise of the tide. Also, pieces of plant material torn from the sea-bed below low tide level become deposited on the shore. The land, too, makes a contribution, various organic substances of terrestrial origin being stranded on the beach.

8.6 Zonation

A gradient of environmental conditions extends across the shore, due mainly to the different durations of submergence at each level. The lowest parts of the shore are uncovered only during the lowest spring tides, and then only for brief periods. The highest levels are seldom fully submerged and are mainly wetted by wave splash. Intermediate levels experience intermediate durations of alternating exposure and submersion. Even on non-tidal shores there are gradients between permanently submerged and fully terrestrial conditions, with intermediate levels wetted by splash or irregularly covered and uncovered as wind action alters the water level.

The requirements for life in air and water are so different that no organism is equally well suited to every level of the shore. Different levels are therefore occupied by different assemblages of plants and animals, each species having its main abundance within a particular zone where conditions are most favourable for it. Above and below this zone it occurs in reduced numbers, or is absent, because environmental conditions are less suitable: physical conditions may be too difficult to allow its survival; it may be ousted by competition with other species better suited to these levels; it may be eliminated by predation; or some biotic factor essential for its life, such as suitable food, may be lacking. Some examples of these types of interactions are given in papers by Barnett, 1979; Connell, 1961; Coombs, 1973; and Schonbeck and Norton, 1978, 1979.

Alternations of uncovering and submergence present somewhat different problems to different organisms, depending on the type and level of shore they inhabit, and their mode of life. For surface-living forms, drying, wave action, illumination and temperature are major factors influencing zonation but their relative importance must vary with level. The occupants of the upper shore are exposed to air for long periods between spring tides, and must therefore be able to withstand conditions of prolonged drying, extremes of temperature and strong illumination. Direct wave impact is a relatively brief and infrequent hazard, and strong wave action favours the upward spread of this population by increasing the height to which spray regularly wets the shore.

Lower on the shore, surface populations are never uncovered for more than a few hours at a time, and here the problems of desiccation, temperature fluctuations and excessive illumination are less severe. These organisms experience longer and more frequent periods of wave action with the attendant risks of damage or dislodgement, and during submergence the illumination may be inadequate for certain species of plants. Other factors which vary with shore level, and influence distribution, include the duration of periods during which feeding is possible, the maximum depth of water, and often the type of substrate. These factors may largely account for the zonation of many of the forms which are not ordinarily left exposed, but hide in crevices or under the algae, or burrow into the rock or sediment. Because the vegetation varies at each level of the shore, the distribution of herbivorous animals and their predators may be influenced by the dependence of grazers on certain preferred algae.

Although the tides exert a major influence on zonation, the distribution of littoral (shore-dwelling) species relative to particular tidal levels is by no means constant. Wide variations occur from place to place due to differences of geography, geology and climate. Factors which modify zonation and vary with locality include the intensity of wave action, the range of temperatures and humidities, the aspect of the shore with respect to the sun and prevailing wind, the type of rock or sediment, the amount of rainfall and fresh water run-off, and the period of day or night when extreme low tides occur. Lewis (1964) has emphasized that the *littoral zone* cannot be satisfactorily defined solely in relation to sea level, but is better described in biological terms as the strip between sea and land inhabited by characteristic communities of organisms which thrive where the shore surface undergoes alternations of air and seawater.

These communities comprise three main groups occupying different levels, and the littoral zone can be correspondingly subdivided (see Section 8.8.1). The three-zone system was first proposed by two biologists in the 1940s (Stephenson and Stephenson, 1972) and was extended by Lewis (1964). Today the system still provides a useful descriptive framework.

The highest zone, with the uppermost communities which require mainly aerial conditions, is called the *littoral fringe*. This zone is submerged only at spring tides or wetted only by wave splash. On very sheltered shores this zone is a narrow belt below EHWS level, but it becomes higher and wider with increasing exposure to wave action, until, on the most wavebeaten rocky coasts, it lies entirely above EHWS level and may extend upwards for 20 m or more. Below the littoral fringe is the broad *eulittoral zone*, occupied by communities tolerant of short periods of exposure to air between tides but requiring regular submersion, or at least thorough wetting, at each tidal cycle. The low-shore zone is called the *sublittoral fringe* and really forms part of the permanently submerged *sublittoral zone*. It is only exposed at spring low tides. At their lowest levels the populations of the eulittoral zone overlap those of the sublittoral fringe (see Figure 8.11). Examples of intertidal zonation around the British Isles are given later (see page 284 ff).

8.7   Fitting the shore environment

The detrimental effects of exposure to air restrict much of the littoral population to sheltered parts of the shore which are not left completely uncovered at low tide. Many algae, anemones, hydroids, bryozoans, prawns and fish occur only in rock pools where they are safe from drying, and relatively protected from wide temperature fluctuations. Shelter can also be found between the fronds and holdfasts of the shore algae, under stones and boulders or in rock crevices. Some of the most numerous shore animals have a flattened shape well suited for hiding in narrow spaces, for example the crab *Porcellana platycheles*, the leptostracan crustacean *Nebalia*, amphipods, isopods and chitons. Others burrow into the shore deposits for protection, and these comprise virtually the entire population of sandy and muddy shores.

Environments with fluctuating conditions favour the evolution of species that exhibit wide variations, both physiological and anatomical. Where an unstable habitat also includes environmental gradients, different forms of a species are likely to occupy different zones. On the seashore it is obvious that many of the dominant species are highly variable, with differences which can to some extent be correlated with differences of zonation. A notable example is *Littorina saxatilis* which shows great variety of shell size, colour, thickness, shape and surface texture (see Section 8.8.1, page 284). Other common polymorphic species on the shore include *Littorina obtusata*, *Nucella lapillus* (Kitching, 1977) and *Mytilus edulis*, with some evidence of different forms in different localities.

8.7.1   Water loss

Organisms which live on the shore surface in situations where they are frequently uncovered to air must have a protective covering to prevent excessive drying. The exposed algae have mucilaginous outer layers which reduce evaporation (Chapman, 1979). The beadlet anemone, *Actinia equina*, often inhabits uncovered rocks and is coated with slimy exudations which presumably help to conserve water. Most of the surface-dwelling animals have a strong shell, the

Table 8.1   Uric acid content of nephridia of *Littorina* spp. (From Nicol, 1967 by courtesy of Pitman)

Species	Dry weight (mg/g)
Melaraphe (Littorina) neritoides	25
Littorina saxatilis	5
Littorina obtusata	2.5
Littorina littorea	1.5

orifice of which can be kept closed while they are uncovered. The limpet, *Patella*, draws down the rim of its shell so close to the rock surface that only a very narrow gap is left, sufficient for oxygen and carbon dioxide to diffuse through without allowing much evaporation. Winkles, top shells, dog whelks and serpulid worms close the shell aperture with an operculum. In barnacles, the movable plates of the shell, the terga and scuta, are kept shut most of the time the animals are uncovered, occasionally opening momentarily for renewal of the air enclosed within the shell. To some extent animals adjust to the drying conditions at different shore levels by metabolic adjustments. For instance, specimens of *Patella vulgata* living at high shore levels during summer have lower respiratory rates, lower rates of water loss and can tolerate greater percentage water losses than those low on the shore (Davies, 1966–70).

In addition to losses by evaporation, animals also lose water by excretion. The majority of marine creatures excrete ammonia as their chief nitrogenous waste product. This is a highly toxic substance which has to be eliminated in a very dilute urine, involving the passage out of the body of a copious amount of water. On the seashore, this must present a difficulty to animals already in danger of desiccation, and some of the littoral gastropods reduce their excretory water-loss by excreting appreciable amounts of uric acid, a less soluble and less toxic substance than ammonia which can be excreted as a semi-solid sludge, thereby conserving water. The different species of *Littorina* form a series, those which live highest on the shore excreting the greatest amount of uric acid in the most concentrated urine (Table 8.1).

8.7.2   Protection from waves
The surface population is also exposed to the dangers of wave impact and dislodgment, and shells are again the chief form of protection. The heavy wear sometimes visible on the shells of shore molluscs indicates the severity of the abrasion to which they are subjected. To resist dislodgment, some of the surface forms have great powers of adhesion. The large algae are anchored to the rock by strong holdfasts. Barnacles and serpulid worms have shells firmly cemented to the rock. The common mussel, *Mytilus edulis*, attaches itself to rocks and stones

by strong byssus threads. The remarkable adhesion of the foot of *Patella* has given rise to the expression 'sticking like a limpet'. Four different genera of fish found on British coasts, *Gobius, Liparis, Cyclopterus* and *Lepadogaster*, have the pelvic fins specialized to form a ventral sucker by which they can cling to a firm surface.

8.7.3 Respiratory adaptations

Intermittent submergence presents problems in connection with respiration because no respiratory organs function equally well in both air and water. The majority of shore-dwelling animals perform aquatic respiration, but within the littoral fringe the infrequency of immersion calls for the ability to breathe air. Some of the inhabitants of this fringe zone are essentially marine forms which have become adapted for aerial respiration. *Talitrus saltator* and *Orchestia gammarella* can live in moist air. In *Melaraphe (Littorina) neritoides* and *L. saxatilis* the gill (ctenidium) is reduced and the mantle cavity is modified to function as a lung. There are also a few animals which have colonized the shore from the land, and several of these show adaptations for storing air during the periods they are submerged. The collembolan insect, *Anurida maritima*, widespread among rocks above mid-tide level, carries a layer of air among its surface bristles. The intertidal beetle, *Aepus marinus*, has internal air sacs for air storage.

8.7.4 Reproductive adaptations

The difficulties of survival on the shore have their effects on all phases of life, including reproductive processes and larval and juvenile stages. The majority of benthic organisms start life as floating or swimming forms in the plankton, and may become widely dispersed in the water before they settle on the sea bottom. Shore creatures face special risks of great losses of pelagic eggs and larvae during this phase if they drift far from the shore and settle outside the zone in which survival is possible. For some inhabitants of the shore the chances of successful settlement in suitable areas are enhanced by certain aspects of the behaviour of their larvae. Some produce larvae which are at first strongly attracted by light, and presumably rise close to the sea surface during the day. Wind direction is often landward during the day time, driving the surface water towards the coast, and so it is likely that positive phototaxis improves the chances of pelagic larvae returning to the shore. At night, offshore winds tend to predominate, and surface water is moved away from the shore with replacement water coming in shorewards along the bottom. Under these circumstances, descent of larvae to deeper levels away from the surface must then have a similar effect of keeping them concentrated along the shoreline.

It has been mentioned earlier that the larvae of many benthic species can discriminate between substrates and can for a time delay settlement until favourable conditions are encountered (see Section 6.2.2) (Rainbow, 1984). Some tend to settle gregariously, often in response to the presence of successfully metamorphosed members of the species, for example barnacles, the reef-building worm *Sabellaria*, serpulid worms, mussels and some bryozoa. Settlement in shallow water may also be favoured by larval response to wave action; for example, cyprids of *Semibalanus balanoides* are reported to settle more readily under fluctuating water pressure. When settlement occurs sublittorally, or at lower levels of the shore than are occupied by the adults, for example *Melaraphe neritoides* (see Section 8.7.5), the responses of the juveniles to various environmental stimuli may cause migration upshore towards the appropriate zone.

In many shore animals the planktonic phase is abbreviated or omitted, and this simplifies the problem of finding the correct shore level. For instance, the lugworm *Arenicola marina*, one of the most successful littoral worms burrowing in muddy sand, has only a brief pelagic period. In late autumn or early winter the gametes are shed from the burrow onto the surface of the sand during low spring tides, and here fertilization occurs. The fertilized eggs may be dispersed upshore to a limited extent by the rising tide, but being heavier than water and slightly sticky, they tend to adhere to the surface of the sand. They hatch after about 4–5 days and the larvae, although capable of swimming, seem from the outset to burrow into the deposit wherever the substrate is suitable.

Direct development

Other shore animals completely eliminate pelagic stages by developing directly from egg to miniature adult form. These eggs are well charged with yolk to enable the young to hatch in an advanced state. *Littorina obtusata* and *L. mariae* deposit eggs in gelatinous masses on the surface of seaweed. The young occasionally emerge as advanced veliger larvae, but probably more often do not hatch until after the velum has been resorbed and then appear as tiny crawling winkles. The dogwhelk, *Nucella lapillus*, lays vase-shaped egg capsules which often occur in large numbers stuck to the underside of stones or sheltered rock surfaces. Each capsule contains several hundred eggs, but eventually only a dozen or so whelks crawl out of each capsule, the rest of the eggs serving as food for the first few to hatch.

Parental protection

Eggs laid on the seashore are exposed to all the vicissitudes of this environment, and certain shore creatures contrive in various ways to give their eggs some protection. For example, several species of inshore fish guard their eggs, such as the butterfish (*Pholis gunnellus*) and the shanny (*Lipophrys pholis*). They lay sticky masses of large eggs from which advanced young are born, and one of the parent fish, usually the male, remains close to the eggs until they hatch, courageously protecting them against marauders. Female pipefish lay their eggs in a pouch on the male's belly. Here they remain until they hatch as miniature adults. Amphipods and isopods, often very numerous on the shore, also retain their eggs within a brood-pouch from which fully formed young emerge. In *Littorina saxatilis*, *L. rudis* and *L. neglecta* the eggs remain in the mantle cavity until they hatch as minute winkles. The tiny bivalve *Lasaea rubra*, often abundant in rock crevices and empty barnacle shells, incubates its eggs and young within the gills until they are sufficiently developed to crawl out and maintain themselves near the parent. The viviparous blenny, *Zoarces viviparus*, gives birth to well-developed young about 4 cm in length.

8.7.5 Behavioural and activity responses of mobile animals

Physiological and behavioural adaptations are necessary to withstand the fluctuating nature of the shore environment (Gibson, 1969). The wide and rapid changes of temperature and salinity that occur on the shore surface during low tide require wide eurythermy and euryhalinity in the exposed population (Cornelius, 1972; Southward, 1958). They must also be capable of making appropriate adjustments of behaviour in response to changes in their surroundings. The limpet (*Patella*), if wetted with freshwater, pulls its shell hard down and remains still; but if repeatedly splashed with seawater it begins to wander about (Arnold, 1972). The lugworm, *Arenicola marina*, before commencing the intermittent irrigation cycles which replace the water in its burrow, moves backwards up the rear end of the burrow and appears to test the quality of the surface water with its tail, modifying the subsequent sequence of irrigation activity accordingly (Wells, 1949). Animals that live among algal fronds are usually coloured to match their surroundings, and some, such as the sea scorpion (*Taurulus bubalis*) have considerable powers of colour change to match different backgrounds.

Appropriate changes of activity are required to meet the profoundly different conditions of submergence and exposure to air. Movement, feeding or reproduction is only possible for many littoral species during the periods when they are covered by water. When uncovered, the heart rate slows and they become more or less inactive. In this quiescent state their respiratory needs are reduced, water is conserved and there is less danger of attracting the attention of seabirds and terrestrial predators. Restriction of movement as the shore dries during low tide may also help to confine free-living forms to their appropriate zones. Some regulate their activity according to wind conditions, only moving about over the shore surface in calm or light winds but in strong, drying winds sheltering in crevices or under stones, e.g. the top shell *Monodonta lineata* (Courtney, 1972). Many free-living forms show seasonal changes of level, usually moving slightly downshore during the coldest part of the winter and ascending in spring.

The tiny estuarine mud snail *Hydrobia ulvae* is very common on the surface of mudflats, usually mainly above mid-tide level (see references in Graham, 1988). When the snails are first uncovered by the tide the majority are found crawling on the surface of the mud. In areas where the mud surface remains wet, the animals continue their active browsing on the surface throughout the tidal cycle; but in places where the surface becomes dry the majority of the population then burrow just below the surface, remaining buried until they are covered by the returning tide. Newell (1962) describes a behaviour cycle related to feeding. As the tide ebbs, the snails follow the water down, feeding as they go. When the tide starts to flow back in, they float to the surface and are carried back to their starting level. At high water, numbers of floating *Hydrobia* can sometimes be skimmed from the sea surface, but this may not be a regular cycle of activity (see Graham, 1988).

Endogenous rhythms

Free-living shore creatures are in some danger that their own movements may carry them out of their proper zones into levels too high or too low on the shore, or into positions too exposed to wave, wind or sun. Studies of their behaviour reveal some of the mechanisms whereby they find and keep within suitable parts of the shore. Many shore animals, e.g. the crab *Carcinus maenas*, the polychaete *Eurydice pulchra*, the blenny *Blennius pholis*, and the prawn *Palaemon elegans*, display cyclical changes of activity having a tidal frequency even when removed from the shore, apparently controlled by endogenous rhythms (Alheit and Naylor, 1976; Naylor, 1958, 1976; Palmer, 1973; Rodriguez and Naylor, 1972). In various small crustaceans which burrow in intertidal sand (*Corophium volutator*, *Synchelidium*, *Eurydice pulchra*) an endogenous rhythm evokes emergence from the sand for swimming mainly during the ebb tide. This pattern of activity avoids the danger of stranding too high up the shore but promotes swimming at the stage of the tide when food and oxygen have just been replenished.

Homing and responses to physical factors

A notable feature of the movements of the adult limpet (*Patella vulgata*) is its 'homing' behaviour, usually returning to a particular site or 'home' on the rock after foraging for food, often over a distance of several feet. *Patella* feeds chiefly by scraping the surface with its long, toothed radula, rasping off the microscopic film of algae which forms a slimy coating on the rocks. During the day it seldom moves about unless submerged, but at night *Patella* can often be found crawling on moist surfaces while uncovered. Homing behaviour is most strongly developed in individuals living high on the shore. On returning to its home, *Patella* always settles in the same position and the margin of the shell grows to fit the rock surface very accurately. On soft rocks the home is often marked by a ring-shaped groove, the 'limpet scar', conforming to the shell margin and presumably worn into the rock by slight movements of the shell. Homing obviously ensures that the animal maintains its zonational position, and the exact fit between shell and rock reduces water loss during exposure, and also lessens the danger of the shell being prised off the rock by predators.

It is uncertain how *Patella* finds its way back to its home (see references in Graham, 1988), but in some animals it is evident that the direction of their movements is related to factors such as light (phototaxes), gravity (geotaxes), lateral contact (thigmotaxes), humidity (hydrotaxes) or direction of flow of water (rheotaxes) (Fraenkel and Gunn, 1961). This is a complex field of study because animal behaviour is seldom altogether consistent, varying from one individual to another and sometimes changing at different stages of the life-history or reproductive cycle. It can also be modified or even reversed by alterations in the condition of the animal or the environment; for example, by changes of temperature or salinity, the animal's need for food, its state of desiccation or its previous experiences. Nevertheless, it has been demonstrated that some shore creatures display patterns of movement which carry them into situations for which they are well suited and enable them to remain in appropriate zones despite their need to move about over the shore in search of food or for mating.

The movements of *Lasaea rubra*, for instance, are influenced by light, gravity, and contact (Morton, 1960; Morton *et al.*, 1957). This tiny bivalve is widely distributed throughout the littoral zone, extending to a high level and occurring mainly in the protection of crevices, empty barnacle shells and tufts of *Lichina*. It makes temporary attachment by means of byssus but is capable of moving freely over the surface by using its extensible foot to crawl on a mucus film. On a level surface *Lasaea* moves away from light, but on a sloping surface it climbs even against the light. Its response to lateral contact overrides the effects of both light and gravity, causing the animal to move into crevices and small holes. Laboratory experiments have shown that *Lasaea* will crawl into a narrow hole even downwards towards bright light.

The periwinkle *Littorina obtusata* is numerous under cover of middle-shore algae, where it usually matches the weed in colour and often also with respect to size of air bladders (Gill *et al.*, 1976). On level surfaces it usually moves away from light. *L. saxatilis* extends high in the littoral fringe, showing a strong tendency to move towards light and to climb. *L. littorea* is widely distributed across the middle shore in both sheltered and moderately exposed situations. The direction of wave action is one clue whereby this snail, if displaced, moves towards its original level (Gendron, 1977; Williams and Ellis, 1975). Its movements have been studied on the Whitstable mud flats (Newell, 1958a,b), and found there to be related to the direction of the sun. In this locality the feeding excursions occur mainly during the periods shortly after the winkles are uncovered by the receding tide, or submerged when the tide returns. The majority of winkles on the mud move at first towards the general direction of the sun, but later they reverse their direction. They therefore tend to retrace their course, their overall movement following a roughly U-shaped path which brings them back approximately to their starting-point. These animals, accustomed to a horizontal surface, show no geotaxic responses, but experiments with other specimens collected from the vertical faces of groins demonstrate responses to both light and gravity, these too moving over a U-shaped track, at first downwards and later upwards. Looped tracks have been reported for a number of other shore creatures, in some cases orientated to light, in others to gravity, and such behaviour has obvious advantages in enabling free-living animals to range about over the shore without moving far out of the levels in which they find favourable conditions.

The adults of *Melaraphe (Littorina) neritoides* occur high in the littoral fringe, but their eggs and larvae are planktonic, and settlement is mainly on the lower shore. The attainment of adult zonation is brought about by a combination of responses to light and gravity, modified by immersion (Fraenkel and Gunn, 1961). This winkle is negatively geotactic, and climbs upwards on rock surfaces. It is also negatively phototactic, and so will move into dark crevices. But the reaction to light reverses if the animal is immersed in water while it is upside down. If its crevice becomes submerged, *Littorina neritoides* therefore tends to crawl out along the ceiling towards the light, and then climb higher on the shore.

Vision

Some shore creatures have sufficiently well-developed vision to be able to see nearby objects and direct their movements by sight. An example is the sand-hopper *Talitrus saltator*, which burrows in upper-shore sand in the daytime and emerges at night during low water to feed on the surface. These feeding explorations carry it well down the shore, but it eventually finds its way back to high-water level. If removed from its burrow during daytime and released lower on the shore on a firm, unbroken sand surface, irrespective of slope of the beach or direction of sun or wind, it tends to move over the surface towards the back of the beach, where it burrows on reaching the drier, looser sand. If both eyes are covered, the movements of *Talitrus* released low on the shore are haphazard and show no tendency to carry it back upshore. Experiments both on the beach and in the laboratory suggest that *Talitrus* is capable of seeing shapes (form vision), that certain shapes attract it, and that its movements towards the top of the beach are probably associated with its ability to see the line of the backshore, even in dim, night-time illumination.

Some experiments with periwinkles on a rocky shore have demonstrated that they too direct their movements in relation to what they see of their surroundings (Evans, 1961). From an examination of the structure of the eye of *Littorina littorea* (Newell, 1965), it appears that this can form sharp images of distant objects in air, and can probably accommodate to bring near objects into focus.

Pardi (1960) and others have studied the movements of *Talitrus* from various parts of the coastline when placed above high-water level. They have demonstrated that the animals generally move in a direction which would carry them towards the sea in the localities from which they were taken, even when removed to areas remote from the sea. Their orientation is based on their sight of the sun or, in shade, on their perception of the polarization of light from the sky. Even in animals bred in captivity in uniform light, which have never had previous sight of the sun or contact with the shore, the orientation shows once they are exposed to sunlight.

8.8   Rocky shores

Rocky shores exist where the effect of waves on the coastline is mainly erosive, wearing down the softer materials and carrying them away, leaving the hardest rocks exposed. Most of the substrate is therefore stable and permanent, forming a secure surface upon which can grow a variety of organisms requiring attachment; for example, large algae, barnacles, mussels and limpets. The appearance of the shore depends largely upon the type of rock exposed. Horizontal strata often erode to a stepped series of fairly uniform level platforms which provide little shelter from the waves. Tilted strata running across the shore usually produce a very varied shore with numerous protruding rock ledges and overhangs, and deep pools in the gullies between them. Certain types of rock erode to a smooth surface, while some laminated rocks readily gape to form deep narrow fissures.

Many rocky shores are heavily populated. The agitation of the water keeps it well oxygenated, and favours plant growth by continually replenishing the supply of nutrients. The plants provide a primary food supply for animals, and copious additional food is available from the plankton.

Rocks present a variety of habitable environments – exposed rock faces, sheltered overhangs, crevices, deep or shallow pools, silt within fissures or under boulders, in the shelter of algae or in their ramifying holdfasts – each offers a domain which some species can occupy.

The size and composition of rocky shore communities are profoundly influenced by the intensity of wave action because this is one of the major factors determining the amount and type of algal growth on the rocks (Moyse and Nelson-Smith, 1963; Stephenson and Stephenson, 1972). Where wave intensity is moderate, large algae cover the shore and give shelter to many small animals which cannot tolerate complete exposure to air and sun, for example coelenterates, sponges, bryozoans and small crustaceans. Stronger waves prevent the growth of plants, and the rock surface then becomes covered mainly with barnacles and limpets, or sometimes at the lower levels by mussels. In extreme conditions of wave exposure, rock faces are swept virtually bare and the population is restricted to fissures and crevices. Because of the wetting effects of splash, heavy wave action tends to raise the levels to which sublittoral and littoral populations extend up the shore, and greatly increases the width and height of the littoral fringe (Figure 8.11). Where sand is deposited between rocks, their lower parts may be kept bare by the scouring effects of wave-tossed sand.

The intensity of wave action is a difficult parameter to evaluate. Attempts have been made to measure it with dynamometers (Jones and Demetropoulos, 1968), but an alternative approach has been to define degrees of wave exposure in terms of their biological effects. The most obvious contrast to be made is between shores where the rock surface is mainly encrusted with barnacles or mussels, i.e. *barnacle-dominated* and *mussel-dominated shores*, and those where the rocks are covered by a copious growth of seaweeds, i.e. *algal-dominated shores*. The former occur where the force of wave action is too fierce to allow the survival of large plants; the latter where the intensity of wave action is much gentler. By studying the differences of population between exposed headlands and sheltered inlets, numerical values for wave intensity can be assigned to particular patterns of population on the assumption that the differences are due mainly to wave effects, i.e. a biologically defined scale of exposure to wave action (Ballantine, 1961; Lewis, 1964).

8.8.1   Zonation patterns

The zonation of plants and animals is often clearly visible on rocky coastlines of the British Isles. An excellent summary of 'typical' patterns of zonation on sheltered, moderately exposed and exposed shores is given in Hawkins and Jones (1992). For a fuller picture, students should refer to Lewis (1964) where a full account is given of distribution patterns on many different shores throughout the British Isles. Other useful references include Ballantine (1961) and Moyse and Nelson-Smith (1963).

Zonation is a feature of shore populations everywhere. In the bibliography at the end of this chapter are a few references giving information on the zonation of intertidal communities in some other parts of the world (Dakin, 1987; Knox, 1960, 1963; Lawson, 1966; Morton and Miller, 1968; Ricketts and Calvin, 1962; Stephenson and Stephenson, 1972).

In the following pages, Figure 8.11 indicates in a general way the zonation of the more conspicuous fauna and flora on an algal-covered rocky shore in the south-west. Figure 8.12 indicates the distribution of certain species in various intensities of wave action. Some components of rocky shore food webs are shown in Figure 8.13.

Plant zonation
Where waves are not too violent to allow the growth of seaweeds, the rocks across much of the eulittoral zone of the shore have a brownish-green covering of large

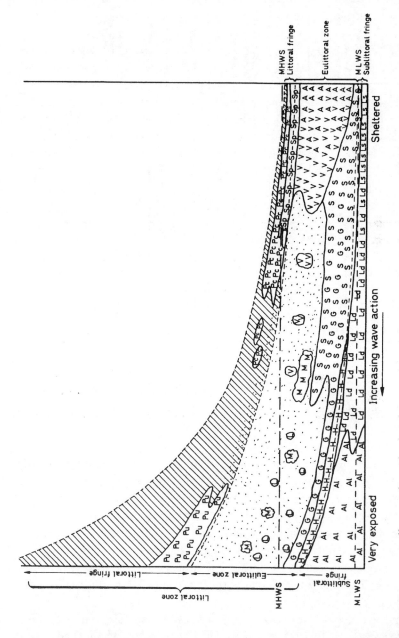

Figure 8.11 Diagram to illustrate how the distribution of some dominant plants and animals of rocky shores varies with wave intensity. ▨ Verrucaria with Littorina saxatilis and Melaraphe neritoides, ▣ Barnacle zone. A – Ascophyllum nodosum, Al – Alaria esculenta, G – Gigartina stallaris, H – Himanthalia elongata, L – Fucus vesiculosus f. linearis, Ld – Laminaria digitata, Ls – Laminaria saccharina, M – Mytilus edulis, Pc – Pelvetia caniculata, Pu – Porphyra umbilicalis, S – F. serratus, Sp – F. spiralis, V – F. vesiculosus.

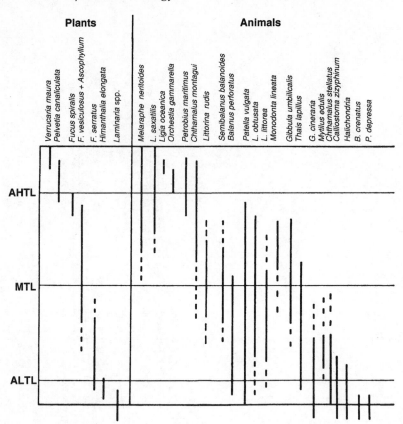

Figure 8.12   Zonation of some common plants and animals of rocky shores exposed to moderate wave intensities along south-west peninsulas of England and Wales.

fucoids, with bands of different colour at different levels indicating the dominance of particular species. Above the level of these large algae there is often a strip of relatively bare rock on which there are few plants except the black tufts of the lichen *Lichina pygmaea*. Higher still, in the upper part of the littoral fringe, there is a black, tar-like streak of blue-green algae and the encrusting lichen *Verrucaria maura*. Above this the terrestrial vegetation begins, sometimes as a light-coloured lichen zone tinged with yellow or orange by encrustations of vivid species such as *Xanthoria* and *Caloplaca*, and where there are spray-resistant angiosperms, notably sea thrift (*Armeria maritima*).

Examining the fucoids more closely, we find certain species restricted within fairly close levels. Usually at the highest level there is the channelled wrack, *Pelvetia canaliculata*, forming a narrow band in the lower part of the littoral fringe where it is wetted by salt spray but only occasionally submerged. Just below

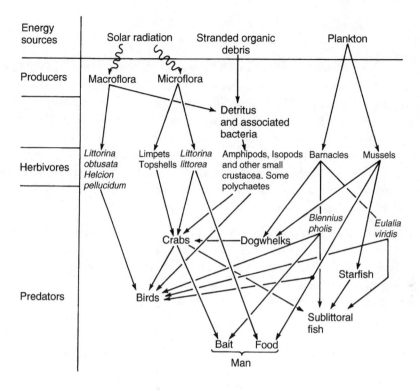

Figure 8.13 Simplified food web of a rocky shore.

Pelvetia canaliculata, in the lower limit of the littoral fringe, there is often a narrow zone of *Fucus spiralis*. The greater part of the eulittoral zone is more or less covered with a mixed growth of *Ascophyllum nodosum* and *Fucus vesiculosus*, the former predominating in sheltered areas and the latter in regions more open to waves. Overlapping with *A. nodosum* and *F. vesiculosus*, but extending below them into the lower part of the eulittoral zone, is a belt of toothed wrack, *Fucus serratus*. On many shores the lower limit of the eulittoral zone is indicated by a narrow band of the thong weed, *Himanthalia elongata*. Conspicuous plants in the sublittoral fringe, usually uncovered only at spring low-tides, are species of *Laminaria*, or, on the most wave-beaten rocks, *Alaria esculenta*. Since 1973 another conspicuous weed, *Sargassum muticum*, has been found along parts of the south coast (Boalch and Potts, 1977; Farnham *et al.*, 1973). Being native to Japan, it is commonly referred to as Japweed and was probably introduced with imported Japanese oysters, *Crassostrea gigas*. It was first seen at Bembridge, Isle of Wight, and was soon discovered on the Hampshire coast. It now extends along the English Channel coast from Cornwall to Kent.

Animal zonation

Barnacles

The zonation of littoral animals is well illustrated by some of the species that live on the surface, for example, the barnacles that cover the rocks where wave action is too strong for plants. The biology of British littoral barnacles is described in full in Rainbow (1984) and their identification in Hawkins and Jones (1992). The most diverse populations of shore barnacles in the British Isles occur on the south-west peninsulas of England and Wales where at least seven species are to be found, namely, *Chthamalus montagui*, *C. stellatus*, *Semibalanus (Balanus) balanoides*, *Balanus perforatus*, *B. crenatus*, *Verruca stroemia* and *Elminius modestus*. Their distribution can be summarized as follows:

- *Chthamalus montagui* reaches the highest level, sometimes extending above *Pelvetia* into the littoral fringe (Southward, 1976). It is a southern species ranging northwards along Atlantic shores from Morocco, the British Isles being the northernmost limit of its distribution. Here it is the dominant barnacle of the upper eulittoral zone in the south-west. It occurs in reduced numbers further north up the west coast to Shetland and eastwards along the English Channel as far as Dorset and the Isle of Wight. *C. stellatus* is found in the Mediterranean and on Atlantic shores between the tropics and the British Isles. In Britain it is found as far north as the west coast of Shetland (Rainbow, 1984) but is absent from the north Irish Sea. In most British localities it is limited to lower parts of the shore than *C. montagui*, mainly below mid-tide level. It is usually most abundant on the most heavily wave-beaten coasts whilst *C. montagui* favours embayed sheltered situations.

- *Semibalanus balanoides* is a boreal form flourishing in colder water than *Chthamalus* spp. It is found on all British coasts except the tip of Cornwall and is the dominant eulittoral barnacle on north and east coasts. It does not reach as high a level as *C. montagui*, and scarcely enters the littoral fringe. In the south and west, it occurs only in the mid- and low-eulittoral where it competes for space with *C. montagui*. Where conditions of moderate shelter or lower temperature favour *Semibalanus balanoides*, *Chthamalus montagui*, if present, tends to be restricted to a narrow fringe zone above the level of *S. balanoides*. Stronger wave action broadens the *Chthamalus* zone at the expense of *Semibalanus*, and favours *C. stellatus* more than *C. montagui*. On the Devon, Cornwall and Pembrokeshire coasts these species all compete for space in the lower eulittoral with the larger *Balanus perforatus*, which in some localities virtually ousts them. This south-western form extends from the lowest levels of the shore up to about the middle of the eulittoral zone.

- *Balanus crenatus* is essentially a sublittoral species, intolerant of exposure to air. It may be found in rock pools or under algal cover very low on the shore all round the British Isles. *Verruca stroemia* is a shallow-water species extending to the lower shore, where it is restricted mainly to pools or the underside of stones.

- *Elminius modestus* is an introduction to this country which appeared in Chichester harbour and the Thames estuary during World War II. It is an Australasian species and is thought to have been brought to Britain on ships' hulls. It has now spread around much of the British coastline (Crisp, 1958), and has crossed the seas to Ireland and the mainland of Europe, extending northwards from the French coast along the shores of Belgium, Holland and the North Sea coasts of Germany and Denmark. The northernmost population yet recorded is in the Shetland Isles. This is over 400 km north of the nearest known population, and it is likely that its spread to the Shetland Isles has again been effected by shipping. In sheltered areas it tends to oust the native population of *S. balanoides*, being able to survive at shore levels at least as high and as low as *Semibalanus* and having advantages in its rapid growth rate and extended breeding season through the summer months. It also has a wide tolerance of desiccation, silting, and variations in temperature and salinity.

Periwinkles

Periwinkles, *Littorina* spp., are very common on almost all rocky shorelines. Several species occur on the coast of the British Isles but there are still taxonomic problems with a number of them. Hawkins and Jones (1992) give basic and advanced identification tables for all species. Graham (1988) gives detailed descriptions and a key. A recent review of the ecology and taxonomy is given by Raffaelli (1982). The following is a summary of their distribution.

- Adults of the small periwinkle, *Melaraphe (Littorina) neritoides* live mainly in the upper littoral fringe, extending almost to the highest levels of wave splash. They are most numerous on shores exposed to heavy wave action and are often absent from sheltered regions. They are found mainly in crevices on hard rocks, and are consequently absent from most of the eastern part of the English Channel and southern part of the North Sea where suitable substrates are lacking.
- *L. obtusata*, *L. mariae* and *L. littorea* are eulittoral species found on all British coasts but more abundant in sheltered regions. *L. obtusata* and *L. mariae* are very similar species frequently referred to together as *L. littoralis*, the flat periwinkle. Yellow, black, brown and olive morphs are all common. *L. obtusata* eats the fucoid algae under which it lives. *L. mariae* feeds on epiphytes on the fucoids. They occur together on algal-dominated shores. On sheltered shores, it is reasonably certain that flat periwinkles in the *Ascophyllum* zone are *L. obtusata* and in the *F. serratus* zone are *L. mariae* (Hawkins and Jones, 1992). *L. littorea*, the common periwinkle, feeds by

scraping the microflora from the rock surface and on macro-algae. It is often found on bare rocks and is especially numerous in gulleys or on the sheltered faces of boulders.

- The *Littorina saxatilis* (= *L. rudis*) group, often called the rough periwinkle, includes several species with much confusion of nomenclature. Species falling within this group are *L. saxatilis*, *L. rudis*, *L. arcana*, *L. nigrolineata* and *L. neglecta*. Identification of the different species in the group is difficult in the field and requires some practice and experience. Students are advised first to learn to recognize the group. They can be distinguished from *Melaraphe neritoides* and *Littorina littorea* by looking carefully at the lip of the opening which meets the body whorl at right-angles. In the other two species, it meets tangentially. For more detailed studies, students should refer to the references given above and the papers referred to in them.

    The species now called *L. saxatilis* is the highest littorinid apart from *M. neritoides* and is mostly found in crevices at and above the *Pelvetia* zone. *L. arcana* extends down into the upper eulittoral and prefers more exposed situations. Most workers treat *L. rudis* as a variant of *L. saxatilis* though Smith (1981) regards it as a distinct species. *L. neglecta* is a very small eulittoral form that lives in small crevices, among mussel byssus and in *Laminaria* holdfasts, but is most reliably found in empty barnacle shells. It is most common on shores exposed to strong wave action. *L. nigrolineata* is a eulittoral species of rather local distribution on moderately exposed shores to very exposed shores. The shell is distinctively ridged, light coloured and often patterned with dark spiral lines within the grooves.

Reproductive differences between these littorinids are as follows. *Melaraphne neritoides* and *L. littorea* are oviparous with planktonic eggs and larvae; *L. obtusata* and *L. mariae* are oviparous and lay non-planktonic eggs in jelly-like egg-masses from which the young snails hatch directly; the *L. saxatilis* group are generally ovoviviparous bearing live young, with the exception of *L. arcana* and *L. nigrolineata* which lay egg masses as in *L. obtusata*.

Limpets

Three species of *Patella* occur on the British shoreline. The common limpet, *Patella vulgata*, extends throughout the eulittoral zone on virtually all rocky shores around the British Isles. In the south-west its distribution in the lower eulittoral overlaps with the black-footed limpet, *P. depressa* (= *P. intermedia*) on moderately exposed to exposed shores (Isle of Wight to Anglesey, absent in Ireland). *P. aspera* (= *P. ulyssopensis*) occurs on most suitable coasts around the British Isles but not between the Thames and the Humber on the east coast. It lives mainly below MLWN but on exposed shores it can be found in mid-shore rock pools. *P. depressa* and *P. aspersa* are rarely found on sheltered shores.

British rocky shore zones
Some of the common species comprising the communities of British rocky shores are listed below.

The littoral fringe
Verrucaria maura, *Lichina confinis*, *L. pygmaea*, Myxophyceae, *Pelvetia canaliculata*, *Fucus spiralis*, *Porphyra umbilicalis*, *Melaraphe* (*Littorina*) *neritoides*, *Littorina saxatilis*, *L. rudis*, *Ligia oceanica*, *Orchestia gammarella*, *Petrobius maritimus*.

Eulittoral zone
Ascophyllum nodosum, *Fucus vesiculosus*, *F. serratus*, *Himanthalia elongata*, *Osmundea pinnatifida*, *Corallina officinalis*, *Mastocarpus stellatus*, *Palmaria palmata*, *Semibalanus balanoides*, *Balanus perforatus* (SW), *Chthamalus montagui* and *C. stellatus* (W and SW), *Elminius modestus*, *Littorina neglecta*, *L. obtusata*, *L. littorea*, *Patella vulgata*, *Nucella lapillus*, *Gibbula cineraria*, *G. umbilicalis* (W and SW), *Monodonta lineata* (W and SW), *Actinia equina*, *Anemonia sulcata* (W and SW), *Mytilus edulis*, *Porcellana platycheles*, *Carcinus maenas*, *Blennius pholis*

Sublittoral fringe
Laminaria spp., *Alaria esculenta*, *Corallina officinalis*, *Chondrus crispus*, *Lithothamnia*, *Balanus crenatus*, *Verruca stroemia*, *Gibbula cineraria*, *Patella aspera*, *Helcion* (*Patina*) *pellucidum*, *Anemonia sulcata*, *Asterias rubens*, *Psammechinus miliaris*, *Galathea* spp., *Pisidia longicornis*, *Halichondria panicea*, *Alcyonidium* spp., *Archidoris pseudoargus*, *Acanthodoris pilosa*, *Aeolidia papillosa*.

Other animals on rocky shores
Inaccessible cliffs, rocky stacks and islands provide breeding grounds for innumerable seabirds (Lloyd *et al.*, 1991). The following are some common British species which nest mainly on rock ledges: the herring gull (*Larus argentatus*), the greater black-back gull (*L. marinus*), the kittiwake (*Rissa tridactyla*), the guillemot (*Uria aalge*), the razorbill (*Alca torda*), the shag (*Phalocrocorax aristotelis*), the cormorant (*P. carbo*) and the gannet (*Sula bassana*). The puffin (*Fratercula arctica*) is a bird of the open sea and forms breeding colonies on certain rocky islands, such as the Farnes and Shetlands, where it makes its nest in burrows in the turf.

At the beginning of the breeding season, the grey seal (*Halichoerus grypus*) comes ashore onto rock ledges and inaccessible beaches where the pups can be born out of reach of the sea. During the first two weeks of life the young avoid the water and may drown if they fall into the sea. The adult seals usually stay ashore for several weeks at this time.

8.9   Sandy shores

Detailed accounts of the ecology of sandy and muddy shores and their inhabitants can be found in Bassindale and Clark (1960), Brafield (1978), Brown and McLachlan (1990), Eltringham (1971), Evans and Hardy (1970), Hails and Carr (1975), Hayward (1994), Ranwell (1972) and Swedmark (1964).

The sizes of particles to which the name 'sand' is applied has been given as: coarse sand 2.0–0.5 mm; medium sand 0.5–0.25 mm; fine sand 0.25–0.062 mm (see Section 6.1.1).

Seashore sands contain particles of many types and sizes, often including silt and clay, deposited from many sources (see Section 8.2.1). The main constituent of sand on British coasts is silica fragments. Our yellow beaches often consist almost entirely of coarse siliceous sand. Grey, muddy beaches contain silica particles mixed with silt, clay and organic debris. Various other substances also contribute to sandy deposits; for example, fragments of shell, diatoms, calcareous algae, foraminifera, and, in low latitudes, coral. The beautiful white sand beaches on the west coasts of the Outer Hebrides in Scotland are formed mostly from shells and calcareous algae, and give rise to extensive lime-rich coastal grasslands, the machair.

8.9.1   Plant life

The surface of a sandy beach is liable to disturbance by wind and waves, and provides no firm anchorage for superficially attached plants and animals. There is consequently seldom much obvious surface-living flora and fauna. Where the sand contains embedded stones, certain algae can grow attached to these, for example, *Chorda filum*. In sheltered areas the deposit may be sufficiently stable to allow rooting plants such as eel-grass (*Zostera*) or cord grass (*Spartina*) to become established.

Usually the plant population consists only of a microscopic vegetation of diatoms and coloured flagellates existing in the interstices between the surface sand grains. Probably the majority are attached to the sand particles, brought to the surface or buried by movements of the sand by wind and waves. Others are motile and appear capable of rhythmical movements, forming green or brown patches on the shore surface at low tide and retreating below the surface as the tide advances. Mucilaginous secretions produced by microscopic algae are probably important in helping to stabilize fine sediments. Along the backshore, accumulations of wind-blown sand may form sand dunes, and when these are sufficiently stable they become consolidated by a characteristic dune vegetation, for example marram grass (*Ammophila arenaria*), sand couch grass (*Agropyron junceiforme*) and sand sedge (*Carex arenaria*).

8.9.2   Animal life

The animal population of sandy shores mainly dwells below the surface, but includes some species which at times emerge to crawl or swim. Along the upper parts of some sandy beaches is a zone where the sand dries out and air penetrates during intervals between periods of submergence. This is the littoral fringe of the sandy shore, which in the British Isles is often inhabited by the burrowing amphipod *Talitrus saltator*, occurring in great numbers where there is much organic debris deposited along the strand line. It is an air-breather, usually restricted within a narrow strip of upper shore except when it moves over the surface for feeding at night (see Section 8.7.5, Vision). The strand line and adjacent sand also house a diverse population of insects, for example *Bembidion laterale*, *Chersodromia arenaria* and larvae of *Coelopa frigida*.

Lower on the shore the superficial layers of sand may dry briefly during low tide, but capillary forces hold a water table some height above sea level depending on the size of spaces between the sand grains. In sand containing much fine material the deposit remains waterlogged throughout the tidal cycle, but in coarse sands the water table may drop considerably as the tide recedes. Even where there is air in the sand its humidity is high, and burrowing creatures are not in appreciable danger of desiccation. The deeper-dwelling forms are also well insulated from surface fluctuations of temperature and salinity, which seldom produce much effect below a depth of a few inches. However, many burrowing animals descend deeper in the sand during winter than in summer. Populations of sandy beaches also exhibit considerable seasonal changes of biomass and composition associated with breeding and growth periods (Harris, 1972).

Factors which vary with shore level, and influence the zonation of burrowing populations, include the duration of submergence which for many species determines the time available for feeding and respiratory exchange; also the particle size of the substrate (Longbottom, 1970) (see Section 6.2.1) and the range of hydrostatic pressure. A general indication of the zonation of some sand-dwelling species is given in Figure 8.14 and some interconnections of the food web are illustrated in Figure 8.15.

Coarse sand

Wave action exerts a major control over the distribution of sand populations because it influences, directly or indirectly, many important features of the substrate, including its stability, particle size, gradient, drainage, oxygenation and organic content. Beyond a limit, strong wave action produces a virtually barren shore because no sizable organisms can withstand the crushing effects of deep churning in a coarse shifting beach. In less wave-beaten conditions the sand is more stable, and survival becomes possible first for creatures that can burrow in fairly coarse material and do not require any permanent tube or burrow. These forms must be sufficiently robustly built to withstand pressures from sand movement, or must be able to avoid them by burrowing at a safe depth, or

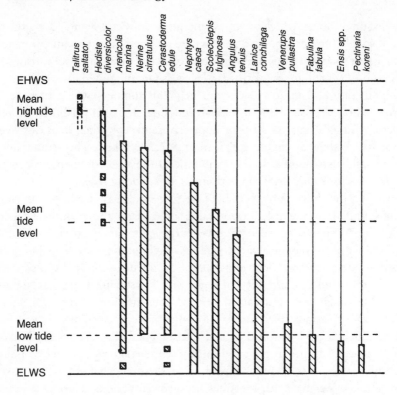

Figure 8.14   Zonation of some common inhabitants of intertidal deposits.

escaping by swimming. Coarse sand is also found in calmer conditions where the water is fairly fast-moving, for example the shores in tidal channels.

Among the inhabitants of the coarser sands of British shores are the molluscs *Donax vittatus*, *Fabulina fabula*, *Lucinoma borealis*, *Ensis siliqua* and *Lunatia (Natica) alderi*, the crustaceans *Portumnus latipes*, *Nototropis swammerdami*, *Eurydice* spp., *Bathyporeia* spp., *Urothoe* spp. and *Haustorius arenarius*, the echinoderms *Astropecten irregularis* and *Echinocardium cordatum*, and sand eels *Ammodytes* spp. Sandy shore pools often contain large numbers of shrimps (*Crangon vulgaris*) and small flatfish. Some sandy shores are important nursery areas for young plaice (*Pleuronectes platessa*), dabs (*Limanda limanda*) and soles (*Solea solea*), also sand gobies (*Pomatoschistus microps*) and whiting (*Merlangius merlangius*). The venomous weever fish (*Echiichthys vipera*), whose poison spines on its gillcovers and first dorsal fins can give the unwary an agonizing sting, burrows superficially in clean sand. It is an inhabitant of shallow water which sometimes occurs on the lower shore where there are abundant shrimps and small fish on which they feed.

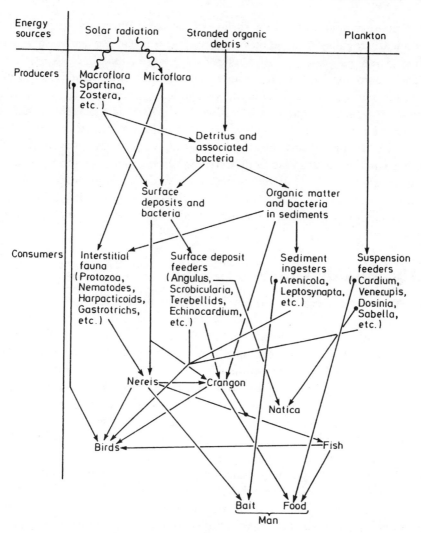

Figure 8.15 Simplified food web of a depositing shore.

Muddy sand

With increasing shelter, movements of sand by wave action extend less deeply until only a superficial layer is disturbed. Below this the substrate is quite stable. Sheltered conditions permit the settlement of smaller particles and also favour the deposition of organic material, the latter providing food for a more numerous and diverse population than occurs in coarser, less stable sand. Sulphide-darkening of the subsurface layers indicates the activity of anaerobic bacteria in a deoxygenated medium. The burrowing population is then dependent upon oxygen obtained from the overlying water by drawing a current through the burrow, or extruding the respiratory organs above the surface of the sand. Where the surface dries at low tide, there must be some means of storing sufficient oxygen to survive this period.

The macrofauna of muddy sands of British shores includes the worms *Arenicola marina*, *Lanice conchilega*, *Neoamphitrite figulus*, *A. edwardsi*, *Nerine foliosa*, *Notomastus latericeus*, *Scolecolepis fulginosa*, *Glycera convoluta*, *Perinereis cultrifera* and *Nephtys* spp., the molluscs *Cerastoderma* (*Cardium*) *edule*, *Venus* spp., *Venerupis decussata* and *V. pullastra*, the holothurian *Leptosynapta inhaerens*, the crab *Carcinus maenas* and the shrimp *Crangon crangon*. There is also a diverse microfauna and meiofauna of flattened and threadlike forms living in the interstices of the sand (Harris, 1972; Swedmark, 1964) including the smallest known species from most of the invertebrate phyla.

8.10   Muddy shores

The term 'mud' is loosely applied to deposits containing a high proportion of silt or clay particles (see Section 6.1.1). The fine particles settle only from still water, and shores of mud are therefore found where conditions are normally calm and without strong currents, for example within sheltered bays, at the landward end of deep inlets and in river mouths where conditions grade from marine to estuarine. The upper levels of a muddy shore often merge into salt marsh where cord grass (*Spartina*) and other halophytes become established. Except for drainage channels, these shores have little slope, and extensive areas of 'mud flat' may be exposed as the tide recedes. The mud surface seldom dries appreciably, and a layer of standing water may be left at low tide. In these conditions organic debris readily settles, and the organic content of the mud is correspondingly high, often about 5 per cent of the total dry weight. The deposit usually compacts into a soft, stable medium, easy for burrowing, in which permanent burrows and tubes remain undisturbed. Because there is little exchange of interstitial water the mud beneath the extreme surface layer becomes completely deoxygenated and its sulphide content is high. The infauna therefore encounter problems of deficient oxygenation and clogging of respiratory and feeding organs by fine particles. The high organic content of the substrate is a potentially rich source of food. In comparison with sands, the populations of intertidal muds are usually less diverse but often of much greater biomass.

The population of seashore mud is made up of forms which can readily tolerate silt, among which are the worms *Neanthes* (*Nereis*) *virens*, *Sabella pavonina* and *Capitella capitata*, the clam *Mya truncata*, the anemone *Cereus pedunculatus* and the crustaceans *Upogebia deltaura* and *Callianassa subterranea*. The populations of muddy sand merge gradually into those of finer mud, and the muddy shore population often includes the worms *Arenicola marina*, *Amphitrite* spp., and

Marphysa sanguinea and the bivalves *Cerastoderma edule, Mya arenaria* and *Venerupis* spp. The gastropods, *Nassarius reticulatus* and *Ocenebra erinacea*, both to be found on rocky shores, occur also in soft mud, ploughing their way through the superficial layers with their long siphons extruding above the surface. *Hydrobia ulvae, Hediste (Nereis) diversicolor* and *Macoma baltica* are essentially estuarine forms which are also widely distributed in intertidal mud.

Sand and mud shores have a special ecological importance as feeding grounds for a great variety of wading birds. Also certain remote areas of mud and sand flats, such as occur in the Wash, form breeding grounds for the common seal, *Phoca vitulina*. Unlike the grey seal (see Section 8.8.2), the pups of this species can take to water within a few hours of birth. A gravid female can land on an exposed mud bank to calve, and the young seal is able to swim off with its mother by the time the tide returns.

8.11 Estuaries

8.11.1 Salinity
At a river mouth the salinity gradient between fresh water and the sea fluctuates continuously with the state of the tide and varies with the amount of fresh water coming downstream. The salinity regime in many British estuaries can be described as 'partially mixed' and the salt content varies both vertically and horizontally. Dense salt water flowing upstream along the estuary bottom mixes with the lighter fresh water flowing downstream. In this way, an irregular and often indistinct, vertical salinity gradient forms between surface and bottom (see isohalines in Figure 8.16), and mixed water flows seaward at the surface.

Exactly where and how much mixing occurs will vary with, for instance, the state of the tide and the amount of freshwater flow. In estuaries such as the Severn and the Thames, which have very strong tidal flows, mixing may be almost complete so that there is little or no vertical stratification. At the other end of the scale, 'salt-wedge' estuaries have almost no vertical mixing and a marked halocline exists because a high river flow rate holds back the lesser flow of salt water. Usually the estuary bottom and lower shore experience wider variations of salinity than the higher shore levels.

Where the salinity is vertically stratified, the water is prone to develop deoxygenated layers (Figure 8.16), especially in summer when river flow is less. Water of intermediate density may ebb to and fro as the tide rises and falls without much mixing with well-oxygenated surface water or deeper salt water.

8.11.2 Chemical composition
Estuary water is not a simple dilution of seawater. Many subtle changes in composition are involved, varying with local conditions. The relative proportions

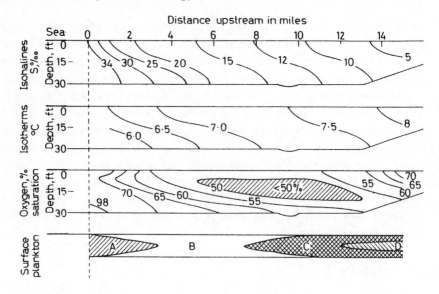

Figure 8.16   Section of Tyne Estuary showing isohalines, isotherms, percentage oxygen saturation and distribution of dominant planktonts at 2 m depth. Zone A – marine plankton containing Sagitta elegans, Nyctiphanes couchi, Calanus, Paracalanus, Pseudo-calanus, *nauplii of* Semibalanus balanoides, *polychaete larvae, etc. Zone B – very sparse plankton; water contains many fragments of organic debris. Zone C – abundant estuarine copepod,* Eurytemora hirundoides, *with elvers of* Anguilla. *Zone D – filamentous freshwater algae, mainly* Ulothrix.
(Data obtained during a one-day students' field trip 18 March 1965.)

of the various constituents in river water are always widely different from those of seawater. Some minor constituents of seawater, e.g. Fe, Si, PO_4, are often present in river water in much greater concentration than in the sea. Depending largely on the geology of the drainage area, some rivers are relatively rich in Mn, Cu or Zn. Rivers may also transport large amounts of solid matter, and both suspended particles and sediments may absorb or release solutes and colloidal micelles. Between river and sea there are sometimes irregular changes of pH due to the effects of varying salinity on the dissociation of bicarbonate. Consequently, estuary water differs from seawater not only in concentration but also in the relative quantities of many constituents.

8.11.3   Substrata

River water often carries large amounts of suspended silt which become deposited in the estuary where tidal and river currents slow down. Settlement of the fine particles carried down by rivers is enhanced in the estuary by salt flocculation. The influx of salt water results in an increase of electrolyte concentration, which in turn causes fine clay particles to stick together. These clumped particles settle out more quickly. The inflow of seawater along the floor of an estuary also tends to carry sediments into the estuary mouth, adding to the material deposited by the river. This continual trapping and settlement of sediments means that most estuaries are muddy with shores of fine unconsolidated mud. However, if the inflowing river drains mainly hard ground and if the offshore areas are also rock, sand or shell, a predominantly sandy estuary will result (e.g. many of those in north Devon). In narrow parts of an estuary, tidal ebb and flow of great volumes of water can produce powerful currents and heavy scouring of the bottom, particularly when the water carries much silt.

The prevalence of fine silts in estuarine deposits leads to severe deoxygenation of the mud, and silting conditions often change seasonally or even daily with resultant effects on the surface of the substrate.

The intensity of wave action, and the speed and direction of water movement, vary continuously in estuaries through each tidal cycle. At the seaward end of some estuaries, the interaction of waves and currents piles up the bottom sediments forming a zone of shallow water across the river mouth, termed a bar. This prevents large waves penetrating up-river, except at the seaward end at high water when there is sufficient depth over the bar to allow waves to cross without breaking. Throughout the rest of the tidal cycle the water is calm. In these circumstances only the higher levels of the shore receive appreciable wave action and may in consequence be rocky in places. Lower levels of the shore do not experience waves and are therefore likely to be covered with soft sediments, the grade varying at each level depending on the flow rate of the water. The fastest rates of tidal flow occur around the mid-tide periods. Therefore, only the estuary bottom and the shore below mid-tide level encounter the full scouring of fast flow. Consequently an estuary shore often displays considerable changes in substrate between upper and lower tidal levels. At the seaward end there may be rocks or stones at high-water level where wavebreak occurs; muddy sands or mudflats on the upper parts of the middle shore; coarse, unstable sands and often shifting sandbanks on the lower parts of the middle shore; and gravelly deposits or even exposed rocks on the estuary bottom. Depending on the geography of the estuary, tidal flow rates usually reduce upstream, and fine deposits correspondingly predominate until, towards the landward end, the substrate changes to reflect mainly the flow rate and silt burden of the river.

8.11.4 Water temperature

Areas of mud flat are also formed subtidally, beneath expanses of calm shallow water where there is little movement beyond the rise and fall of the tide. In these shallows, cooling and warming of the water produce wider changes of temperature than occur in river or sea. The lowest temperatures frequently coincide with the lowest salinities because in winter the outflow of fresh water may be greatly increased by rainfall or the rapid melting of snow. In summer, high temperatures and high salinities exist together when drought periods reduce the amount of fresh water at the same time as sunshine is heating and evaporating the water. Where industrial plant is sited near a river mouth, cooling installations may pour out quantities of warm water, so that the temperature of the estuary becomes continuously raised.

8.11.5   Pollution

Pollution by sewage or industrial effluents produces various effects which are discussed in Chapter 10. The carrying capacity of an estuary for wastes depends on the 'flushing time'. This is the time it takes an estuary to exchange most of its water with the adjacent sea. The flushing time can be worked out approximately by dividing the estuary volume by the net seaward flow rate.

8.11.6   Fauna and flora

Pelagic organisms in estuaries face difficulties of maintaining position in the ebb and flow of water. The net transport of water is downstream, tending to flush pelagic organisms out to sea. On the rising tide there is some danger that they might be carried up into lethally low salinities. Some pelagic animals or larvae adjust their behaviour to reduce these dangers. Some swim mainly on the flood tide and sink to the bottom on the ebb, thereby avoiding being washed too far seawards, e.g. oyster larvae. Some are retained in the estuary because they swim only within a particular range of salinities, being otherwise on the bottom. At the landward end some avoid being carried into fresh water by settling on the bottom if salinity falls below a safe level.

The population of estuaries therefore encounters peculiar and difficult conditions. Nevertheless, a restricted range of species can successfully cope with these problems, and population densities and biomass are often very high (Barnes, 1974; Barnes and Green, 1972; Green, 1968; Perkins, 1974). Shore species which occur both within estuaries and on the open coast, especially the herbivorous fauna, e.g. *Littorina littorea, Mytilus edulis* and *Cerastoderma (Cardium) edule,* often grow faster, mature more rapidly and sometimes spawn earlier in the year in estuaries than in more exposed sites. This is probably the result of a combination of shelter, abundance of food and sometimes raised temperature. The end-products of organic decomposition enrich the water with plant nutrients and the rate of primary production in estuaries is often very high. This derives from several sources. The phytoplankton is often rich; there may be an abundant growth of salt marshes; and the surface of mudflats, despite the absence of visible vegetation, is usually highly productive by the photosynthetic activity of micro-organisms (Joint, 1978).

At the seaward end there is always some penetration by marine species, and estuarine shores near the mouth are commonly inhabited by ordinary littoral forms, for example, *Semibalanus balanoides, Chthamalus montagui, Elminius modestus, Patella vulgata, Littorina littorea, L. rudis, Mytilus edulis, Cerastoderma (Cardium) edule, Nucella lapillus, Crangon crangon, Arenicola marina* and *Carcinus maenas*. The distance these extend upstream depends partly upon their powers of osmotic adjustment or osmoregulation, and partly upon the protection afforded by shells, tubes or deep burrows into which some species retire during periods when the salinity falls below a safe level. In mid-estuary, where the widest and most rapid fluctuations of salinity occur, only extremely euryhaline forms can survive. This typically estuarine community (Figure 8.17) often includes *Enteromorpha intestinalis, Fucus ceranoides, Corophium volutator, Hydrobia ulvae, Hediste (Nereis) diversicolor, Scrobicularia plana, Macoma baltica, Carcinus maenas, Sphaeroma rugicauda, Gammarus zaddachi, G. duebeni, Balanus improvisus* and *Pomatoschistus microps*. Certain mysids are numerous in brackish water, for example *Neomysis integer*, and in some estuaries there are often shoals of whitebait, i.e. small herrings and sprats. The plankton sometimes contains dense patches of the copepod *Eurytemora hirundoides* (Figure 8.16).

In the upper parts of estuaries the gymnoblast hydroid *Cordylophora caspia* occurs on stones or wooden piers. The prawn *Palaemonetes varians* is sometimes very numerous in salt-marsh pools. At these levels, part of the population is of freshwater or terrestrial origin. There are often many larvae of midges (*Chironomus*) and mosquitoes (*Aedes*), also oligochaetes (*Tubifex*) and a variety of beetles (*Colymbetes, Ochthebius*) and bugs (*Notonecta, Sigaria*).

The distribution of many estuarine species changes seasonally. The flounder *Platichthys flesus*, a common estuarine fish, migrates upstream in summer but returns to the sea during the colder months for spawning. *Carcinus maenas* and *Crangon vulgaris* also move up estuaries in summer and seawards in winter, their movements being influenced by changes of both temperature and salinity because these species osmoregulate less effectively as the temperature falls. Others make the reverse movements. *Pandulus montagui* goes out to sea in summer and into estuaries in winter.

River mouths where there are wide expanses of sand and mud are important feeding grounds for many species of birds, especially in winter. Among these are the heron (*Ardea cinerea*), the oystercatcher (*Haematopus ostralegus*), the curlew (*Numenius arquata*), the dunlin (*Calidris alpina*) and a variety of plovers, sandpipers, ducks and geese. However, estuaries are being increasingly subjected to industrial developments, barrage schemes and recreational developments. Over the past few years there has been an increasing awareness of the importance of estuaries to wildlife. In Britain, this is reflected in many new 'estuary initiatives' by statutory and voluntary conservation bodies (Davidson *et al.*, 1991).

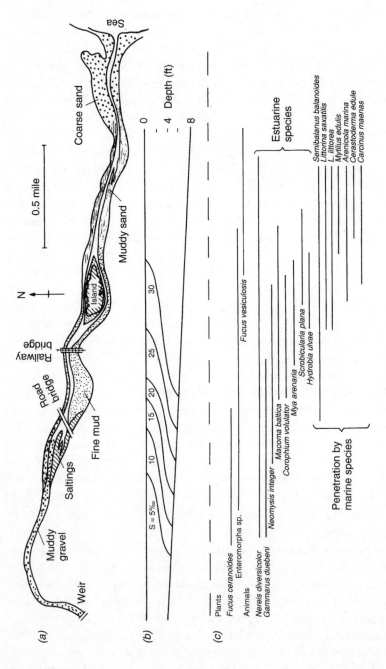

Figure 8.17 Distribution of the benthos of the Wansbeck Estuary, Northumberland. (a) Distribution of substrate. (b) Estimated positions of isohalines at high water. At low water the estuary drains and the shallow outflow is virtually fresh (S = 0.6–4.6‰). (c) Horizontal distribution of organisms.
(Data from a one-day students' field trip 1 March 1967.)

8.12   Tropical shores and coral reefs

8.12.1   Mangrove swamps

In tropical and subtropical areas a characteristic feature of the upper levels of estuaries and sheltered parts of the seashore is the mangrove swamp forest. This develops on mud flats which are exposed at low tide. The genus *Rhizophora* is a common mangrove tree which grows to large size, supported above the surface of the mud on a number of downcurving prop-roots resembling flying buttresses. These prop-roots contain air spaces which provide oxygen for the underground root system embedded in the waterlogged, oxygen-deficient mud. In some genera there are also aerial roots growing vertically out of the mud as slender, erect structures. There may also be 'pillar-roots' which support the branches. This elaborate rooting system reduces water movement and entraps and stabilizes the mud so that the mangrove forest tends to increase in extent, forming broad flat areas of swamp cut by drainage channels through which the sea flows with rise and fall of the tide. The spread and consolidation of the forest are assisted by a reproductive peculiarity of some mangrove species. These have seeds which germinate while still within the fruit borne on the tree, producing a long slender radicle growing downwards from the branch sometimes as much as 60 cm before dropping from the tree to stick into the mud. As the forest advances, the vegetation becomes zoned between land and sea with different species of mangrove at each level and different communities of smaller salt-marsh plants between the trees.

The mangrove swamp harbours a complicated community of animals. The roots of the trees provide a secure substrate for a variety of attached animals, especially barnacles, bivalves, serpulid worms and tunicates. Fish and free-living molluscs and crustaceans find shelter in the crannies between the roots. In the mud are large numbers of burrowing crabs, molluscs and fish, and the branches of the trees contain insects, lizards, snakes, birds and monkeys. Details of some shores of this type, common at low latitudes, can be found in references at the end of this chapter (Macnae, 1968; Sasekumar, 1974; Teas, 1983).

8.12.2   Coral reefs

Coral reefs have always fascinated man. Their richness and diversity have provided a livelihood for coastal communities for thousands of years. Now their spectacularly colourful beauty can be seen by millions through television and increasing numbers of tourists visit reefs to scuba dive, snorkel or view the reef from glass bottomed boats.

The major part of most coral reefs lies below the low tide mark since corals are killed by prolonged exposure to air. So the coral reef is not truly a seashore habitat but is included here because it is such an important ecosystem. Many coral reefs do have a type of shore in the form of extensive 'reef flats' (see below) which are exposed at low tide. Such flats usually remain at least partially covered by a few inches of water. Only the hardier corals can thrive in these areas because although they rarely dry out completely, the temperature of the shallow water can rise dramatically when the tide is out and the corals will also be subjected to intense UV radiation. Coral reef animals such as shrimps, crabs and small fish will be subjected to increased predation from birds such as reef herons. Many reef flats are now being unwittingly damaged by tourists walking over the exposed corals and hunting for souvenirs. The reef flats also provide a source of food and materials for local people, and in many communities the women are sent out onto the reefs to collect clams, crustaceans, fish, seaweed and sea cucumbers. Valuable species such as *Tridacna* clams have become very scarce as both their flesh and shells can be sold.

The following brief account of coral reefs as a habitat is intended as an introduction to the subject. The geological history, formation, structure and functioning of coral reefs and their natural history are described in a number of well-illustrated recent texts to which the student should refer for more information and which are given in the references at the end of this chapter. Coral reefs in the context of climate change and global warming are referred to in Section 10.2.

Distribution

Coral reefs are found in tropical and sub-tropical areas where the average minimum temperature is not less than 20°C and so are found mainly between the Tropics of Cancer and Capricorn where water temperatures remain steady all year round. They are not able to withstand any great fluctuations in temperature and recent episodes of coral bleaching have been associated with rises in water temperature (see Section 10.2.3). The approximate worldwide distribution of coral reefs is shown in Figure 8.18, from which it can be seen that no reefs are found up the west coast of Africa, although this coast lies within the tropics. The reason for this is the occurrence of the Benguela current (see Figure 1.6) which carries cold water northwards from Antarctic waters. Wherever cold currents intrude into the tropics, corals will not thrive. Likewise, where significant amounts of fresh water flow into the sea from major rivers such as the Amazon, growth of coral reefs is also severely limited. Reefs cannot develop without a suitable hard substratum and so are not found in extensive areas of sediment.

Formation and structure

The physical basis of a coral reef is laid down by the corals themselves in their hard calcium carbonate skeletons. The deep underlying structure consists mostly of old corals that have been broken down and re-consolidated into calcite rock by many microbial and chemical processes. The living coral forms a layer on top

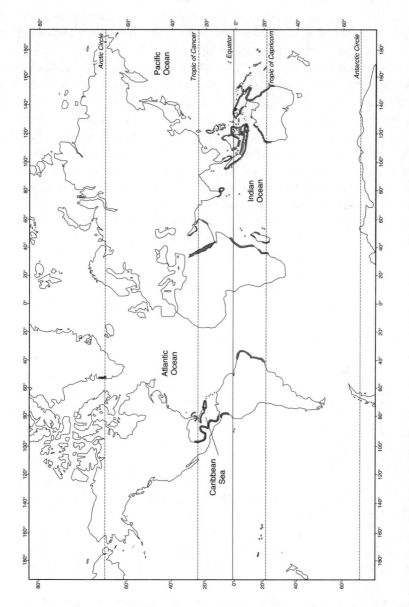

Figure 8.18   Worldwide distribution of coral reefs. Thick lines indicate where coral reefs are present.

of this. Massive amounts of material can be laid down over centuries and exploratory drilling through reefs in some areas has revealed coral rock deposits thousands of feet thick. Two boreholes drilled in the Enawetak Atoll in the Marshall Islands passed through about 1000 m (3280 ft) of coral rock before reaching volcanic rock. Living coral reefs need light (see below) and these deep deposits were formed in response to changes in land levels. Over geological time, the volcanic island with its surrounding reefs sank due to movements of the seafloor itself. As the island subsided, the corals maintained their upward growth in order to remain in the sunlit upper layers. When the land finally disappeared, a circular reef or atoll remained with a central lagoon inside. This theory of atoll formation was first put forward by Darwin and is still generally accepted.

Coral reefs are often classified according to the way in which they have been formed and where they are growing (Wood, 1983). Atolls have already been described. The shallow continental shelf provides a suitable base for the growth of many reefs known as *fringing reefs*. Fringing reefs grow close to the shore and as their name suggests, form a line along the coast or around an island or island group. *Barrier reefs* also form a line along the coast but further out on the continental edge between the waters of the continental shelf and the open sea. The Great Barrier Reef of Australia (Reader's Digest, 1984) is so large it can be seen from outer space and is known to have been in existence for at least 2 million years. *Platform reefs* and *bank reefs* occur away from land where irregularities in the continental shelf bring the sea-bed close enough to the surface for reef-building corals to grow. In the former, the reef is close to or breaks the surface whilst in the latter the reef top is up to 40 m below the surface.

Zonation

Within most coral reefs, a series of zones can be distinguished where bottom topography, depth (and therefore light), wave exposure and temperature vary. The patterns differ according to the type of reef and the location but the main divisions are illustrated in Figure 8.19. A typical fringing reef will have a shallow back reef or reef flat on the landward side, the seaward edge of which may be raised out of the water forming a reef crest. Beyond this, where the topography slopes downwards is the strongly growing fore reef. This is usually divided into the shallow fore reef (reef rim, reef front) and the deep fore reef (reef slope). The reef crest is exposed to the maximum wave action and here where the waves break, there is often extensive dead coral rock and rubble and sometimes extensive growths of calcareous algae (see below).

Corals living on the reef back are subjected to extreme conditions when the tide goes out. The corals may be exposed or if a shallow lagoon remains, then water temperatures can be very high. Consequently this area is often impoverished in terms of coral growth. However, areas of sand often support dense growths of

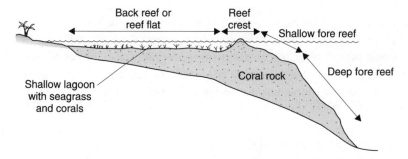

Figure 8.19   Main zones and coral habitats of a typical fringing reef.

seagrass and the back reef may be very rich in terms of invertebrate life. Molluscs, worms, crustaceans and echinoderms are often abundant. On the fore reef, coral growth is most vigorous below the depth at which wave pounding might cause damage but shallow enough for good light penetration and for a good supply of oxygen and nutrients brought in by water movements. At deeper depths, the species composition of the corals changes mostly in response to changing light intensity.

Apart from the corals themselves, many other organisms, especially red algae, help to build and consolidate the reef. A number of genera of red algae grow as heavily calcified crusts or nodules that help to bind the reef frame together. These algae are especially common on reefs exposed to considerable wave action, growing at the junction of the fore reef and back reef where the waves crash down. On some reefs their growth results in a broad algal ridge running parallel to the edge of the reef. Other material is added from the lime-containing skeletons of molluscs, echinoderms and crustaceans. Some reef fish, in particular parrotfish, eat the coral and excrete the hard material as sand. This can fill crevices and cracks and again adds to the strength of the reef.

Living coral

Reef-building stony corals are found only in well-lit, relatively shallow water (to about 50 m). Since corals are animals, this requirement for light puzzled biologists for many years. Microscopic studies of coral polyps finally revealed the answer. Coral polyps contain large numbers of microscopic dinoflagellate algae known as *zooxanthellae* in their tissues and it is the algae that require the light. Many details of this symbiotic relationship have still to be worked out but in simple terms the algae provide extra food for the corals, through photosynthesis, allowing them to build their massive skeletons, whilst the corals provide a safe haven for the algae and a supply of nutrients in the form of the corals' waste products. The corals also feed in a normal animal-like way by catching plankton with their tentacles. Many corals feed at night and retract their tentacles by day necessitating a night visit to see their true splendour. The vivid colours of corals are mostly imparted by their zooxanthellae and as the corals can expel their algal cells when under stress, the coral then appears white or bleached (see Section 10.2.3). Sometimes this results in the death of the coral but under the right conditions the zooxanthellae can be regained and the coral recovers.

Productivity and diversity

Apart from areas of back reef in shallow lagoons, algal growth on coral reefs is conspicuously absent and it requires careful searching to find the small clumps that are present. However, numerous algae are present and form a vital food source without which the reefs could not support the large numbers of fish that many do. A careful search of the reef around Sipadan Island in Sabah by one of the authors (Dipper), revealed at least 43 species on this relatively small reef (Wood, 1994). Part of the reason for the apparent lack of algae lies in the intensity with which some reefs may be grazed. Fish such as surgeonfish and parrotfish and also turtles eat the algal growths as rapidly as they grow. Thus although the seaweed is potentially abundant, it is never allowed to build up into large standing crops. Exclusion experiments have shown that on well-lit reefs as much as 1 to 5 kilograms of seaweed can grow per metre per year. At any one moment, however, there will be only a few grams present (Sheppard, 1983). Tough encrusting calcareous and coralline red algae and green algae with calcareous deposits such as *Halimeda* resist grazing and so may be more conspicuous.

Primary production in the reef ecosystem is carried out by benthic algae, suspended phytoplankton and especially by the zooxanthellae living in the coral tissues. It is therefore very difficult accurately to estimate primary productivity and few figures are available. However, those attempts that have been made suggest figures of around 1500 to 5000 $gC\,m^{-2}\,yr^{-1}$. This is very high but most of the production is recycled particularly with respect to the corals and their zooxanthellae.

The diversity and complexity of the coral reef ecosystem is astonishing and rivalled only by that of tropical rain forests. Estimates suggest that a single reef may contain up to 3000 different species with more if the meiofauna and microfauna are included. The most diverse reefs with the highest number of coral genera are located in the Indo-Pacific where at least 500 reef-building species have been identified. This is thought to be due to the long evolutionary history, with reef ages measured in millions of years, stable environmental conditions and wide variety of suitable habitats present in this region. Atlantic reefs, in contrast, are 10 000 to 15 000 years old and support only around 60 reef-building corals. The diversity of the coral faunas is reflected in the diversity of the fish and invertebrates associated with the reef. There are around 2000 fish species in Indo-Pacific reef areas compared to 500 or so in Atlantic reef areas. This is further illustrated by a recent coral reef survey in which one of the authors (Dipper) took part. The reef around the tiny island of Sipadan off the coast of Sabah in Borneo was found to support 386 species of fish in spite of the fact that the reef is only a few kilometres in circumference (Wood, 1994).

Coral reefs are immensely important in terms of productivity and in their role as fixers of atmospheric carbon dioxide. We utilize them for food, profit and pleasure. Like the tropical rain forests, their inhabitants are a potential and as yet largely untapped source of an endless variety of medicines and drugs. They are also a wonder and delight, a place most of us would like to visit at some time in our lives. Unfortunately coral reefs are increasingly coming under threat from man's activities (see Section 10.2.3) and considerable effort will be needed to prevent the tragic loss and degradation of large areas of coral reef.

References and further reading

Student texts

Barnes, R.S.K. (1974). *Estuarine Biology*. London, Edward Arnold.
Brafield, A.E. (1978). *Life in sandy shores*. London, Edward Arnold.
Clayton, K. (1979). *Coastal geomorphology*. Macmillan Education.
King, C.A.M. (1972). *Beaches and Coasts*. 2nd edition. London, Arnold.
Little, C. and Kitching, J.A. (1996). *The Biology of Rocky Shores*. Oxford, Oxford University Press.
Newell, R. (1970). *The Biology of Intertidal Animals*. London, Logos Press.

References

Alheit, J. and Naylor, E. (1976). Behavioural Basis of Intertidal Zonation of *Eurydice pulchra*. *J. Exp. Mar. Biol. Ecol.*, **23**, 135–44.
Arnold, D.C. (1972). Salinity tolerances of some common prosobranchs. *J. Mar. Biol. Ass. UK*, **52**, 475–86.
Ballantine, W.J. (1961). A Biologically Defined Exposure Scale, for the Comparative Description of Rocky Shores. *Fld. Stud.*, **I** (3), 1.
Barnes, R.S.K. (1974). *Estuarine Biology*. London, Edward Arnold.
Barnes, R.S.K. (ed.) (1977). *The Coastline*. Chichester, Wiley.
Barnes, R.S.K. and Green, J. (eds) (1972). *The Estuarine Environment*. Chichester, Wiley.
Barnett, B. E. (1979). Predation by *Nucella* on Barnacles. *J. Mar. Biol. Ass. UK*, **59**, 299–306.
Bassindale, R. and Clark, R.B. (1960). The Gann Flat, Dale: Studies on the Ecology of a Muddy Beach. *Fld. Stud.*, **I** (2), 1.
Bemert, G. and Ormond, R. (1981). *Red Sea Coral Reefs*. London, Kegan Paul International.
Boalch, G.T. and Potts, G.W. (1977). *Sargassum* in the Plymouth area. *J. Mar. Biol. Ass. UK*, **57**, 29–31.
Brafield, A.E. (1978). *Life in sandy shores*. London, Edward Arnold.
Brown, A.C. and McLachlan, A. (1990). *Ecology of sandy shores*. Amsterdam, Elsevier Science Publishers B.V.

Chapman, A.R.O. (1979). *Biology of Seaweeds*. London, Edward Arnold.

Connell, J.H. (1961). Effects of Competition, Predation by *Thais lapillus*, and other Factors on Natural Populations of the Barnacle, *Balanus balanoides*. *Ecol. Monogr.*, **31**, 61.

Coombs, V. (1973). Factors in Vertical Distribution of Dog Whelk. *J. Zool. Lond.*, **171**, 57–66.

Cornelius, P.F.S. (1972). Thermal Acclimation in some Intertidal Invertebrates. *J. Exp. Mar. Biol. Ecol.*, **9**, 43–53.

Courtney, W.A.M. (1972). Effect of Wind on Shore Gastropods. *J. Zool. Lond.*, **166**, 133–9.

Crisp, D.J. (1958). The Spread of *Eliminius modestus* in North-West Europe. *J. Mar. Biol. Ass. UK*, **37**, 483.

Dakin, W.J. (1987). W.J. Dakin's classic study *Australian Seashores*. A guide to the temperate shores for the beach lover, the naturalist, the shore-fisherman and the student. Fully revised and illustrated by Isobel Bennett. London, Angus and Robertson.

Davidson, N.C. *et al.* (1991). Nature conservation and estuaries in Great Britain. Peterborough, Nature Conservancy Council.

Davies, P.S. (1966–70). Physiological Ecology of *Patella. J. Mar. Biol. Ass. UK*, **46**, 647–58; **47**, 61–74; **49**, 291–304; **50**, 1069–77.

Day, J.W., Hall, C.A.S., Kemp, M.W. and Yanezarancibia, A. (1989). *Estuarine ecology*. New York, John Wiley and Sons.

Doodson, A.T. and Warburg, H.D. (1941). *Admiralty Manual of Tides*. London, HMSO (reprinted 1973 with minor amendments).

Eltringham, S.K. (1971). *Life in Mud and Sand*. London, English University Press.

Evans, F. (1961). Responses to Disturbance of the Periwinkle *Littorina punctata* on a Shore in Ghana. *Proc. Zool. Soc. Lond.*, **137**, 393–402.

Evans, S.M. and Hardy, J.M. (1970). *Seashore and Sand Dunes*. London, Heinemann.

Farnham, W.F., Fletcher, R.L. and Irvine, L.M. (1973). Attached *Sargassum* found in Britain. *Nature*, London, **243**, 231–2.

Fraenkel, G.S. and Gunn, D.L. (1961). *The Orientation of Animals*. Dover edition.

Gendron, R.P. (1977). Habitat Selection and Migratory Behaviour of *Littorina littorea. J. Anim. Ecol.*, **46**, 79–92.

Gibson, R.N. (1969). The Biology and Behaviour of Littoral Fish. *Oceanogr. Mar. Biol. Ann. Rev.*, **7**, 367.

Gill, J., Ramsay, R. and Smith, S. (1976). Ecology of Flat Periwinkles. *J. Biol. Educ.*, **10**, 237–41.

Graham, A. (1988). *Molluscs: Prosobranch and Pyramidellid gastropods*. Synopses of the British Fauna (New Series), No. 2 (2nd ed.). E.J. Brill/Dr W. Backhuys.

Gray, W. (1993). *Coral Reefs and Islands. A natural history of a threatened paradise*. David and Charles.

Green, J. (1968). *The Biology of Estuarine Animals*. London, Sidgwick and Jackson.

Hails, J. and Carr. A. (1975). *Nearshore Sediment Dynamics and Sedimentation*. Chichester, Wiley.

Hansom, J.D. (1988). *Coasts*. Cambridge University Press.

Harris, R.P. (1972). Seasonal Changes in Meiofauna of Intertidal Sand. *J. Mar. Biol. Ass. UK*, **52**, 389–403.

Hawkins, S.J. and Jones, H.D. (1992). *Rocky Shores. Marine Field Course Guide 1*. Marine Conservation Society. Immel Publishing.

Hayward, P.J. (1994). *Animals of sandy shores. Naturalists' Handbook 21*. The Richmond Publishing Co. Ltd.

Joint, I.R. (1978). Microbial Production of an Estuarine Mudflat. *Est. Coastal Mar. Sc.*, 7, 185–95.

Jones, O.A. and Endean, R. (eds) (1973–7). *Biology and Geology of Coral Reefs* (4 vols). London, Academic Press.

Jones, W.E. and Demetropoulos, A. (1968). Exposure to Wave Action; Measurements of an Important Ecological Parameter on Rocky Shores in Anglesey. *J. Exp. Mar. Biol. Ecol.*, 2, 46.

Kitching, J.A. (1977). Variability in *Nucella*. *J. Exp. Mar. Biol. Ecol.*, 26, 275–87.

Knox, G.A. (1960). Littoral Ecology and Biogeography of Southern Oceans. *Proc. Roy. Soc. B.*, 152, 577.

Knox, G.A. (1963). Biogeography and Intertidal Ecology of the Australasian Coasts. *Oceanogr. Mar. Biol. Ann. Rev.*, 1, 341.

Lawson, G.W. (1966). Littoral Ecology of West Africa. *Oceanogr. Mar. Biol. Ann. Rev.*, 4, 405.

Lewis, J.R. (1964). *The Ecology of Rocky Shores.* London, English University Press.

Lloyd, C., Tasker, M.L. and Partridge, K. (1991). *The status of seabirds in Britain and Ireland.* T&AD Poyser, London.

Longbottom, M.R. (1970). The Distribution of *Arenicola marina* with Reference to Effects of Particle Size and Organic Matter in Sediments. *J. Exp. Mar. Biol. Ecol.*, 5, 138.

MacGarvin, M. (1990). *The North Sea.* Greenpeace. The seas of Europe. Collins and Brown.

Macnae, W. (1968). A General Account of the Fauna and Flora of Mangrove Swamps and Forests in the Indo-West Pacific Region. *Adv. Mar. Biol.*, 6, 73.

Morton, J. and Miller, M. (1968). *The New Zealand Sea Shore.* London and Auckland, Collins.

Morton, J.E. (1960). Responses and Orientation of the Bivalve, *Lasaea rubra*. *J. Mar. Biol. Ass. UK*, 39, 5.

Morton, J.E., Boney, A.D. and Corner, E.D.S. (1957). The Adaptations of *Lasaea rubra*, a Small Intertidal Lamellibranch. *J. Mar. Biol. Ass. UK*, 6, 383.

Moyse, J. and Nelson-Smith, A. (1963). Zonation of Animals and Plants on Rocky Shores around Dale, Pembrokeshire. *Fld. Stud.*, I (5), 1.

Muir, R. (1993). *The coastlines of Britain.* Macmillan.

Naylor, E. (1958). Tidal and Diurnal Rhythms of Locomotory Activity in *Carcinus maenas. J. Exp. Biol.*, 35, 602.

Naylor, E. (1976). Rhythmic Behaviour and Reproduction in Marine Animals. In *Adaptation to Environment.* R.C. Newell, ed. London, Butterworths.

Newell, G.E. (1958a). The Behaviour of *Littorina littorea* under Natural Conditions and its Relation to Position on the Shore. *J. Mar. Biol. Ass. UK*, 37, 229.

Newell, G.E. (1958b). An Experimental Analysis of the Behaviour of *Littorina littorea* under Natural Conditions and in the Laboratory. *J. Mar. Biol. Ass. UK*, 37, 241.

Newell, G.E. (1965). The eye of *Littorina littorea. Proc. Zool. Soc. Lond.*, 144, 75.

Newell, R. (1962). Behavioural Aspects of the Ecology of *Peringia ulvae. Proc. Zool. Soc. Lond.*, 138, 49.

Palmer, J. D. (1973). Tidal Rhythms: The Clock Control of Rhythmic Physiology of Marine Organisms. *Biol. Rev.*, 48, 377–418.

Pardi, L. (1960). Innate Components in the Solar Orientation of Littoral Amphipods. *Cold Spring Harb. Symp. Quant. Biol.*, 25, 395.

Perkins, E.J. (1974). *Biology of Estuaries and Coastal Waters.* London, Academic.

Raffaelli, D.G. (1982). Recent ecological research on some European species of *Littorina.*

J. Moll. Stud., **48**, 342–54.

Rainbow, P.S. (1984). An introduction to the biology of British littoral barnacles. Reprinted from: *Field Studies*, **6**, 1–51.

Ranwell, D.S. (1972). *Ecology of Salt Marshes and Sand Dunes.* London, Chapman and Hall.

Reader's Digest (1984). *Reader's Digest Book of the Great Barrier Reef.*

Ricketts, E. F. and Calvin, J. (1962). Revised by J. W. Hedgpeth. *Between Pacific Tides.* Stanford, California, Stanford University Press.

Rodriguez, G. and Naylor, E. (1972). Behavioural Rhythms in Littoral Prawns. *J. Mar. Biol. Ass. UK*, **52**, 81–95.

Royal Yachting Association (1981). *Navigation. An RYA Manual.* London, RYA and David and Charles Publishers.

Sasekumar, A. (1974). Distribution of Macrofauna on a Malayan Mangrove Shore. *J. Anim. Ecol.*, **43**, 51–69.

Schonbeck, M. and Norton, T.A. (1978). Factors Controlling the Upper Limits of Fucoid Algae on the Shore. *J. Exp. Mar. Biol. Ecol.*, **31**, 303–13.

Schonbeck, M. and Norton, T.A. (1979). Effects of Brief Periodic Submergence on Intertidal Fucoid Algae. *Est. Coastal Mar. Sc.*, **8**, 205–11.

Sheppard, R.C. (1983). *A Natural History of the Coral Reef.* Blandford Press.

Smith, J.R. (1981). The Natural History and Taxonomy of Shell Variation in Periwinkles of the *Littorina 'saxatilis' Complex. J. Mar. Biol. Ass. UK*, **61**, 215–41.

Southward, A.J. (1958). Temperature Tolerances of some Intertidal Animals in Relation to Environmental Temperatures and Geographical distribution. *J. Mar. Biol. Ass. UK*, **37**, 49.

Southward, A.J. (1976). Taxonomy and distribution of *Chthamalus. J. Mar. Biol. Ass. UK*, **56**, 1007–28.

Steers, J.A. (1969). *The Sea Coast.* 4th edition. London, Collins/New Naturalist.

Stephenson, T.A. and Stephenson, A. (1972). *Life between Tidemarks on Rocky Shores.* San Francisco, Freeman.

Swedmark, B. (1964). The Interstitial Fauna of Marine Sand. *Biol. Rev.*, **39**, 1.

Teas, H.J. (ed.) (1983). *Biology and ecology of mangroves.* The Hague, Dr W. Junk Publishers.

Underwood, A.J. (1979). Ecology of Intertidal Gastropods. *Adv. Mar. Biol.*, **16**, 111–210.

Veron, J.E.N. (1993). *Corals of Australia and the Indo-Pacific.* Hawaii University Press.

Wells, G.P. (1949). The Behaviour of *Arenicola marina* in Sand. *J. Mar. Biol. Ass. UK*, **28**, 465.

Wells, S.M. (1988). *Coral Reefs of the World.* Vols. 1–3. IUCN, Cambridge, UK and Gland, Switzerland.

Williams, I.C. and Ellis, C. (1975). Movements of *Littorina littorea. J. Exp. Mar. Biol. Ecol.*, **17**, 47–58.

Wood, E.M. (1983). *Corals of the World.* Biology and field guide. T.F.H. Publications.

Wood, E.M. (ed.) (1994). Pulau Sipadan: Reef life and ecology. Unpublished report to the World Wide Fund for Nature.

9 Sea fisheries

9.1 Fishing methods

Even with the refinements of modern science and technology, fishing remains an essentially primitive method of obtaining food. In our exploitation of the fish stocks of the sea, we still behave mainly as nomadic hunters or trappers of natural populations of animals living in the wild state. There are broadly three ways of capturing fish: they may be scooped out of the water by means of a net bag such as a trawl or seine; they may be enticed to bite upon a baited hook attached to a fishing line; or they may be snared or entangled in some form of trap such as a drift net. In whaling the quarry is pursued and speared with a harpoon. The main gears and vessels used in commercial fishing worldwide are described in Sainsbury (1996).

Each method is modified in innumerable ways to suit local conditions, and the following brief account refers mainly to the fishing techniques used commercially in the north-east Atlantic area, though large commercial fisheries worldwide use much the same techniques. These fisheries comprise two major groups, the *demersal* and the *pelagic* fisheries. Demersal fishing takes place on the sea floor for species which live mainly close to the bottom, for example cod, haddock, saithe, hake, plaice and sole. The principal method of demersal fishing is by trawl. Danish seines and long lines are also used. Pelagic fisheries seek shoals of fish which, while they may roam throughout a considerable depth of water, are caught near the surface, for example herring, mackerel, pilchard and sprat. The chief pelagic fisheries make use of drift nets, ring nets and seines, and pelagic line fishing. The areas in which various species are sought are indicated in Figure 9.1.

9.1.1 Keeping the catch

Keeping the catch fresh once the fish have been caught has always been a problem and the length of time iced fish will remain in reasonable condition effectively limits the duration of fishing for distant-water trawlers if they rely solely on ice for preservation. The development in recent years of efficient and cost-effective refrigeration and storage systems on board larger fishing boats, has made much longer fishing trips possible. If fish are stored directly in the hold, then dry refrigeration or refrigerated brine spray can be used to lower the hold

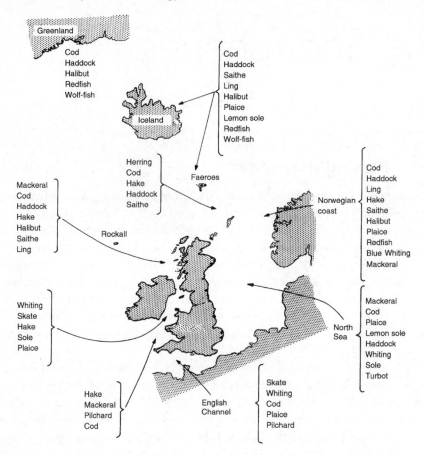

Figure 9.1   Important commercial species of fish sought in north-east Atlantic areas.

temperature. For very high quality, circulating refrigerated seawater tanks may be used.

The most notable advance in fish preservation has been the development of deep-freezing techniques. At temperatures around −30°C fish can be kept without serious deterioration for long periods, but for good results they must be frozen as soon as possible after capture, certainly within three days. This presents an obvious problem where the fishing grounds are several days' voyage from port unless it is possible to deep-freeze the fish at sea. To do this, some trawlers now carry refrigeration equipment so that some or all of their catch can be frozen immediately, and the duration of their voyage can thus be extended. It is not essential to freeze the entire catch because the later-caught fish can be brought to market in good condition on ice. When fish are frozen aboard, they are often preserved as whole fish which may be thawed later and dealt with through the normal wet-fish marketing channels. These fish are also suitable for smoke curing.

In recent years, various types of factory ship have been developed. Factory vessels can process their catch directly so that the packaged and frozen products are ready for distribution upon arrival back at port. High-value fish such as tuna for the raw fish market, can even be individually and rapidly frozen in low-temperature liquid tanks before storage. Factory ships may be enlarged trawlers, equipped both to catch and to process the fish, or larger ships equipped solely for processing and storing great quantities of fish, and operating in conjunction with a fleet of trawlers which bring their catches to it.

Preservation by freezing has advantages in addition to making possible longer fishing voyages in distant waters. Because frozen fish can be kept for long periods without deterioration, the market supply can be regulated independently of seasonal variations in the quantities of fish caught. Freezing has also opened the market to a wider range of species, because blocks of frozen filleted fish, from species not ordinarily bought by the public, can be successfully marketed.

9.1.2 Demersal fishing

The beam trawl

The beam trawl (Figure 9.2) is a tapering bag of netting which can be towed over the sea-bed. The mouth of the bag is held open by a long beam, the ends of which are supported above the sea-bed by a pair of strong metal runners. The upper leading edge of the bag is attached to a strong headrope lashed to the beam. The under part of the bag, which drags over the bottom, is attached to a considerably longer groundrope. As the trawl is towed, the ground rope trails behind the headrope so that fish disturbed on the bottom by the groundrope are already enclosed under the upper part of the net.

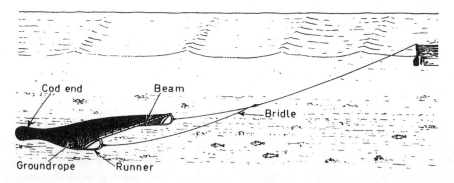

Figure 9.2 The beam trawl.

The hind part of the trawl net tapers to a narrow sleeve of stronger, finer-mesh netting known as the cod end, within which the captured fish accumulate. The rear of the cod end is tightly closed by a rope, the cod line. When the trawl is hauled up onto the vessel the catch can be released by untying the cod line, letting the fish out of the back of the bag.

Beam trawls have been in use for several hundred years, and were the chief method of demersal fishing in the days when fishing vessels were sail-driven. Although less used in modern commercial fisheries, beam trawls are still operated in Europe from small boats by inshore fisheries where the modern otter trawl is too large. They are used mainly for flatfish and in the north-east Pacific (Alaska) for shrimp. Being easy to handle from small craft, small beam trawls also have applications in biological work.

The size of a beam trawl is limited by the length of beam which it is practicable to handle. The largest nets are sometimes as much as 15 m wide. During the latter years of the nineteenth century, the introduction of steam trawlers capable of towing much heavier nets caused the beam trawl to become almost entirely superseded by the larger, more easily handled and more efficient otter trawl except for inshore use. Recently there has been an increase in the use of beam trawls made more effective by the addition of heavy 'tickler chains' in front of the groundrope. These chains plough up the bottom immediately ahead of the net so that fish buried in the sediment cannot escape by letting the net ride over them. Electrified ticklers are also used, which cause burrowing animals to jump out of the sand, increasing the likelihood that they will be enclosed in the net. These electrified chains, being lighter, do less damage to the bottom.

The otter trawl

This is now the chief method of demersal fishing. The otter trawl (Figure 9.3) has a bag of netting resembling that of the beam trawl in general shape, but considerably larger. The sides of the bag are extended outwards by the addition of wings of netting attached to large, rectangular, wooden 'otterboards'. These otterboards are towed by a pair of very strong steel cables, the warps, which are attached to the otterboards in such a way that the pressure of water causes the otterboards to diverge as they move, pulling the mouth of the net wide open horizontally. The under-edges of the otterboards slide over the sea-bed, and are shod with steel for protection.

The headrope, to which the upper lip of the trawl net is laced, is usually some 30–40 m in length, and bears numerous hollow metal floats which keep it a few metres above the bottom. Sometimes, elevator boards known as 'kites' are fitted to the headrope to increase the gape of the net. The lower lip and groundrope are considerably longer, about 40–60 m, and trail well behind the headrope during trawling. The groundrope is a heavy, steel-wire rope, carrying on its central part

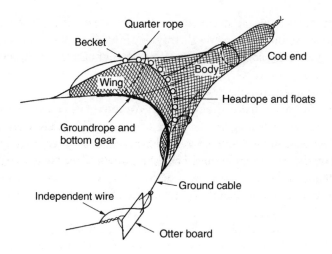

Figure 9.3 The otter trawl.

a number of large steel bobbins, about 60 cm in diameter, and on its lateral parts several large rubber discs. These help the trawl to ride over obstructions on the sea-bed. A bottom trawl cannot be used, however, on very rough ground or where wrecks may snag the net. Loss or damage is very expensive.

At the junction of the body of the trawl and the cod end, there is a vertical flap of netting, the flopper, which hangs down and acts as a valve to prevent the escape of fish from the cod end if the trawl should stop moving. The under part of the cod end is protected from chafing on the bottom, and a cod line is tied to close the free end. The trawl wings may be joined directly to the otterboards, but nowadays trawls of the Vigneron Dahl type are commonly used, in which long cables about 50–60 m in length are inserted between the wings and the otterboards. This simple modification gives the trawl a greater sweep, disturbing more fish and considerably increasing the catching power of the gear.

The deck of a trawler carries a powerful winch for winding the warps, and the net may be operated either from the side or the stern depending on the design of the vessel. The older method is side trawling. For shooting and hauling the trawl, the sides of the trawler carry pairs of strong gallows supporting the blocks over which the trawl warps run. On British vessels, starboard was the traditional working side for the trawl; but later, trawlers were fitted with gallows on both sides, and carried two trawls, working each side alternately. Stern trawling is a more recent development that is superseding side trawling. Built over the stern there is usually a gantry for operating the net and warps, and many stern trawlers have a slipway in the stern up which the net is hauled. This means the cod end does not have to be lifted over the stern or side. This technique allows greater mechanization, speeding up the hauling of the net and making the trawlerman's life a little less arduous. Although side trawling is likely to remain in use in various parts of the world for some years, few, if any, new boats are being built for side trawling.

When shooting the trawl, the ship moves ahead as the warps are paid out until the otterboards are submerged and the net overboard. The two warps must be of equal length to ensure correct opening of the net. When the trawl strikes the bottom, the drag of the net slows the vessel and the subsequent speed of trawling depends upon the power of the ship. The length of warps is adjusted to about three times the depth of the water, and on side trawlers they are braced in at the stern to keep them clear of the propeller. The strain of towing is taken on the winch brakes.

When the net is to be hauled, the trawler continues ahead as the warps are wound-in by the winch. The cod end usually comes up to the surface with a rush, due to the distension of the swimbladders of fish brought up quickly from the bottom. When the cod end comes alongside a side trawler, it is brought aboard by a hoist from the mast-head and suspended about 1 m above the deck. On a stern trawler the net is usually drawn up a slipway to the deck. The cod line is then untied and the fish discharged on to the deck.

The fish are then sorted and trash fish and bycatch are discarded. The fish are gutted and the heads removed from the larger species such as cod, saithe and haddock before being thoroughly washed by hosing. On factory ships the fish are processed directly. Otherwise they are stored on ice, refrigerated or frozen, before storage.

The Spanish trawl (pareja)

The Spanish trawl (Figure 9.4) is a very large net, similar in general principle to other bottom trawls, but towed by two vessels working together. This trawl has no need of otterboards, the mouth of the net being held open horizontally by the lateral pull of the warps, one of which is attached to each vessel. The technique of using two boats is known as pair trawling and is more commonly used in mid-water trawling.

The net has a wide sweep and great catching power, the headrope sometimes being as much as 100 m in length. The mesh is constructed of lighter material than the otter trawl, and generally produces a catch in rather better condition. In spite of its great size, this net can be towed by a pair of relatively low-powered vessels which individually could not undertake otter trawling. The Spanish trawl can be operated in deep water down to 600 m on the continental slope.

The Danish seine

The Danish seine (Figure 9.5) is a light-weight net for taking fish from the bottom, used in shallow water by small motor vessels. The net consists of a central bag

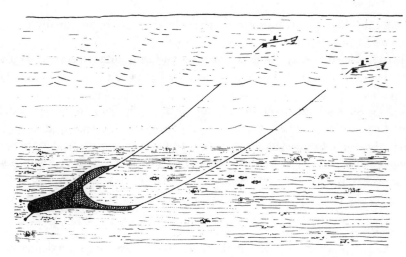

Figure 9.4 The Spanish trawl.

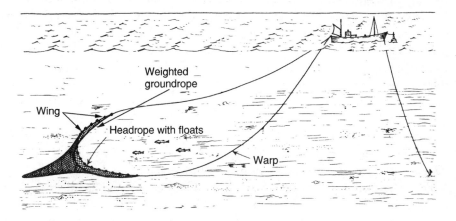

Figure 9.5 The Danish seine.

with lateral wings extending 25–40 m on either side. The groundrope is weighted, and the headrope buoyed by small floats.

There are several ways in which this net may be operated. Usually one end of a warp is attached to an anchored buoy, and the vessel steams downtide letting out a great length of warp, sometimes as much as two miles. The vessel now turns at right-angles and shoots the net across the tide. A second warp is attached to the net, and the vessel turns and steams back to the buoy running out the second warp. Both warps are then winched in, dragging the net over the sea bottom. The converging warps sweep the fish between them into the path of the net, gradually drawing the wings of the net together to enclose the catch, and finally the net is hauled up to the vessel.

Alternatively, the warps and net may be laid in a circular course, and then towed behind the vessel for some distance before hauling. This is termed fly-dragging.

The Danish seine carries no devices for clearing obstacles on the sea-bed, and is therefore only used where the bottom is fairly smooth. The condition of the catch is usually better than that obtained by trawl, and because many of the boats make only short voyages the catch is fresh. Fish may be brought up alive and can be carried live to port in seawater tanks.

Line fishing

Line fishing captures fish on baited hooks and can be used for both demersal and pelagic fishes. The latter is described in Section 9.1.3. Both horizontal and vertical lines can be used. The technique is often laborious because usually each hook has to be individually baited, and each captured fish removed from the hook. If the line becomes caught or entangled on the sea-bed, its recovery may take many hours; or the line and its catch may be lost. Line fishing is, however, widely practised throughout the world. The gear is simple and relatively inexpensive, and can be operated from any size of boat which need not be specially designed for this purpose. The capital costs can be kept low by conversion to 'lining' of various obsolescent craft. It is therefore well suited to the needs of communities where little capital is available for investment in fishing. In addition, automated mechanized systems are now available for use by larger boats. Line fishing is very adaptable to local circumstances and to the species sought. To some extent it can be made selective by choice of bait and size or shape of hook. Line-caught fish often fetch a high price because, being individually caught and handled, they can be brought to market in specially good condition.

For long-line fishing on the bottom the fishing gear comprises a very long length of strong line bearing at intervals numerous short lengths of lighter line, the snoods, which carry baited hooks (Figure 9.6). In laying the line, one end is anchored to the sea bottom and its position marked by a buoy. The liner steams along a straight course running out the full length of the line along the bottom, and the other end is marked by a second buoy. As the line is laid, the hooks are baited with pieces of fish or squid. A 10 m or so long vessel with one crew may fish 2500 hooks a day and double that with two crew. Mechanized long-lining systems may fish up to 40 000 hooks a day. Each line is left on the sea bottom for a few hours before hauling. As the line is brought aboard, the fish are removed from the hooks and the line carefully coiled for use again.

This technique does not generally compete directly with trawlers, because it can particularly exploit fishing grounds where trawls cannot be operated and it

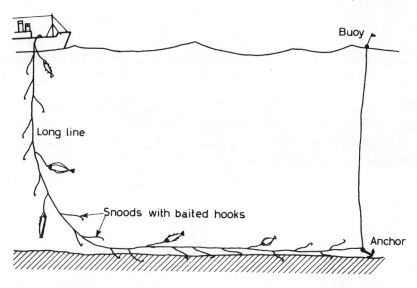

Figure 9.6   Long-line fishing.

concentrates on valuable species such as halibut, which are not caught in quantity by trawl. In continental shelf areas and offshore banks, species targeted include cod, haddock, whiting, dogfish, skates and especially halibut. Its use by British vessels is confined mainly to areas where the sea bottom is too deep or too rough to be suitable for trawling. Line fishing vessels, termed 'liners', operate in the northern part of the North Sea and down the continental slope to the north and west of the British Isles.

Various miniature versions of long-line fishing, for example haddock lines, are used in inshore waters around the British Isles, using shorter lines and small hooks suited to smaller fish.

9.1.3   Pelagic fishing

Drift nets (gill nets)

Drift nets can be used to catch fish which form shoals near the surface, for example herring, mackerel, pilchards and sometimes sprat. A drift net consists of a series of rectangular, light-weight nets joined end-to-end to form a very long vertical curtain of netting which hangs loosely in the water. The top edge of the curtain bears floats; the lower edge is weighted by a heavy rope, the messenger, by which the net is attached to the vessel. At the junctions between the individual pieces of netting are buoy-ropes, or strops, by which the net is suspended from a series of surface floats (Figure 9.7). The size and mesh of net, and the depth at which it hangs in the sea, are chosen to suit the particular fish sought.

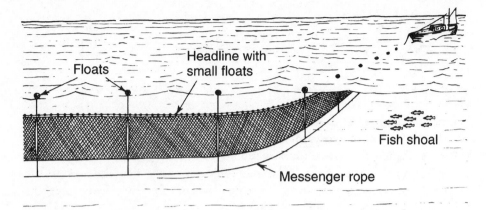

Figure 9.7   Drift nets.

For most pelagic species, drift nets are usually shot shortly before darkness. The fishing vessel, known as a drifter, moves slowly down wind while the net is paid out until the full length has been put into the water. The messenger rope from the foot of the net is brought to the bow of the drifter, which swings bow to wind and simply drifts attached to its long curtain of netting. During darkness, shoals of fish ascend into the surface layers, and become entangled as they attempt to swim through the net. The mesh size is selected so that the head of the fish passes easily through the net but the larger middle part of the body will not go through. When the fish try to wriggle out backwards, the net catches behind the gillcovers so that the fish are unable to escape. This method of capture is fairly selective, retaining only fish within a particular range of sizes. Smaller fish can swim right through, while larger fish may not pass sufficiently far into the mesh for the net to slip behind the gillcovers.

The net is hauled a few hours before dawn. The strain of hauling is taken on the strong messenger rope, which is wound in by a capstan or winch on the deck of the drifter. The net is carefully drawn in by hand over special rollers fitted on the side of the drifter. As the net comes aboard, the fish are shaken out and fall through hatches into the hold.

Drift nets have largely been superseded by purse seines, ring nets and pelagic trawls for the capture of herring and most other shoaling fish. In the North Sea the use of herring drift nets had ceased by 1970. They are still used for certain species which follow particular migration routes at particular times, notably for tuna. Salmon are also caught in this way, both on their feeding grounds at sea and in inshore waters as they approach the rivers for spawning. Modern drift nets are mostly nylon and monofilament which means they are very light and relatively simple to handle even though they may reach lengths of several kilometres. The maximum permitted length in the EC is 2.5 km.

There is now considerable concern over the use of monofilament drift nets in the open ocean. This is discussed further in Section 10.4.2. They are effectively invisible to animals using sonar and are responsible for the deaths of many dolphins and porpoises each year. In the eastern tropical Pacific, yellowfin tuna live alongside various dolphin species. Fishermen use spotter planes to locate the schools of dolphin on the surface, and, knowing the tuna will be below, they set their nets in the area. This leads to many dolphin deaths. Turtles and diving birds can also become ensnared earning drift nets the nickname 'walls of death'. A further problem is the virtual indestructibility of nylon nets. Lost nets may carry on 'ghost fishing'. The nets ensnare and entangle marine life, and sink to the bottom where the catch rots away or is eaten. The net then rises up and starts to fish again.

Ring nets and purse seines

A ring net is a curtain of fine-mesh net hung vertically in the water to encircle surface shoals of fish. Around the British Isles it has been used mainly for capturing herring, although the technique is applicable to many other species which form shoals near the surface. It is specially suitable for inshore and enclosed waters such as sea lochs and estuaries where the enormous lengths of net used by drifters are unmanageable.

The size and mesh of ring nets are selected to suit local conditions and the species sought. Herring ring nets are up to 200 m in length and 30–40 m deep. The top of the net is buoyed by corks and floats at the surface. The lower edge of the net is attached to a weighted sole rope (Figure 9.8). A fine-mesh net is used because the fish are not entangled, as in a drift net, but are trapped by being surrounded by netting.

Ring nets are often laid by a pair of vessels operating together. When a shoal

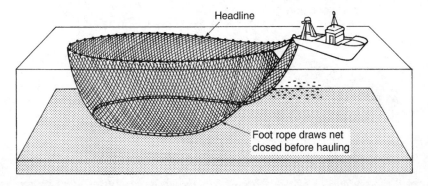

Figure 9.8 A purse seine net.

of fish has been detected, nowadays often by echo-location, one end of the net is secured to one vessel while the other steers a semicircular course paying out the net. The two vessels then steam on parallel courses, towing the net, and finally turn to meet and enclose the shoal. A pair of messenger ropes attached to the lower edge of the net are winched in, closing the bottom of the net below the shoal so that the fish cannot escape by swimming underneath. The net is gradually hauled on board until the captured fish are densely crowded within the central part of the net. They are then scooped out of the net and brought aboard by a dip-net known as a brailer, or by a suction hose.

Purse seines are similar in principle to ring nets but much larger, sometimes as much as 400 m in length and 90 m in depth. In the sheltered conditions of the Norwegian fiords they have been used for capturing herring. Mackerel shoals are caught by purse seines in the English Channel and North Sea.

In warm seas purse seines are extensively used to encircle shoals of tuna. As with the drift nets, dolphins are often incidentally caught and eventually drown in the nets. New tuna nets have now been developed to help avoid this problem. The nets have a fine-mesh area, the 'Medina panel', which is visible to dolphins. So as the net tightens, the dolphins can jump over the net rim and may even be given a helping hand by the fishermen.

Pelagic trawling (mid-water trawls)

Mid-water trawls are used for capturing pelagic fish shoals and also for demersal species during periods when they leave the sea-bed. A pelagic trawl is usually a conical net kept open by floats on the headrope, a weighted footrope, and by

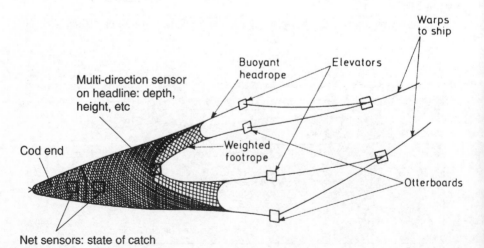

Figure 9.9 A pelagic trawl. Modern systems may have acoustic sensors on the net indicating the position of the net and the state of the catch.

various otterboards, elevators and depressors attached to the mouth (Figure 9. 9). The depth of trawl can be regulated by the length of the warps and the speed of the vessel, but its control requires considerable skill. Modern nets may be under computer control which may greatly increase their efficiency. Sensors can show height above the bottom, depth below the surface, vertical mouth opening and data on fish entering the net. Accurate location of the fish shoals is obviously essential. Sonic techniques are used to enable the position of the net and its relation to the fish shoals to be accurately known and this too can be fed into the computer system. Hybrid semi-pelagic nets which can be used either on the bottom or in mid-water have also been developed.

Pelagic long lines
For capturing pelagic species, long lines can be floated a variable distance below the surface suspended at intervals from a series of surface buoys. The long line lies more or less horizontally in a series of sagging connections between each buoy, with numerous short hooked lines attached and hanging below (Figure 9.10). This method is often more effective than nets over deep water for catching certain large species which do not form dense surface shoals. It is widely used in all oceans by Japanese and Korean fishermen for catching tuna, especially in areas where the thermocline lies deep. A single ship may operate over 100 km of line.

Pole lining
Large pelagic fish such as tuna and sharks are also caught on single hooked lines dangled or trolled through the water from strong poles which are either manhandled or may be hinged to the side of the boat and mechanically raised and lowered. The hooks are baited either with artificial lures or more often with live fish. Sardines or anchovies are favoured bait, and numbers of these are first pitched overboard to attract groups of tuna to the fishing boat. The method is used by Breton and Basque fishermen for tuna in the Bay of Biscay, and by the Japanese in parts of the Pacific where the thermocline is shallow. In the western English Channel, Cornish fishermen use hand lines carrying numerous feathered hooks for catching mackerel.

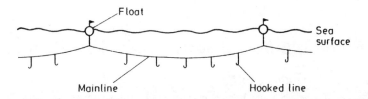

Figure 9.10 Pelagic long line for tuna. Length of mainline between floats is usually 300 m.

9.1.4 Other fishing methods

The fish pump

This method of fishing attempts to capture fish by sucking them out of the sea through a wide hosepipe in a powerful stream of water pumped aboard the fishing vessel. This is not as simple as it may appear. The disturbance of the water caused by pumping frightens the fish away, and only those quite close to the orifice of the hosepipe are likely to be drawn in. The fish must therefore be attracted in some way towards the hose so as to concentrate them near the opening before the pump is switched on. The method has been applied commercially in the Caspian Sea where there are species of *Clupeonella* which can be attracted close to the pump by light, but many fish will not come near to a powerful lamp. Another possibility is to attract fish to the pump electrically (see below).

Electric fishing

When direct electric current is passed through water, a fish within the electrical field will turn and swim towards the anode. This is known as the *anodic effect*. The intensity of this effect depends upon the potential gradient to which the fish is exposed, and large fish are therefore influenced more than small ones. If the strength of the electric field is progressively increased, the fish eventually becomes paralysed and finally electrocuted.

These effects have been applied in various ways to the capture or enclosure of fish in fresh waters. In the sea, it is difficult to maintain electric fields of sufficient strength because of the high conductivity of the water. None the less, various experiments in marine electrical fishing have been conducted, and some commercial applications may ensue. For example, the anodic effect might be used to attract fish towards nets or fish pumps. As they approach the anode, the fish are likely to become paralysed by the increasing field strength and consequently unable to avoid capture.

Electrocution has been applied commercially to tuna fishing to cut short the struggles of fish caught by electrified hand lines, and electric harpoons have been used in whaling. Electrodes attached to trawls increase the catch of animals that burrow beneath the surface (shrimps, Norway lobsters, plaice, soles, etc.), pulses of current stimulating reflex muscle contractions which cause them to leap out of the sand or mud and become caught in the net.

9.2 The biology of some european food fishes

9.2.1 Cod (*Gadus morhua*)

No other single species has been of such importance as cod for human consumption. Many millions are still taken each year although there is now growing concern over the state of many stocks. For example the Canadian Grand

Banks cod fishery in the NW Atlantic finally collapsed by 1992. Cod are mainly caught by trawl, and some are also captured by long line.

Distribution

The cod (Figure 9.11) has an extensive range over the continental shelf and slope to a depth of about 600 m throughout the Arctic and the northern part of the north Atlantic. Although the isotherms do not set firm limits to its distribution, cod is most abundant in seas within the temperature range 0–10°C. It is found around Greenland (mainly on the west coast) and Iceland, in the Barents Sea and around Nova Zemlya. It occurs around the Faroes, along the Norwegian coast, in the North Sea and Baltic, the Irish Sea, and the English Channel. On the eastern side of the Altantic, the Bay of Biscay is as far south as cod extend in any numbers. It is also found along the coasts of Labrador, Newfoundland and south along the North American coast as far as Virginia. In the northern part of the Pacific, a closely related form, *G. macrocephalus*, occurs over a wide area.

Although tagging experiments (see Section 9.4.2) have revealed extensive migrations by individual cod throughout the North Atlantic and into the North Sea and Barents Sea, there does not appear to be any large-scale movement of populations between different areas apart from the tendency to congregate for spawning. Results from tagging experiments in the North Sea showed a maximum distance travelled from the release point of about 200 miles. Occasionally much longer migrations have been reported. For example, a tagging experiment in the central North Sea resulted in two fish being recaptured off the Faroe Islands and one from Newfoundland (Macer and Easey, 1988).

The cod population consequently comprises a number of fairly distinct stocks; notably those of the Arcto-Norwegian (Cushing, 1966), North Sea, Faroe,

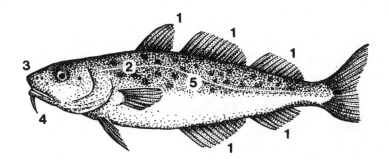

Figure 9.11   The cod, Gadus morhua. *Identification features: 1 = Three dorsal and two anal fins; 2 = Light coloured lateral line; 3 = Overhanging upper jaw; 4 = Long chin barbel; 5 = mottled colour.*
(Modified from *British Sea Fishes*, F. Dipper (1987), by kind permission of Robert Irving and Underwater World Publications Ltd.)

Iceland, East Greenland, West Greenland, Newfoundland and Labrador regions. There appears to be little interchange of stocks by movement of cod across intervening areas.

Life history

During winter, mature cod move towards particular areas for spawning in late winter and spring. In the northern part of their range there is a general tendency to move southwards in winter prior to spawning, returning northwards for feeding in summer.

The following information relates chiefly to cod in the North Sea (Macer and Easey, 1988). Spawning occurs between January and April, the peak spawning period in the North Sea usually being March to April in water temperatures of 4–6°C. The fish collect in shoals close to the sea-bed, and the female sheds between 3 and 7 million pelagic eggs, diameter about 1.4 mm. The male cod sheds its milt (sperm) into the water and fertilization is external. The fertilized eggs are buoyant and gradually float to the surface.

The major spawning areas around the British Isles are shown in Figure 9.12. There are probably many other subsidiary spawning areas. Further afield, cod are known to spawn around Iceland, Faroes and the Lofotens, also on the west coast of Greenland, on the Newfoundland Banks and along the Atlantic coast of North America.

After spawning the shoals disperse. Time to hatching depends on the water temperature but is typically 2 to 3 weeks. At the time of hatching the larva is about 4 mm long, the mouth has not yet formed and the animal is at first entirely dependent for food on the ventrally attached yolk-sac beneath which it floats upside down. About a week later, the yolk-sac has become completely resorbed, and the mouth has perforated and the young fish begin to feed for themselves in the surface waters. At this early stage the nauplius larvae of copepods are a major part of their food.

The planktonic phase lasts for about ten weeks, by the end of which time the young cod has grown to about 2 cm in length and increased in weight about forty times. Throughout this period copepods remain the chief food. In the North Sea, *Calanus*, *Paracalanus*, *Pseudocalanus* and *Temora* are important foods for the cod fry, which in their turn are preyed upon by carnivorous zooplankton, particularly ctenophores and chaetognaths.

At the end of the planktonic phase the cod fry disappear from the surface layers and go down to the sea-bed. In the North Sea there are nursery areas to the south-east of the Dogger Bank and around the Fisher Banks. Maps showing the distribution of cod at various stages in their life history in the North Sea are given in Macer and Easey (1988). The fish are not easy to find at this stage because they are quite small and occur mainly in areas where the sea bottom is rocky, making

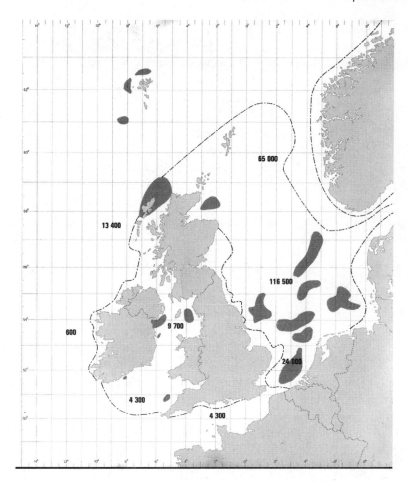

Figure 9.12   Main areas of cod spawning around British Isles as identified from egg surveys. Numbers in sea areas refer to the average international catch in tonnes during the period 1973–77.
(From Lee, A.J. and Ramster, J.W. (eds). (1981). *Atlas of the Seas around the British Isles.* MAFF, London. © Crown Copyright, 1981.)

it difficult to operate nets. Young fish sometimes occur in rock pools on the shore. At this stage they are often an orange-brown with a distinctly chequered pattern in contrast to the more usual adult sandy brown.

The change from planktonic to demersal life involves a change of diet. The young demersal cod feed at first on small benthic crustacea such as amphipods, isopods and small crabs. As the fish increase in size, they take larger and faster-moving prey. Shoals of adult cod actively hunt, chasing pelagic prey through the middle depths. They feed mainly on other fish such as sand eels, whiting and

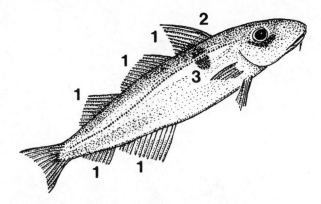

Figure 9.13   The haddock, Melanogrammus aeglefinus. *Identification features: 1 = three dorsal and two anal fins; 2 = Pointed triangular first dorsal fin; 3 = Black thumbprint mark on sides.*
(Modified from *British Sea Fishes*, F. Dipper (1987), by kind permission of Robert Irving and Underwater World Publications Ltd.)

haddock, and also squid. A variety of benthic annelids, crustacea and molluscs are also included in the diet, when the fish are feeding on the bottom.

The growth rate of cod varies in different areas. In the North Sea they have reached approximately 8 cm in length at the end of their first six months, 14–18 cm by the end of the first year, and 25–35 cm by the end of the second year. Further north the growth rate is slower. Off the Norwegian coast, cod attain only about 8 cm during their first year, reaching 30–35 cm by the end of the third year. The fish begin to be taken in trawl nets once they exceed about 25 cm in length.

In the North Sea, cod reach maturity when about 50 cm long at 3–4 years of age. Given the chance, they grow to considerable size, sometimes reaching about 1.5 m in length and weighing 30 kg or more.

9.2.2   Haddock (*Melanogrammus aeglefinus*)

Distribution
The distribution of haddock is similar to that of cod, but with slightly narrower temperature limits, not extending quite so far as cod to the north or south. It is most abundant in the northern part of the North Sea, but is also widely distributed around Britain and Ireland including the Orkneys, Shetlands and Rockall. It also occurs around the Faroes, Iceland, the west coast of Greenland, around Newfoundland and on the Atlantic coast of North America.

Life history
The spawning period is February to June, the peak spawning in the North Sea being between March and April. The female haddock deposits up to three million eggs very like those of cod. The spawning shoals usually congregate in rather

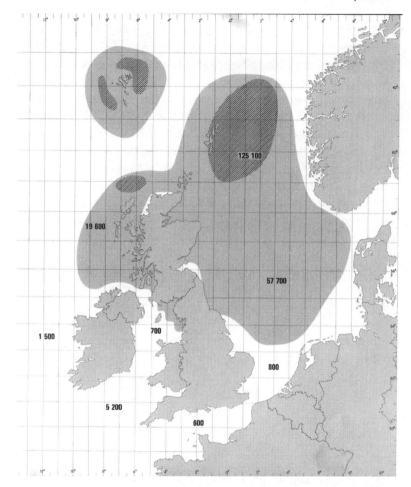

Figure 9.14 Main spawning areas of haddock around British Isles as determined by egg surveys. Dark shading indicates spawning areas; all shaded areas indicate fishing ground. Numbers in sea areas refer to the average international catch in tonnes during the period 1973–77.

(From Lee, A.J. and Ramster, J.W. (eds). (1981). *Atlas of the Seas around the British Isles*. MAFF, London. © Crown Copyright, 1981.)

deeper water than cod, mainly 80–120 m depth. This preference for deeper water restricts the haddock spawning areas of the North Sea to a more northerly distribution than those of cod, the main area extending from the east coast of Shetland to off the coast of Norway (Figure 9.14). When preparing to spawn, the North Sea haddock migrate northwards towards the deeper water to the north of the Fisher Banks.

After spawning the fish return to shallower water and form feeding shoals in the central part of the North Sea. This movement of the fish has its effect on the

North Sea haddock fishery, which generally shows two peaks of landings (a) during February to May, when the best catches are made in the spawning areas in the northern part of the North Sea; (b) in September to October, when the main fishery occurs amongst the feeding shoals in the central part of the North Sea.

The buoyant eggs rise to the surface layers, and in the northern North Sea the eggs take some 14–20 days to hatch, a 4 mm larva with attached yolk-sac emerging. The miniature adult form is reached when about 2–2.5 cm in length. Haddock tend to remain in the surface and mid-depth waters rather longer than cod, sometimes until 5 cm or more in length. At night Group 0 fish (less than 1 year old) can be caught in surface and mid-water trawls, as well as on the bottom, indicating a tendency to make diurnal vertical migrations. As they grow larger the fish become demersal and show a greater tendency than cod to congregate in shoals. Young haddock prefer rather deeper water than young cod, which sometimes occur close inshore.

During their pelagic phase, haddock feed mainly on copepods. Once they have become demersal, haddock obtain much of their food by searching about in the deposit for crustacea, molluscs, annelids and echinoderms. Although they feed to some extent on fish, they do so far less than cod, spending more of their time foraging on the bottom, and there is thought to be no strong competition for food between these two closely related species.

Growth rates vary considerably in different areas, the southern part of the North Sea being a rapid growth area where haddock average about 45 cm in length at 4–5 years of age. In the northern part of the North Sea, haddock of the same age average only about 30 cm in length. They can grow to about 90 cm. Probably most haddock reach maturity during their third year.

9.2.3   European hake (*Merluccius merluccius*)

Distribution

Hake (Figure 9.15) extend along the eastern side of the Atlantic from Norway and Iceland (where it is only seasonally common) to Morocco in North Africa. Hake are also found in the Mediterranean. It is a fish of deep water, occurring mainly at depths below 200 m, but throughout adult life it performs seasonal migrations into shallower water for spawning during spring and summer, returning to deep water in the autumn. Further south, along African and American coasts, extending down to South Africa and Patagonia, there are other closely related species of hake.

Around the British Isles, the fish are mainly found off the west coast but may enter the northern part of the North Sea during the summer months when a migration occurs from beyond the Orkneys and Shetlands. The commercial fishery mainly operates over shelf regions, but the fish can be trawled from depths well below the continental edge.

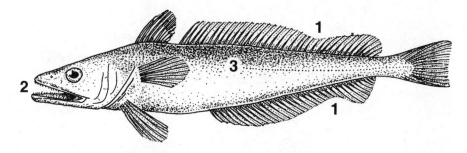

Figure 9.15   The hake, Merluccius merluccius. *Identification features: 1 = Shallow dip in second dorsal, and anal fins: 2 = Large jaws and teeth; 3 = Straight lateral line.*
(Illustrated by Robert Irving.)

Life history

Hake spawn in numerous areas off the west coast of the British Isles between April and October (Figure 9.16). During the early part of the season, spawning mainly occurs close to the continental edge, but as the season proceeds the spawning fish are found in progressively shallower water. The fish also spawn in the Mediterranean, off the Atlantic coast of Morocco and in various areas in the western part of the North Atlantic.

The female probably deposits between a half million to 2 million eggs on or near the bottom. The eggs are small, approximately 1 mm in diameter, and buoyant. They float up to the surface and hatch after about seven days, the newly hatched larva being about 3 mm long. From spawning areas to the west of the British Isles, the eggs and developing larvae are carried mainly eastwards by the surface drift towards shallower coastal water.

Little is known of the biology of the early stages of hake. It seems that the fish remain pelagic for their first two years of life, and then descend to the sea-bed. They grow to about 10 cm by the end of their first year, 20 cm by the end of the second year and after that probably add about 7–8 cm per year, eventually reaching over 1 m in length.

There is a difference between the sexes in the age of onset of maturity. The majority of male hake mature in groups III to VI (see Section 9.4.4) when between 28 and 50 cm long, but females not until they are Group VI to VIII fish at about 65–75 cm.

Throughout life, hake feed almost entirely on pelagic prey. The young fish feed at first on copepods, later on larger planktonic crustacea such as euphausids, and also on small fish and small cephalopods. The diet of older fish consists almost entirely of fish and cephalopods. The hake is cannibalistic, and sometimes small hake comprise as much as 20 per cent of the food of the larger adults. In addition a large range of shoaling fish is taken. Bottom-feeding fish are seldom eaten and hake appear to feed mainly at night, making diurnal feeding migrations from the

Figure 9.16   Main area and times of hake spawning around British Isles. Arrows indicate summer drift of currents transporting pelagic eggs and larvae.

bottom to mid-depth water during darkness. Trawling for hake must therefore be performed during daylight while the fish are on the sea-bed.

9.2.4   Plaice (*Pleuronectes platessa*)

Distribution

Plaice are commercially important flatfish (Figure 9.17) occurring in the north-east Atlantic area. They range from the Barents Sea and Iceland, south to south Spain and the western Mediterranean but are most common throughout the North Sea, English Channel and Irish Sea. Plaice are most commonly found on sandy bottoms on the shallow continental shelf in depths of less than 80 m. They

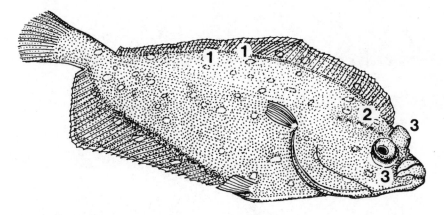

Figure 9.17 *The plaice,* Pleuronectes platessa. *Identification features: 1 = Bright orange or red spots; 2 = Row of bony knobs on head; 3 = Right-eyed.*
(Modified from *British Sea Fishes*, F. Dipper (1987), by kind permission of Robert Irving and Underwater World Publications Ltd.)

occasionally extend down to 120 m or more and can be found on mud and gravel as well as sand. Plaice are caught mainly by trawl or Danish seine, and make up about 7 per cent of the landings of British fishing vessels. They are taken in shallow water all around the British coasts but a large part of the commercial catch comes from the southern North Sea. They are fished for all year round although the quality varies and is rather poor around spawning time.

Life history

Although not great long-distance swimmers, adult plaice from the main Irish Sea, English Channel and North Sea populations, make regular migrations between known feeding grounds and spawning areas. Spawning (Simpson, 1959) occurs all round the British Isles but the main well-defined areas are in the southern North Sea and the Irish Sea and here large numbers congregate (Figure 9.18). The largest area is in the Flemish Bight between the Thames estuary and the Dutch and Belgian coasts. A large part of the southern North Sea plaice population migrates southward during the early winter months towards the tongue of slightly warmer and more saline water which flows into the North Sea through the Strait of Dover. Here, spawning occurs between January and March. After spawning, the spent fish return northwards to their main feeding grounds in the central North Sea. Other spawning concentrations in the North Sea occur to the east of the Dogger Bank, off Flamborough Head and further north in the Moray Firth and around the Shetland Isles.

Spawning also occurs in the Irish Sea, the English Channel and the Baltic, mainly during February and March. In more northerly latitudes, plaice spawn a little later in the year, for example, in the Barents Sea in April, and around Iceland

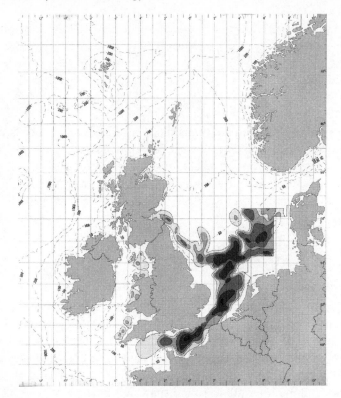

Figure 9.18   Main spawning areas of plaice around the British Isles as determined by egg surveys. The darker the shaded areas, the more eggs per square metre (darkest = >27 m²; lightest = 1 m²).
(From Lee, A.J. and Ramster, J.W. (eds). (1981). *Atlas of the Seas around the British Isles*. MAFF, London. © Crown Copyright, 1981.)

between March and June. There is no special uniformity of depth, temperature or salinity in the spawning areas, but they are never far from the coast, and always in areas where water movements will carry eggs and larvae towards sandy coasts for nursery grounds.

Sexual maturity depends on size rather than age. In the North Sea most females mature and first lay eggs when between 30 and 40 cm in length and 4–5 years of age, and males mature at 20–30 cm when most are four years old. A female plaice spawns between 10 000 and 600 000 eggs in a season. The eggs average about 1.9 mm in diameter. They are spawned close to the bottom but are buoyant, and are mostly fertilized in mid-water as they float up to the surface, especially at night. The eggs increase slightly in density as the embryos develop, so that by the time the larvae hatch they are drifting in near-surface waters. They hatch after about 2 to 3 weeks into a symmetrical larva, pigmented canary yellow,

6.0–7.5 mm in length with a ventral yolk-sac which is resorbed within about eight days. When active feeding commences, the larvae take chiefly flagellates and small diatoms. As the larvae grow, larger diatoms, molluscan larvae, early stages of copepods and the larvaceans *Fritillaria* and *Oikopleura* are taken, the larvaceans being a particularly important component of the diet of plaice larvae, often forming virtually the entire food.

Normal symmetry and a planktonic mode of life continue until about 4 to 6 weeks after hatching, sometimes longer. At this stage, when the plaice larva is about 10 mm long and 2 mm in height, its metamorphosis begins, and within the next 17 days the larva becomes gradually converted into a 'flatfish'. It becomes laterally flattened and acquires a new swimming position with its left side downwards. The skull is progressively transformed by the movement of the left eye to a new position dorsal and slightly anterior to the right eye on what now becomes the uppermost side of the body. The swimbladder which is present in the planktonic larva is gradually lost. Colour disappears from the new underside, and the upper parts develop the characteristic pigmentation and spots of the plaice. During this period of metamorphosis, the fish becomes demersal and grows to about 14 mm in length and 7 mm in 'height', i.e. the maximum distance between the bases of the dorsal and ventral fins.

The chief nursery areas for young plaice are along sandy stretches of coastline. Eggs and larvae spawned in the Flemish Bight are carried by the drift of the surface waters in a north-easterly direction towards the Dutch, German and Danish coasts, usually travelling some 1.5–3 miles per day. By the time their metamorphosis is complete, the young fish have drifted close to the coast where they settle upon the sea-bed. The extensive sandy expanses along these shores, particularly the west coast of Denmark and the Waddensee, form a major nursery area for young plaice. Larvae spawned in various other places near the British Isles also drift towards regions of shallow sandy bottom, and the Wash and the greater part of the Lancashire and Cheshire coasts are other plaice nursery areas. Considerable fluctuations occur from year to year in the numbers of young plaice which successfully complete their metamorphosis. Of several factors which may influence survival at this stage, an important one may be the strength and direction of the wind, upon which depend the speed and direction of drift of the surface water. Poor survival may occur in years when a large proportion of the larvae fail to reach, or are carried beyond, the sandy regions required by the young fish.

Once the fish have become demersal, they feed at first on a variety of small benthic organisms including annelids, harpacticoid copepods, amphipods, small decapods, small molluscs, etc. While very small, much of their food is obtained by biting the siphons off molluscs such as *Tellina* or the palps off spionid worms, which then regenerate. As they grow larger they begin to take whole polychaetes and small crustacea, and molluscs form an increasingly important part of the food. The diet varies from place to place according to the nature of the food

available but molluscs often account for about 25 per cent of the total. *Venus,*
Cardium, Spisula, Solen, Mactra and *Tellina* are important plaice foods, and they
also take the tectibranch *Philine* and numerous annelids, crustacea, echinoderms,
actinians and sand-eels.

Tagging experiments (see Section 9.4.2) have shown that during their first year
of life, the majority of plaice are to be found close inshore, mainly in sandy areas
in water of less than 5 m depth. With increasing size they migrate into deeper
water. This movement usually occurs during the latter part of each summer. At
the end of their first summer, the fish move into the 5–10 m zone and are thought
to spend much of their first winter buried in the deposit. They are now 6–8 cm
in length. In the following spring, they return to shallower water and, during the
summer, range to and fro along the coast in search of food. As autumn
approaches, they again migrate into deeper water, this time to about the 20 m line.
Thus, until their third summer, by which time they are about 20 cm in length, most
plaice are found near the coast in water of less than 20 m depth. Thereafter, as
they increase in age and size, they inhabit increasingly deep water further from
the shore. The size reached by a particular age depends partly on the food supply.
North Sea plaice reach a size of between 35 and 45 cm by about their sixth year
of life. In most areas the females are slightly larger than males of the same age.

Plaice occasionally hybridize with both flounder and dab. Various races of
plaice can be distinguished to some extent by differences in numbers of fin rays
or vertebrae.

9.2.5   Lemon sole (*Microstomus kitt*)

Distribution
Like plaice, the lemon sole (Figure 9.19) is a flatfish of shelf areas of the north-east
Atlantic ranging from the Arctic to the Bay of Biscay. It does not extend as far
south as plaice and generally favours a rougher sea bottom, but the two species
often occur together. Lemon sole are specially abundant in the north-west part
of the North Sea off the east coast of Scotland, also around the Faroes and along
the south-west coast of Iceland and these are the most important fishing grounds
(Figure 9.20). It is most common on gravelly bottoms but may occur on any
substrate. It lives mainly between 40 and 100 m but is found in reduced numbers
to nearly 200 m. The fish are caught mainly by trawls and Danish seines.

Life history
Unlike the plaice, the lemon sole does not have well-defined spawning grounds,
but simply spawns widely throughout its range, gathering in small local
concentrations wherever the fish are normally found. Tagging experiments have
indicated a tendency for the fish to swim against the current during the period

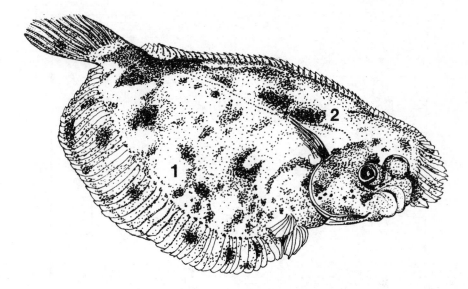

Figure 9.19 The lemon sole, Microstomus kitt. *Identification features: 1 = Colourful speckles on smooth slimy skin with no bony tubercles; 2 = Only slightly curved lateral line.*
(Modified from *British Sea Fishes*, F. Dipper (1987), by kind permission of Robert Irving and Underwater World Publications Ltd.)

preceding spawning. In the north-west North Sea this results in a limited northerly movement towards the Orkneys and Shetlands or westward around the north of Scotland, but there is no great spawning migration. The fish do not appear to require very precise conditions for spawning. In the North Sea it takes place mainly at depths between 50 and 100 m when the bottom temperature is not lower than 6.5°C. Around the British Isles the earliest spawners are usually found in the English Channel in February or March, with a maximum abundance of eggs in April to June. Spawning off the west of Scotland extends from March or early April until late July with the peak in April to May. In the North Sea, spawning begins in the north in early May and a little later further south, with a maximum in June to August, and lasting into November in the central North Sea. Around the Faroes and Iceland spawning lasts from May to August with the peak in June to July.

Lemon sole eggs are smaller than plaice, 1.13–1.45 mm in diameter, and produced in rather greater numbers per female, some 80 000 to 700 000 depending on the size of fish. They are probably shed on the bottom, but float to the surface for two or three days before sinking back to middle depths. Hatching takes place after some six to ten days depending on temperature, and the emerging larva and yolk-sac are symmetrical and about 5 mm long. The mouth and gullet are narrow, allowing only small food objects to be taken, such as eggs,

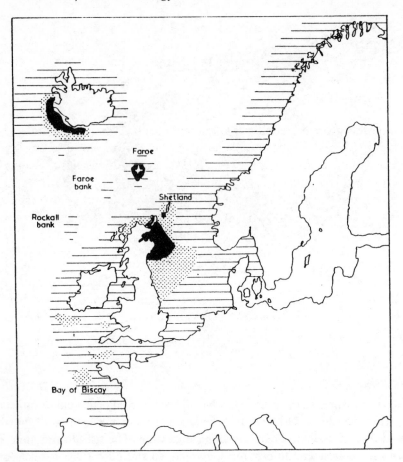

Figure 9.20   Distribution and main concentrations of lemon soles, ☰ *light,* ▨ *moderate,*
■ *heavy.*
(From Rae (1965), by courtesy of Fishing News (Books) Ltd.)

peridinians, diatoms, *Podon* and copepod nauplii. Later larvae also eat *Limacina*
and *Oikopleura*.

Although the larvae can be taken in tow nets at all depths, the majority are near
the bottom and few are found close to the surface. In the north-west North Sea
the drift of eggs and larvae is southwards towards the central North Sea. The
pelagic phase lasts some two to three months, terminating with a metamorphosis
similar to plaice, after which the fish live on the bottom.

Lemon soles take a wide variety of food from the sea floor, but the major part
of their diet is almost always polychaete worms, especially the eunicids *Onuphis
conchylega* and *Hyalinoecia tubicola*, the terebellids *Lanice conchilega* and
·*Thelepus cincinnatus* and several serpulid species. Their diet is restricted by the
small size of the mouth. The anemone *Cerianthus* is an important food in some

localities. A variety of small benthic crustacea (mainly amphipods and eupagurids), molluscs (mainly chitons and small gastropods) and some ophiuroids are also taken. Small bivalves are eaten and the fish bite the siphons off larger lamellibranchs; but compared to plaice, bivalves form a much smaller part of the diet of lemon soles and it is unlikely that the two species compete much for food.

Lemon sole grow rather slowly attaining about 15 cm in length at the end of their third year and 25 cm by the end of the sixth year. However, the western North Sea off the British coast between Yorkshire and Aberdeen is a region where they grow relatively quickly. Here the lengths at these ages are nearer 25 and 30–35 cm. The fish can grow to nearly 50 cm and rare specimens have been caught between 55 and 67 cm, probably at ages between 15 and 20 years. The majority age of first spawning is four years for males and five years for females. Like several other species, after maturity the males have a slightly greater mortality rate than females so that in the older age groups the females progressively predominate.

9.2.6   Herring (*Clupea harengus*)

Distribution
Herring (Figure 9.21) are widely distributed across the north-eastern Atlantic shelf between Newfoundland and the British Isles, and also in the Arctic. They extend south to the area of Gibraltar on European coasts, and south to Cape Hatteras on the North American coast. However, it is only in northern areas that they are sufficiently abundant to be commercially valuable. They have now become scarce in some areas, such as the North Sea where they were once immensely abundant and they now no longer support the great fisheries they once did. This is discussed further below. With such a wide distribution, it is not surprising that the herring exhibits a number of local races, distinguished by features such as the number of vertebrae and by differences in spawning times and growth rates. A very similar form, *Clupea pallasii*, extends in the North Pacific from Japan to the coast of British Columbia, and also occurs in the North Atlantic. Herring-like fish are found in many other areas, including fresh water.

Herring are pelagic fish of mainly offshore areas, making inshore migrations in great shoals for spawning in certain coastal localities, after which the shoals disperse and the spent fish move out to deeper water, reassembling offshore as feeding shoals. They are therefore caught mainly using ring nets and purse seines.

Life history
When herring are preparing to spawn, they congregate in huge shoals as they approach their spawning grounds. They are the only commercially important marine teleosts in British waters to lay demersal eggs. A spawning female deposits

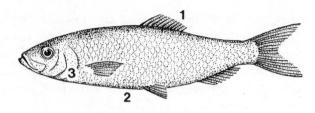

Figure 9.21   The herring, Clupea harengus. Identification features: 1 = Single dorsal fin in middle of back; 2 = Rounded belly (no keel); 3 = Smooth gill covers (not ridged). (Modified from British Sea Fishes, F. Dipper (1987), by kind permission of Robert Irving and Underwater World Publications Ltd.)

10 000–60 000 eggs on inshore banks of small stones and gravel. The eggs are 0.9–1.5 mm in diameter, heavier than water, and form sticky masses which readily adhere to the stones, seaweeds, maerl and to each other. They are often densely packed in layers several eggs deep. Most races spawn in depths from 15 to 40 m but some oceanic races spawn on offshore banks as deep as 200 m (Hodgson, 1967; Parrish and Saville, 1965, 1967).

The precise location and limits of herring spawning grounds have been difficult to determine because herring eggs are not easily found by dredging. It seems that the selected sites are quite patchily spread with eggs. Apart from dredging, the situation of spawning grounds is approximately known from observing the position of ripe and newly spent herring, the occurrence of herring larvae in tow net hauls, and by noting the places from which trawlers bring up 'spawny' haddock, i.e. haddock engorged with herring eggs, on which they prey. The distribution of spawning grounds in the North Sea can also be very roughly equated with the distribution of predominantly gravelly areas.

Around the British Isles, different races of herring can be found spawning at almost any time of year, though there is a peak in the warmer months. The major spawning shoals are of two principal races, oceanic (Atlanto-Scandian) and shelf herring. Oceanic herring have an extensive range in deep water in the North Atlantic, Arctic and Norwegian seas. There is a major movement by spring-spawning shoals of oceanic herring towards the Norwegian coast. Those which approach the British coasts form winter–spring spawning shoals between February and April around the Hebrides, Orkney and Shetland Isles and along the Irish and Scottish coasts. These shoals occur mainly in water of oceanic-neritic type (see Section 4.7.1) where the temperature lies between 5 and 8°C.

The shelf herring of the North Sea, English Channel, Minch and Irish Sea form mainly summer–autumn spawning shoals. The North Sea shoals spawn from the Shetlands to the east coast of Scotland from July to September; off Northumberland, Yorkshire and Lincolnshire between August and October; off the East Anglian coast between October and December; and in the eastern English Channel

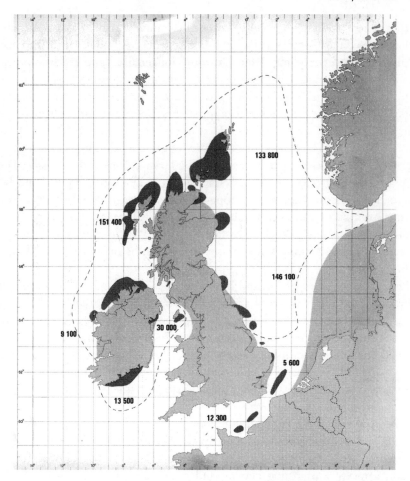

Figure 9.22   Main areas of herring spawning around the British Isles as determined by egg surveys. All the spawning areas shown belong to autumn or winter spawning herring. Spring spawner numbers are relatively very small. The previously very important Dogger Bank spawning grounds are now of little significance due to the collapse of the stocks. Dark areas are spawning grounds; paler areas are nursery grounds; -------- indicates feeding area. Numbers in sea areas refer to the average international catch in tonnes during the period 1973–77.
(From Lee, A.J. and Ramster, J.W. (eds) (1981). *Atlas of the Seas around the British Isles.* MAFF, London. © Crown Copyright, 1981.)

near the north French coast from December to January. These shoals are found in neritic water at 8–12°C.

Slight morphological differences have been noted between the two groups. Both show variation in the number of vertebrae between 54 and 59, with a mean vertebral count slightly above 57 for the oceanic herring and below 57 for the shelf herring. There are also slight differences in the number of gill-rakers, number of keeled scales on the ventral surface and in the structure of the otolith. The oceanic herring are slower growing, later maturing, longer lived and reach a larger size than the shelf herring. It is now generally considered that the two groups, both of which can be further subdivided into several more or less distinct stocks, are biologically separate units with no appreciable interbreeding.

In water of 5–6°C, herring hatch in about 22 days, at 11–12°C in 8–10 days. The newly hatched larva is about 6–8 mm long and at first depends on the food reserves of the yolk-sac. After hatching, the larva swims to the surface. During the period of resorption of the yolk-sac the larva develops a mouth and begins to feed at first mainly on diatoms, copepod eggs and early copepod larvae. As the herring larva becomes larger, it takes the later larval stages of copepods and the adults of small copepods such as *Paracalanus* and *Pseudocalanus*. Until it reaches a length of about 45 mm, when about three months old, the larva has a thin, eel-like appearance.

The larval fish are pelagic and drift with the currents. There is some mixing of broods from different areas and a high proportion of larvae from the North Sea and west coast spawning grounds drift considerable distances into nursery grounds situated in the shallow central and southern North Sea (Heath and Richardson, 1989). Extensive nursery areas exist along the shallow coastline of the Netherlands, Germany and Denmark, and in large estuaries such as the Thames Estuary, the Wash, the Moray Firth and the Firth of Forth (Figure 9.22).

When the young fish reach about 5 cm long, they become able to swim more strongly and start to form shoals actively moving into the shallow inshore nursery and feeding areas. Here they form large shoals known as 'whitebait' which often contain a mixture of young herring and young sprats. At this stage, the food consists largely of small crustacea; for example, the larvae of shrimps and prawns, and estuarine copepods such as *Eurytemora hirundoides*. The young herring remain inshore until near the end of their first year of life, and then start to move offshore. Areas such as the Dogger Bank support large numbers of young fish. By now the fish are mostly some 4–8 cm in length and during their second year these fish grow to some 13–18 cm in length but are still extremely thin.

In their third year these herring fatten, and feeding shoals of the immature fish often approach the English coast during the mid-summer. Throughout life the food is predominantly planktonic, mainly copepods, but also includes chaeto-gnaths, pteropods, hyperiid amphipods, appendicularians, decapod larvae and fish eggs. They sometimes feed on other fish, for example *Ammodytes*. The type of food varies seasonally to some extent and also varies from place to place according to the nature of the plankton.

By the end of their third year they have reached the 'fat herring' stage, and the flesh has become rich in oil for which they are particularly prized by man. During their fourth year, i.e. as Group III fish, the majority of southern North Sea herring become sexually mature. The oceanic fish of the northern part of the North Sea and the Norwegian coast mainly reach maturity later, between their fifth and eighth years. The maturing virgin fish leave the young fish shoals and join the adult spawning shoals.

Certain patterns of movement can be discerned in herring. For example, there are diurnal changes of vertical distribution associated with changes of illumination, the fish usually forming compact shoals on or near the sea-bed during daylight, and approaching the surface during darkness to feed on the plankton. There is also the annual cycle of migration of the adult fish, when they congregate in huge shoals and move into shallow water to spawn, followed by a dispersal to deeper water and the formation of feeding shoals (Nichols and Brander, 1989; Rankine, 1986).

Several factors influence the movement of the feeding shoals; in particular, the condition of the plankton. Where there is an abundance of suitable zooplankton food, the feeding shoals tend to amass, and the numbers of herring can sometimes be correlated with the numbers of *Calanus*. On the other hand, herring are seldom found in water heavily loaded with phytoplankton. It has been suggested that this is due to an 'exclusion' effect (see Section 5.3.4) but because such water usually contains very little zooplankton, the absence of herring shoals may simply be due to the scarcity of herring food.

The quality of the fish depends upon the abundance and nature of their food. The valuable oiliness of the herring reflects the fat content of the plankton, which in turn depends to some extent upon climatic conditions. Generally, a warm, sunny summer produces a rich plankton and provides fish in excellent condition for the autumn fishery. A copious diet of copepods produces a specially oily flesh. If the food consists mainly of pteropods, the fat content of the flesh is poorer.

Fishery and over-exploitation

For many years it was thought that the stocks of North Atlantic herring were so large as to be virtually inexhaustible no matter how intensively fished. As long as fisheries were conducted mainly by drift nets exploiting only the shoals approaching the spawning grounds, this may well have been true. Although the size and movement of herring shoals were subject to many fluctuations, appearing and disappearing in unpredictable ways with the consequent rise and decline of particular local fisheries, there was seldom any overall shortage of herring. However, during the years following 1948 an intensive trawl fishery for immature herring developed on the North Sea nursery grounds, capturing huge numbers of herring in age groups I, II and III for conversion to fishmeal for cattle and poultry food. Far more fish were destroyed in this way prior to spawning than had ever previously been taken from the adult shoals by drift nets.

The capturing power of pelagic fishing techniques has also been greatly increased in recent years. Drift nets have been largely replaced by ring nets, purse seines and pelagic trawls which enclose shoals that have been accurately located by sonic fishfinders. In the 1930s, although herring were heavily fished according to the methods of those times, the total annual catch of herring from the north-east Atlantic then averaged under 1.5 million tonnes. By 1965 and 1966 the landings obtained by newer methods were over 3.5 million tonnes per year. These huge catches were followed by a dramatic decline of landings. In 1969 the catch had fallen to 1.4 million tonnes and by 1976 to less than 800 000 tonnes.

The devastating effects of a combination of the industrial fishery for immature herring and the more intensive fishing of adult shoals caused the North Atlantic herring stocks to collapse and in 1977 a total ban was imposed on the fishing of herring over virtually the whole area in the hope of some recovery of stocks. By 1983 the population had partially recovered and fishing began again with limits on size and numbers taken. However, these measures have not been entirely successful and stocks appear once more to have stopped increasing. The recent history of north-east Atlantic herring fishing is a sorry example of failure to utilize rationally a major natural resource which, if properly managed, could provide large quantities of highly nutritious food for direct human consumption. A major part of the excessive landings of the mid-1960s was processed as fishmeal for animal rearing.

9.2.7 Mackerel (*Scomber scombrus*)

Distribution
Mackerel (Figure 9.23) are found in warmish water on both sides of the North Atlantic. Their range extends from the south coast of Norway and northern North Sea, along the west coasts of the British Isles and into the English Channel, and as far south as the Canaries. They also occur in the Mediterranean and on the western side of the Atlantic from south Labrador to North Carolina.

Life history
The life history and migrations of mackerel have been well documented by Lockwood (1976, 1978, 1989). Two stocks of mackerel are found in north-west European waters: a western stock which spawns mainly along the shelf edge west of Britain, in the Celtic Sea and Bay of Biscay; and a North Sea stock which spawns in the central North Sea and off southern Norway.

During their spawning period, mackerel congregate on the spawning grounds in huge shoals. The western stock spawns between February and July, the most intensive spawning occurring during April in a spawning area some 50–80 miles west of the Scilly Isles (Figure 9.24b), mainly along the line of the continental edge extending southwards into the Bay of Biscay. As the season proceeds, spawning

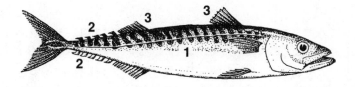

Figure 9.23 The mackerel, Scomber scombrus. *Identification features: 1 = zebra stripes along the back; 2 = Finlets in front of tail; 3 = Dorsal fins widely spaced.*
(Modified from *British Sea Fishes*, F. Dipper (1987), by kind permission of Robert Irving and Underwater World Publications Ltd.)

fish are found in progressively shallower water further to the east, until by July they are spawning in St. George's Channel, the Bristol Channel and the western part of the English Channel. The North Sea stock spawns from May to July, peaking in June.

The females produce about half a million pelagic eggs. These are not all shed together but are produced in successive batches over a period. The eggs are about 1.2 mm in diameter and are laid directly into the upper part of the water column. At first they float to the surface, but after 2 days they lose buoyancy slightly and sink to mid-depth water. They hatch after about 7 days (depending on temperature) into a larva about 2.5 mm long. The yolk-sac is resorbed in about 9 days. Young fish spawned in the Scillies area are presumably carried eastwards by the drift of the water towards the English, Irish and Welsh coasts. In some years, sizable shoals of young mackerel of about 13–17 cm in length are found along the English south-west coast during the summer. Mackerel post-larvae are also abundant in the northern part of the North Sea. In autumn and winter small mackerel have sometimes been brought up from the bottom in fine-mesh trawls, suggesting a pattern of behaviour similar to that of the adult shoals (see below).

Growth is rapid, reaching approximately 23 cm during the first year. Mackerel mature when two years of age at approximately 29 cm, and grow to about 36 cm by their sixth year, but after this period there is no satisfactory method of determining their age (see Section 9.4.4).

Shoals in surface waters feed mainly on copepods, but small fish such as sand eels and pilchards are also taken. Fishermen associate good catches of mackerel with an appearance of the sea which they call 'yellow water', and it has been shown that these areas are particularly well filled with the copepods *Pseudocalanus* and *Calanus*. On the other hand, leaden-coloured water with a slightly unpleasant smell, known to fishermen as 'stinking water' and associated with the poorest catches, was found to be loaded with phytoplankton. Feeding intensity is greatest in summer after spawning and as the shoals approach the coast they tend to disperse, the fish ranging to and fro along the coastline seeking food which

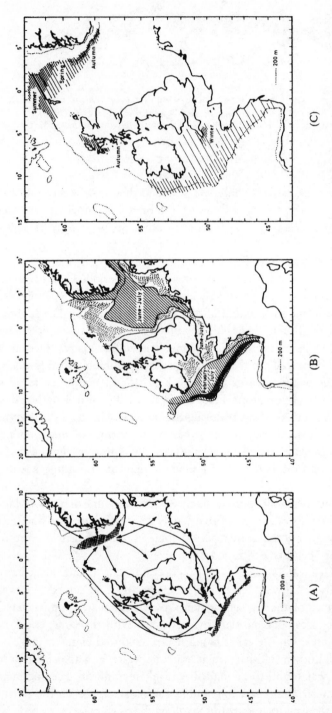

Figure 9.24 (A) The main mackerel overwintering grounds and the annual migration paths to and from the spawning and feeding grounds. (B) The mackerel spawning grounds. Spawning occurs to some extent in all shelf waters around the British isles, but the main areas are in the central-northern North Sea and the western Celtic Sea. (C) The main mackerel fisheries. The Norwegian fisheries are concentrated in the northern North Sea while the UK fisheries are in the Minch, west of Scotland, and off Cornwall. (From Lockwood, S.J. (1978)).

at this time may include various inshore crustacea such as mysids, shrimps and prawns as well as fish and copepods.

Migrations

At the end of the spawning season, part of the western stock of mackerel migrates up the west coast of Britain to the northern North Sea and Norwegian Sea to overwintering grounds (Figure 9.24). The return movement of the western stock down the west coast towards the spawning grounds has become progressively later in recent years and now occurs mainly from January to March (Walsh and Martin, 1986). It is postulated that this may be associated with a change in flow of the North Atlantic Drift at the continental shelf edge. There is also some migration through the English Channel from the west into the southern North Sea and here too the return migration seems to be later in the season than previously. Other major overwintering shoals are thought to extend well down the continental slope to the west of the Scillies and there are also other concentrations along the south coast of Cornwall.

Mackerel migrations off the south-west of the British Isles were investigated by Steven in the late 1940s (Steven 1948, 1949). He found that overwintering fish were demersal congregating on the bottom and feeding on any available food such as small crustacea, polychaetes and small fish. Cascade currents (see Section 5.6) may carry food down from the surface in some areas.

Towards the end of winter, fish begin to leave their overwintering concentrations. At first they still keep to the bottom, but soon begin to perform diurnal vertical movements, ascending during darkness and eventually forming surface shoals. During the early months of the year the surface plankton is sparse and the majority of the fish are still without food, but they readily feed when suitable food is found, for example, on shoals of small fish such as pearlsides (*Maurolicus muelleri*).

During the late winter and spring the surface shoals begin to move towards the spawning areas (Figure 9.24). Those which overwintered in the English Channel and southern North Sea move westwards, those from the Irish Sea move south-westwards, while those which wintered on the continental slope move towards the east, all tending to converge towards the west of the Scilly Isles. From the northern North Sea fish move towards the Norwegian coast, or southwards into the central North Sea.

After spawning the shoals move away from the spawning grounds, some over considerable distances. Many that spawn in the Celtic Sea move eastwards towards the English and French coasts. Some pass up the English Channel into the North Sea and others go northwards west of Ireland or through the Irish Sea to the Hebrides, or even beyond to the Shetlands where there is some mingling of the western and Norwegian stocks. Fish that spawned in the central North Sea return mainly northwards towards the Norwegian coast.

Towards the end of the autumn the fish disappear from coastal regions and return to their overwintering concentrations on the sea-bed. It is thought that the fish tend to return to the same overwintering sites as they left the previous spring.

Fishery and over-exploitation

Mackerel are caught by trawling in the winter, when they are demersal, and through most of the year by various pelagic fishing methods, mainly purse seines and pelagic trawls. Some small boats still use handlines in winter in areas such as along the Cornish coast. The trolled lines carry a number of feathered hooks. The North Sea fishery is exploited mainly by Norwegian purse seiners.

During the mid-1960s the North Sea fishery underwent enormous expansion from normal levels of 100 000 tonnes or less to nearly a million tonnes in 1967. This was partly as a result of the purse seine fleet transferring most of its attention to mackerel after the herring fishery collapsed. The catch was used mainly for fishmeal and oil. During the early 1970s the landings declined rapidly as the stock diminished. The Norwegian government instituted controls to limit mackerel fishing. However, the size of the spawning stock in the North Sea is presently less than 50 000 tonnes and biologists believe the North Sea mackerel will probably become effectively extinct, apart from influxes from Atlantic waters in the north.

The south-western fishery has traditionally been of interest mainly to British, French and Spanish ships, and landings were always less than from the North Sea and remained fairly stable at around 30 000 tonnes per year. The importance of this fishery increased with the decline in herring landings, and in 1968 Soviet ships began to fish the south-western stock. By 1975 annual landings from this fishery had increased to about 500 000 tonnes, with ships from the USSR taking over half this total. In 1977 Soviet ships were excluded from EEC waters. The western spawning stock was still relatively large in 1989, approximately 2 million tonnes, and it is mainly this stock that the international fishery now exploits.

9.3   The overfishing problem

Addressing the International Fishery Exhibition in London in 1883, T.H. Huxley said:

> I believe that it may be affirmed with confidence that, in relation to our present modes of fishing, a number of the most important fisheries, such as the cod fishery, the herring fishery and the mackerel fishery, are inexhaustible. And I base this conviction on two grounds, first, that the multitude of these fishes is so inconceivably great that the number we catch is relatively insignificant; and, secondly, that the magnitude of the destructive agencies at work upon them is so prodigious, that the destruction effected by the fisherman cannot sensibly increase the death-rate.

Shortly afterwards the landings of fish from the seas around north-west Europe increased to an extent that Huxley could not have envisaged. Sailing vessels were superseded by ships with powerful steam engines, enabling the use of much larger nets and the replacement of the old beam trawl by the far more effective otter trawl, and giving fishermen a new independence of wind and tide so that they could fish longer and more often. Within 30 years of Huxley's pronouncement there was already evidence of reduction of stocks of certain favourite demersal species such as cod, haddock and plaice on the more intensively fished areas of the north-east Atlantic. Between the two World Wars the decline became more apparent, and in 1942 E.S. Russell wrote:

> A state of overfishing exists in many of the trawl fisheries in north-west European waters. Two things are wrong. First, there is too much fishing, resulting in catches below the possible steady maximum, and secondly, the incidence of fishing falls too early in the fishes' life resulting in a great destruction of undersized fish which ought to be left in the sea to grow.

Since World War II the catching power of fishing vessels has been further augmented by several innovations. The change from steam to diesel power has raised the power and speed of ships. The development of nets made of stronger, lighter and longer-lasting materials has encouraged the use of even larger nets. The availability of synthetic fibres of various densities, both lighter and heavier than water, facilitates the control and correct orientation of nets in the water by selection of different densities for different parts, and the use of transparent fibres invisible in water has increased the efficiency of certain nets. Sophisticated sonic techniques have been invented for the detection of fish shoals, and for net handling. Modern navigation systems such as GPS (see Section 3.3.5) allow accurate return to good fishing areas and accurate deployment of nets. Refrigeration equipment installed on fishing boats has allowed them to range far afield and continue fishing until full without deterioration of the catch.

The rise in world landings of fish since the end of World War II has been remarkable (Figure 9.25). Over the period 1948–1968 the increase was around 7 per cent per year, bringing the world total annual catch of sea fish from under 20 million tonnes in 1948 to over 60 million tonnes in 1970. Around 1970, the annual catch stopped rising. Between 1980 and 1989 it rose very slowly, and over the past few years appears to have declined slightly. Greater efforts to catch fish have not recently resulted in significantly heavier landings and there is now much concern at the intensity of exploitation of many fish stocks and the risks of diminishing returns through overfishing.

In order to understand what is involved in 'overfishing', we will briefly consider the effects that fishing is likely to have on the size and composition of fish populations. First, we will take the case of a stock of fish which is subjected only to very light fishing. This population can be regarded as having grown to the limits

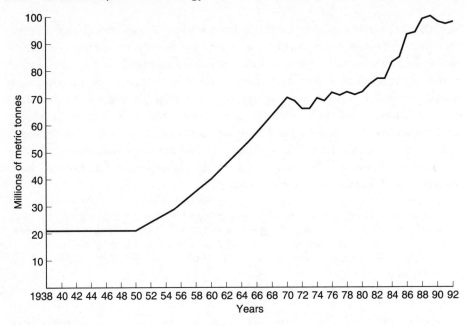

Figure 9.25   World landings of sea fish, 1938–1992.
(Source: FAO yearbooks.)

imposed by the food supply, which restricts both the number of fish surviving and the size to which the individual fish grow. Scarcity of food prevents all the fish from making as much growth as they would if they were better fed, and the slow-growing, older fish compete for food with the rapidly growing younger specimens. The small catches taken by fishermen are likely to consist mainly of the larger, older fish; but due to age, undernourishment or disease, these may not be of good quality, and therefore fetch correspondingly poor prices on the market.

A stock of fish in this condition may be regarded as 'underfished'. The population is overcrowded. An undue proportion of the food is consumed by old fish of poor market quality at the expense of young fish, and none can realize its full growth potential. This stock could support a larger, more profitable fishery of better-quality fish if more fish were caught. A reduction in the size of the population, particularly by the removal of older fish, would promote a better growth rate throughout the remaining stock and improve the condition of the fish, the process being analogous to a gardener thinning out his plants to encourage the best growth and quality of his specimens.

Considering now the opposite case, a stock of fish subject to extremely heavy fishing, this population is likely to consist mainly of young, small specimens because the fish are caught as soon as they reach catchable size. Landings of undersized fish are likely to fetch poor prices because these fish carry little edible meat, being mainly skin and bone. This stock is obviously 'overfished'. Too many fish are caught too early in life. Young fish make rapid growth and, if left longer in the sea, would soon reach a more valuable size, providing heavier landings and better prices. Productivity and profits would eventually improve if the amount of fishing was reduced.

However, in these conditions the fisherman is tempted to try to increase his profits by catching even more fish. This is a vicious spiral which can only lead to a further reduction in the size of the stock, and a further dwindling of the fisherman's income. Indeed a point may be reached at which so many fish are caught before they have lived long enough to spawn that the reproductive capacity of the stock becomes severely impaired, leading to a catastrophic decline in numbers through failure to produce enough young and even a danger of extinction.

We can therefore distinguish two aspects of overfishing. There is what may be termed *growth overfishing*, where catches are poor because too many fish are caught before making optimum growth but recruitment of young fish is not seriously affected. There may also occur *recruitment overfishing* when the stock fails through depressed intake of recruits resulting from a reduction in the numbers of spawners.

9.3.1 Wars and fish stocks

During the first half of the twentieth century the fisheries of the north-east Atlantic were twice interrupted by war. Conditions on some of the major fishing grounds visited by British vessels have fluctuated correspondingly, with symptoms of underfishing during war years and severe overfishing in peacetime. For example, the North Sea provides an important fishery for haddock. During the years prior to 1914, the total landings of haddock from this area showed a fairly steady decline. During the period of the 1914–18 war, fishing in the North Sea was greatly reduced. When normal fishing was resumed after the war, greatly increased yields were at first obtained, amounting to approximately double the pre-war landings, but during subsequent years the catches gradually diminished. By the later 1930s the annual landings of haddock had fallen below their pre-1914 level to about a quarter of their 1920 weight.

During the war of 1939–45, fishing in the North Sea virtually ceased. When normal fishing was re-established, there was at first again a marked increase in haddock landings, but subsequently the yield of the fishery again fell (Figure 9.26).

These diminishing peacetime yields have not been the result of any reduction of fishing effort, rather the reverse. Furthermore, as the yields decline the percentage of the catch consisting of small fish shows a great increase (Figure 9.27).

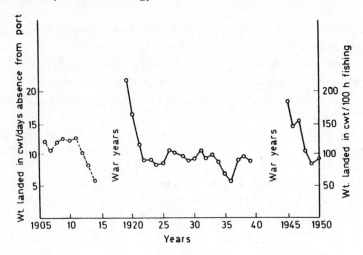

Figure 9.26   Haddock catch per unit of fishing effort by Scottish trawlers in the North Sea.
------ *Landings per days absence from port;* ——— *Landings per 100 hours.*
(From Graham, M. (1956), by courtesy of Edward Arnold.)

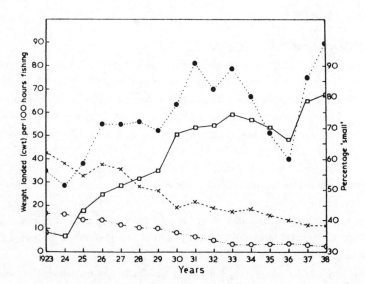

Figure 9.27   Catch, per unit of fishing effort, of North Sea haddock, and percentage of the 'small' category, 1923–38. O----O *large;* ×----× *medium;* ●----● *small;* □——□ *percentage 'small'.*
(From Graham. M. (1956), by courtesy of Edward Arnold.)

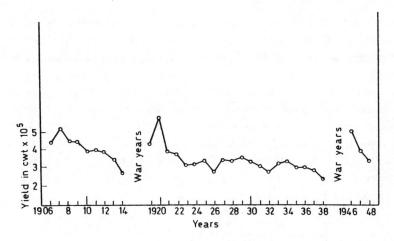

Figure 9.28 Catches of North Sea plaice by first-class English steam trawlers.
(From Wimpenny (1953), by courtesy of The Buckland Foundation.)

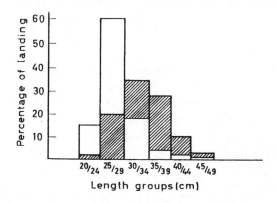

Figure 9.29 Size distribution in plaice landings at Lowestoft before and after World War II. □ *1938* ▨ *1946.*
(From Wimpenny (1953), by courtesy of The Buckland Foundation.)

In the plaice fishery, comparable changes have occurred (Figure 9.28). The largest sizes of plaice in the North Sea have only been abundant in immediate post-war years, and since then the catch has consisted mainly of small fish (Figure 9.29). Although the total weight of plaice landed from the North Sea has remained fairly constant over the first half of this century, the effort and expense of catching them have greatly increased, i.e. the yield per unit of fishing effort has become less. During the 1960s there was some reduction of fishing for plaice in the North Sea because of poor returns. Attention turned to other species and some old vessels were laid up and not replaced. Consequently there was some recovery of plaice stocks in the late 1960s and landings considerably improved.

In normal peacetime operation, plaice and haddock fisheries of the North Sea have therefore shown major features of overfishing, namely, declining yields and a preponderance of small fish. Similar trends have been observed in many areas for such important species as cod, plaice, haddock, halibut and hake.

The war years, on the other hand, produced some symptoms of underfishing. The stock of fish increased greatly, and this was reflected in the heavier catches obtained when fishing was resumed. These early post-war catches contained a high proportion of large fish of the older age groups, but some were diseased and of poor quality. In the case of plaice, there is some evidence that the stock increased to such an extent that the average growth rate of the fish was reduced, presumably by food shortage.

9.3.2   The optimum fishing rate

The weight of a stock of fish tends to increase as the fish grow and young fish join the stock. The effects of fishing and natural mortality operate against this tendency to natural increase. E.S. Russell (1942) combined these factors in a simple equation:

$$S_2 = S_1 + (A + G) - (C + M)$$

S_1 = weight of stock at the beginning of a year,
S_2 = weight of stock at the end of that year,
A = annual increment by recruitment of young fish to the stock,
G = annual increment due to growth of all fish in the stock,
C = total weight of fish removed during the year by fishing,
M = weight of fish lost during the year by death from all the other causes, i.e. the natural mortality.

The amount by which the stock weight would increase if no fishing took place, $A + G - M$, can be termed the *natural yield*. In the special case where the total stock remains unchanged, $C + M$ must equal $A + G$. In these conditions the fishery is said to be 'stabilized', and fishing removes an 'equilibrium catch', i.e. a weight of fish that exactly corresponds with the natural yield of the stock. In theory, equilibrium conditions might be established for any level of stock weight, but the natural yield will vary for different weights and compositions of stock.

Underfishing and overfishing are extremes both of which result in low natural yields. Between the two there must be some intermediate intensity of fishing which would provide, in equilibrium conditions, the maximum possible sustained natural yield, i.e. an *optimum yield*. Such a fishing intensity could be termed the *biological optimum fishing rate*.

There are also economic factors to be considered. Highest profits do not necessarily result from the heaviest landings. If a market is glutted, prices collapse. A biological optimum fishing rate, giving the heaviest possible sustained landings, may not be the same as an *economic optimum fishing rate*, i.e. one giving the greatest financial returns. If a fishing policy is to be acceptable to the industry, it must aim to ensure the maximum output consistent with a fair return to all those engaged in fishing.

At the present time, control of natural factors influencing the yield of commercial sea fisheries cannot be envisaged, and it is therefore only in the regulation of fishing activity that there is a practical possibility of achieving optimum yields. But the relationships between fishing, stock, yields and profitability are by no means simple. The same weight of fish can be taken in innumerable ways; as a small number of large fish, a large number of small ones, or any combination of different sizes. Every variation in the composition of the catch will have a different effect upon the composition of the stock and its natural yield. To achieve anything approaching optimum yields it would therefore be necessary to control not simply the gross weight of fish landed, but also the numbers of each size of fish.

There are broadly two ways of studying the dynamics of fish stocks relative to fishing activities, both of which are used as bases for the formulation of fishery regulations. They are generally termed respectively the *surplus production* approach and the *analytical* or *yield-per-recruit* approach.

The surplus production approach visualizes the stock as a single unit, adopting the simple concept of fisheries already outlined. The effect of fishing is regarded as cropping the natural increase of the stock, thereby reducing the stock to a level below the limit set by the environment and promoting the production on which the fishery depends. Equilibrium yields are considered to be determined mainly by the size of the stock which can be controlled by varying the fishing intensity. The data required are fairly simple, mainly the statistics of fish catches and fishing effort. This information indicates general trends of the fishery from year to year with different intensities of fishing, and enables predictions to be made of the effects of changing the fishing effort.

The alternative analytical, yield-per-recruit method attempts a more fundamental elucidation of all factors producing changes in the size and yield of fish stocks. Instead of regarding the population as a single unit it is analysed in terms of all the individual fish constituting a number of separately recruited units, the annual year-classes of the stock. The objective is to estimate the contribution to the yield of the fishery at various fishing intensities from each year-class, or from one year-class throughout its lifespan in the stock. This requires much more detailed data over a wider range of fish biology than for surplus production studies, especially with respect to the relationships of stock size and composition to growth, mortality and recruitment.

The analytical approach offers possibilities of greater precision of prediction. It separates the effects of changes in the total amount of fishing from those of changes in the selectivity of fishing gear, such as are obtained from alterations in mesh size of nets or hook size on lines. The size or age at which fish first become liable to capture must profoundly influence the stock size, recruitment rate and potential yield of a fish population. With sufficient information it is possible mathematically to compute maximum sustainable yields (MSY) for every combination of fishing effort (boat hours) and mesh size. This leads to the concept of *eumetric fishing*, which may be defined as the optimum combination of effort and mesh giving the maximum sustainable yield.

Theoretical relationships of controlled fishing to fish populations and to fishing industries are discussed in greater depth in references listed at the end of this chapter (Cushing, 1968; Jones, 1974).

9.4   Fishery research

Many interrelated fields of study are included under the general heading of fishery research, including investigations of the distribution and natural history of fish, the size and composition of stocks, and the extent and causes of fluctuations in stocks. One of the chief objectives of this science is to achieve sufficient understanding of the factors influencing fish stocks to be able to predict reliably the long-term effects, on the quantities of fish caught, of varying the methods and intensity of fishing. Much relevant information derives from fishery statistics regarding the quantities of fish landed from each area at various times, the numbers and types of fishing boats, the fishing gear used and the duration of fishing. Additional biological data can be obtained from sampling the catches to determine the numbers of each age and size, and the presence of ripe or spent fish. The following is a brief summary of some of the methods used.

9.4.1   Distribution

Although the general distribution of the most important commercial species has been known for many years from the observations of fishermen, detailed investigations sometimes reveal that the population is not biologically homogeneous, but comprises several more or less separate breeding groups. For the rational exploitation of a stock it is necessary to know of the existence of such subdivisions, the limits of the distribution of each, the extent to which interchanges may occur between them and the contribution that each makes to the fishery in any area.

In some cases, measurements of the sizes of fish may point to a heterogeneity of the population. Normally, the length/frequency curve for fish of equal age from the same stock is unimodal. A curve having an obviously bimodal or polymodal form for fish of the same age group suggests that a mixed population is being sampled. Sometimes anatomical or physiological differences indicate subdivisions of the population. Precise studies of the anatomy of fish from different areas may reveal slight differences in structure; for example in number of vertebrae, the number of fin-rays or the pattern of rings in scales or otoliths. There may also be physiological differences between different parts of the population; for example, in salinity or temperature tolerances, fecundity, breeding season or blood cell counts. Attention has also been given to biochemical and serological studies as evidence of genetic differences between subdivisions of fish stocks (de Ligny, 1969). Fishery research attempts to discover if such differences are characteristic racial features, or are simply variations due to environmental causes.

9.4.2 Tagging

Information on the movements of fish can be gained from marking and tagging experiments (Jakobsson, 1970; Jones, 1977; Earll and Fowler, 1994). In addition to indicating the extent of migrations and interchanges of population between different areas, tagging also provides data on growth rates and for calculations of fishing intensity, stock size and mortality.

Tagging involves the attachment to the fish of some form of label, usually a small metal or plastic disc of which there are various types. For example, in plaice, a tag consisting of two plastic discs engraved with numbers for identification is commonly used. One disc is threaded on a short length of wire which is pushed through the muscles at the base of the dorsal fin at its midpoint and attached to the other disc. This can be done without drawing blood. A small reward is offered for the return of a disc with details of the circumstances of recapture of the fish. In Europe, tagging experiments have been performed on most species of commercial importance, including plaice, lemon sole, halibut, cod, haddock, hake and herring.

Tagging studies of sharks and rays are now providing particularly important information in the light of increasing, and often uncontrolled, fisheries for these groups (Earll and Fowler, 1994). In the USA, a cooperative shark tagging programme, run by the National Marine Fisheries Service, was started in 1962. It is still running and involves recreational and commercial fishermen, scientists and fisheries observers. When a shark is caught on a line, a dart tag with a message inside a capsule is harpooned into the muscle at the base of the dorsal fin with a pole. The shark is then released.

In Europe, an international system has been set up for the exchange of release information and recaptured fish data. Countries that carry out tagging studies circulate a summarized release list to all the countries likely to have these fish recaptured in its fisheries. In most countries, an agreed amount is paid to the finder for each tag returned, irrespective of the tag's origin. The tags and recapture data are then sent to the country of origin, identified by a prefix code on the tags. In the UK, the scheme is coordinated by the Ministry of Agriculture, Fisheries and Food (MAFF).

Recent developments in the use of sonic tags are providing new information on migrations of large species such as basking sharks in the UK and blue sharks in the USA. The tags can record water temperature, depth, speed and position and relay the information via satellites.

9.4.3   Natural history

Fishery biologists seek to accumulate knowledge covering the entire life of a species from egg to adult, including details of spawning areas and seasons, eggs and larval stages, fecundity, the factors that influence brood survival, nursery areas of young fish, their growth and maturation, feeding habits throughout life, changes in distribution at different stages of life and throughout the annual cycles of breeding and feeding, predators, diseases and all causes of mortality. The techniques of investigation vary according to the habits of particular fish, the nature of the area and the research facilities available.

If the fish spawn satisfactorily in aquaria, the early stages of life can usually be investigated in the laboratory. The eggs and larval stages are studied so that they may be readily identified when found at sea, and the effects on development of environmental factors such as salinity or temperature can be experimentally determined.

To discover areas and seasons of spawning, the landings of the commercial fishery may be examined. When the catch includes fish with fully ripe gonads, the regions from which such fish were obtained may be learnt from the fishermen. This gives a general indication of the probable sites of spawning, but for precise information it is usually necessary for research vessels to investigate the area. If the eggs are demersal, an attempt may be made to find them by dredging. If they are pelagic, tow net hauls are made at a large number of stations to determine precisely in which areas and at what depths the main concentration of eggs is found.

Standardized techniques of net hauling and egg counting have been devised for the calculation of the number of eggs in unit volume of water, or beneath unit surface area. The concentration of eggs at each sampling station can be plotted on a map, and lines drawn joining the points where equal numbers of eggs have been found. These 'egg density contours' indicate the zones and limits of the spawning area. As the spawning period proceeds, the areas of main spawning may shift, so these investigations need to be repeated at intervals. Such contour maps, based on the distribution of three-day-old eggs near the peak of the spawning process, are given in Lee and Ramster (1981; see Figure 9.18) for British commercial species.

Studies are also made of the fecundity of the fish of different ages, and in different areas. These may have a bearing on recruitment rates, and also provide some data which may be used for calculations of total population (see page 365).

During the pelagic phase of the life history, distribution and development of larvae and young fish can usually be studied from tow net hauls, but once the fish leave the surface water it is often difficult to find them, especially when they occur in rocky areas where fine-mesh trawls cannot easily be used. The biology of some species is not at all well known during this intermediate period between the end of the pelagic phase and the time when they become large enough to be caught by commercial nets.

Some information on feeding habits may be gained from observation of fish in aquaria, but this may not give a true picture of food preferences under natural conditions. The most reliable information on feeding habits comes from studies of the gut contents of fish captured at sea, but new techniques of underwater observation, such as free diving or underwater television, now open up new possibilities for the study of feeding and many other features of fish behaviour in natural surroundings.

Growth rates can seldom be determined by direct observation. The growth of fish in aquaria is not a reliable guide to growth in natural surroundings. Where tagging is thought to have no detrimental effect on the fish, measurements of tagged fish before release and after recapture provide some information about growth in the sea. In some cases, the width of growth zones between seasonal marks on meristic structures such as scales or otoliths is proportional to the overall growth of the fish. For example, the seasonal rings of herring scales give a close indication of the length of the fish in previous years (Figure 9.30). In most cases, however, average growth rates are determined by correlating size with age in an analysis of the stock composition (see below).

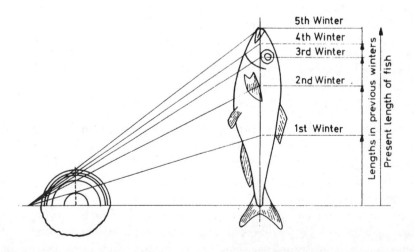

Figure 9.30 Correlarion between scale growth and fish growth in herring.

It is evident that individual fish differ greatly in their rates of growth. Intrinsic differences may be partly responsible for this, but growth rates also vary with locality due to environmental conditions, particularly the water temperature and the availability of food.

9.4.4   Stock analysis

Sampling

The first requirement for any analysis of the composition of a stock of fish is a reliable method of obtaining representative samples. Where there are no market or mesh restrictions leading to the elimination of small fish from the commercial landings, these may include the full range of sizes and provide good samples of the fishable stock. If the commercial catch is not representative of the stock, it may still supply useful data if the sampling error is known. In most cases, however, special fishing gear operated by fishery research vessels is needed to obtain good samples of all age groups. The fishing grounds are explored using trawls with cod ends covered with fine-mesh net capable of retaining the smallest fish. For pelagic fish, drift nets may be used which include a range of mesh sizes suited to the different sizes of fish in the shoals.

Age determination

In many species, age can be determined by examining the periodic markings on meristic structures such as scales, otoliths and opercular bones. The best structure to use varies between species (see Table 9.1). There are many difficulties of interpretation of these markings, and they tend to become less reliable as the age of a fish increases, but the method is useful and widely applied. The age groups of fish are commonly designated as follows:

Group 0   Fish of less than one complete year of life.
Group I   Fish between one and two years of age.
Group II   Fish between two and three years of age.
Group III   Fish between three and four years of age, and so on.

The three parts of the membranous sac in the inner ear of teleost fishes each contain an ear stone or *otolith*. The largest of these, usually the *sagitta*, can often be used for age determination. Otoliths grow by the deposition of lime on the outer surface. These layers are deposited at different rates at different seasons. By cutting thin sections of the otolith, the layers can be seen under the microscope as alternating light and dark concentric rings. Each completed year of life is represented by one darker ring. This method has been extensively used in plaice. Fish of less than one complete year of age, i.e. Group 0 fish, show only an opaque central nucleus.

Table 9.1 Usual methods for age determination in various North Atlantic fish species.

Species	Usual method	Comments
Cod	Scales	Only reliable to 3 years
Haddock	Scales	
Hake	Otoliths	
Plaice	Otoliths and scales	
Lemon sole	Scales	Otoliths are too small
Herring	Scales	
Mackerel	Otoliths	Reliable to 6 years. Scales too easily rubbed off and transferred from fish to fish
Wrasse	Opercular bones	

Scales can be used to age many fish such as cod. The surface of the scale bears a large number of small calcareous plates or sclerites (Figure 9.31), the number of which increases as the scale grows. Sclerites formed during the summer are larger than those formed in winter, and the alternating bands of large and small sclerites indicate the number of seasons through which the fish has lived. It is sometimes difficult to distinguish the zones of summer and winter sclerites clearly. In cod, the method is satisfactory up to three years of age, but as the fish become older the accuracy of the scale age becomes less certain.

Annual growth rings are found in the scales of herring but are not often easy to detect. The rings are due to slight differences in refraction of different regions

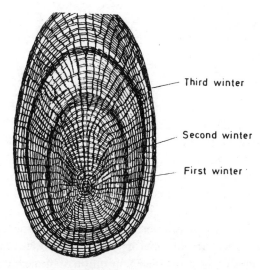

Third winter

Second winter

First winter

Figure 9.31   Growth zones in gadoid scale.

of the scale, and are best viewed under the microscope using a low power objective with dark-ground illumination. The rings are usually most clearly seen in scales taken from the anterior part of the trunk region. The rings indicate the interruption of growth of the scale that occurs during the winter months. Fish spawned in late summer or autumn probably fail to record their first winter as a scale ring, and in these the first ring relates to the second winter. Scale markings may be used to determine growth rates because the width of each zone between the rings is closely proportional to the growth in length of the fish during the period in which that zone was formed (see Figure 9.30, page 361). At sexual maturity the growth rate is reduced, and this is indicated on the scales by a narrowing of the growth zones. The pattern of scale rings therefore varies according to the age at which the fish mature. In shelf herring spawning in the southern North Sea, the first narrow zone is usually to be found between the second and fourth rings. In oceanic herring which mature later, the first narrow zone occurs after the fifth ring. In Baltic herring, most of which mature during their second year, there is usually only one wide zone, the first narrow zone being between the first and second ring.

Age census

Stock samples are examined to determine the number of fish in each age group. Where annual markings are absent, unreadable or of doubtful reliability, Petersen's method of age analysis is used. Measurements are made of the length of each fish in representative samples of stock, and length/frequency graphs are plotted. These polymodal curves can be broken down to their separate modes, each representing an age group within the sample (Figure 9.32). The age assigned to each mode of the curve can be checked against meristic markings of the majority of that group.

The Petersen method is usually applied to the analysis of cod populations because the scale markings are only reliable up to 3 years of age. The age of each length group is taken as the scale age of the majority of fish in each length group.

Studies of the age composition of herring shoals in the southern North Sea have shown that here the adult shoals contain fish from 3 to 11 years old, the maximum age normally reached by herring in this area. Off the Norwegian coast, where the fish mature later, the age range is from 4 to 15 years or older. Great variations occur from year to year in the numbers of fish entering the adult shoals (see 'Recruitment' below), and this is reflected in the relative abundance of each year class (Figure 9.33). Strong year classes, where many young survive and are recruited into the adult population, may dominate the population for many years. Knowledge of the age composition of shoals in a particular year enables a prediction to be made of the probable composition of the next year's shoals, which may be a useful guide to sensible regulation of fisheries.

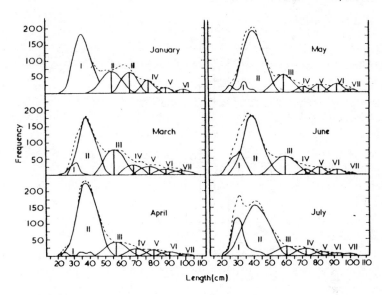

Figure 9.32 Dissection of length/frequency data for cod on Petersen's principle. Polymodal curves are pecked.
(From Graham, M., Buckland Lecture (1948), based on *Fisheries Investigation Ser.* 2, XIII, No. 4, p. 54, Figure 18, by permission of the Controller, HMSO.)

Sizes and growth rates

The modes of the length/frequency curve indicate the length distribution of each age group. From these, average growth rates can be estimated.

Sex ratio

From examination of stock samples, the sex ratio and proportion of mature and immature fish can be determined.

Stock size

There are two principal ways in which estimates of total numbers in fish stocks may be attempted. One is the tagging and recapture method (Cormack, 1968). A known number of tagged fish are released into the sea, and time allowed for dispersal. Records are then kept of the numbers of tagged and untagged fish that are captured. There are several sources of error, but taking the simplest case where mortality can be ignored, where no tags are lost, where experience of capture, tagging and release has not taught the fish to avoid nets and where the tagged fish are evenly distributed throughout a stock unaffected by migrations, the total population could be obtained from the relationship:

$$\text{Total population} = \frac{\text{No. of tagged fish released} \times \text{Total no. of fish caught}}{\text{No. of tagged fish recaptured}}$$

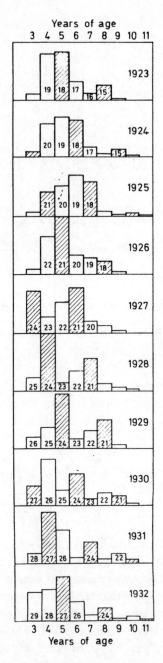

Figure 9.33   Ages of East Anglian herring shoals, 1923–32. The numbers identify the year in which each cohort of fishes was born.
(From Hodgson, W.C. (1934). *The natural history of the herring of the southern North Sea*. London; Arnold. By courtesy of the Buckland Foundation.)

As an illustration, Graham gives the following figures. Over a period of years before 1914 the results of tagging experiments on North Sea plaice indicated a fishing mortality of about 70 per cent of the stock. Average landings of plaice from the area being some 50 000 tonnes, the stock weight was presumably about 70 000 tonnes. Allowing 3000 fish to the tonne gives an average stock of 210 000 000.

Tagging experiments have demonstrated that in some areas fishery for plaice is very intensive. In experiments in the North Sea, over 30 per cent of tagged plaice have been recaptured within 12 months of their release. By relating the percentage recapture of tagged plaice to the total landings of the commercial fishery, an assessment of the total plaice population of the North Sea has been made.

An alternative method is based on egg counts. We have previously referred to the method of plotting egg density contours. From these, an estimate can be made of the total number of eggs laid in a season within a spawning area. If the average number of eggs produced by spawning females is known, the total number of spawning females may be determined. If the sex ratio is known, the total number of spawning males can also be calculated. Taking account of the proportion of the population that are immature, and any fish that may spawn outside the main spawning area, a calculation may be made of total stock numbers.

This method was applied by Wollaston to estimate the plaice population of the southern North Sea. In 1914, an exceptionally good spawning year, the total egg production in the Flemish Bight spawning area was reckoned to be about 3.5×10^{12}. Taking a mean fecundity of 70 000 eggs per female plaice, the number of females spawning in this area would be:

$$\frac{3.5 \times 10^{12}}{70\,000} = 50\,000\,000$$

This figure must be doubled to allow for the number of mature males, and should probably be about doubled again to allow for plaice spawning in other parts of the southern North Sea, making 200 000 000. As approximately half of the fishable stock are immature fish this gives a final figure for the total fishable population of 400 000 000 plaice. These figures (Graham 1956) are a revision of Wollaston's original estimate, which is now regarded as considerably too low, his figure of 200 000 for the mean fecundity of plaice probably being too high by a factor of about three.

Stock growth

The growth potential of the stock may be estimated from data on the composition of the stock and the mean growth rates of each age group, due allowance being made for differences of growth rate of the two sexes and in different areas over which the stock is distributed.

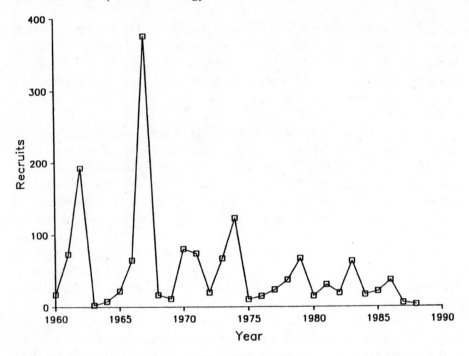

Figure 9.34   North Sea haddock recruitment, 1960–1990.
(From Shepherd, J.G. 1990. 'Stability and the objectives of fisheries management: the scientific background. Lab. Leafl., MAFF Direct. Fish. Res., Lowestoft (64), © Crown Copyright. 1990.)

Recruitment

In the North Atlantic, the number of young fish recruiting to the stocks of various species each year is very variable and so the stocks can fluctuate widely. Haddock, for example, exhibit more variability than most other stocks as shown in Figure 9.34.

This might logically be thought to relate directly to the fecundity of the mature adult females. Fecundity usually varies with size, large fish producing more eggs than small ones (Figure 9.35) so if there are more large adults in the population, the number of eggs produced will be greater. However, only a very small fraction (much less than 1 per cent) of the eggs spawned each year survive the first year of life. So it is actually slight changes in this huge death rate that cause big changes in the number of survivors that are then recruited into the adult population. The major factors causing the changes in death rates and thus controlling recruitment appear to be environmental, chiefly temperature, food supply and movements of the water. The prediction of recruitment rate is therefore very uncertain unless methods are available for sampling the younger age groups before they join the fishable stock. As previously mentioned, this phase of the life history is often the least well known.

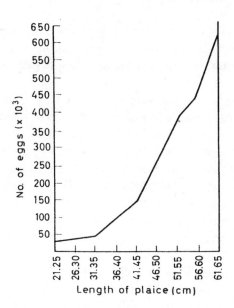

Figure 9.35 Relation between numbers of eggs and length for North Sea plaice.
(From Wimpenny, 1953, by courtesy of The Buckland Foundation.)

Total, natural and fishing mortality

Estimates of total mortality (death from all causes), natural mortality (death from causes other than fishing) and fishing mortality may be made by relating an analysis of stock composition to an analysis of the composition of the catch of the fishery. Total mortality can be derived by noting over a period of years the numbers of fish in each group of the stock. Examination of the commercial catch reveals the numbers of fish removed each year from each age group by fishing, i.e. the fishing mortality. From these, it is possible by subtraction to determine the rate at which the numbers of fish are reduced from causes other than fishing, i.e. the natural mortality.

Mortality rates may also be calculated from the data provided by tagging experiments. If the rate of recapture of tagged fish is known, an estimate may be made of the mortality directly attributable to fishing and the overall mortality from all causes, the natural mortality rate again being the difference between the two.

9.5 The regulation of fisheries

The primary purpose of regulating commercial sea fisheries is an economic one, being to ensure ample supplies of good-quality fish for consumption and to safeguard the profits of fishing (Cushing, 1975; Gulland, 1977; Gulland and Carroz, 1968; Nikolskii, 1969). Fishing is a risky enterprise, and profits must be reasonably assured to attract the large investments needed by an up-to-date industry. Conservation measures for fish stocks are therefore designed principally to preserve the fisherman's livelihood by ensuring that there are plenty of good-sized fish for him to catch, rather than simply to protect the fish. Where it is considered that overfishing is taking place, it may seem obvious that the cure is to catch fewer fish, but the fisherman may not see the matter in such simple terms. The immediate consequence of any reduction of fishing is likely to be some reduction of his earnings. It may be difficult to convince a man whose income is derived from such a precarious occupation that, by reducing his catch, he will eventually be more than reimbursed by larger yields from a recuperated stock. He may feel less confidence in the predictions of fishery scientists than in his own ability to maintain his livelihood by increasing his efforts to catch fish so long as the stocks last. If he does not catch the fish himself, he may well suspect that someone else will; and, unless restrictions can be easily enforced on all fishermen, he is probably right. Regulations to control fisheries are therefore likely to be opposed by the very people they are designed primarily to benefit. Unfortunately, conservation measures avoiding any short-term loss to fishermen are unlikely to be effective. Many of the existing fisheries' management systems, including those applicable in the European Community (see Section 9.5.2), have been unsuccessful or only partly successful. The management of many fisheries on a sustainable basis is still not a reality.

It has been argued that the dangers of overfishing are easily exaggerated because the fishing industry is inherently self-regulating. If overfishing occurs, earnings and profits fall, boats are laid up, men leave the industry, the fishing intensity reduces and stocks are able to recover. Considering that stocks undergo large natural fluctuations (see Section 9.4.4 Recruitment) that cannot be controlled, it might well be thought that the economic checks of *laissez-faire* enterprise would be as effective as any that science can devise for ensuring that fish stocks are not dangerously depleted.

Experience has not supported this argument. When landings decline through overfishing, there is a tendency for the value of the catch to rise due to scarcity. This provides an incentive for even greater fishing effort, especially if there are other concurrent food shortages. Stocks may therefore become over-exploited to such an extent that the general good would be better served by limiting fishing to a level that would allow some stock recovery, and eventually larger quantities of better fish.

9.5.1   Methods for protecting fish stocks

The principal means used to attempt protection of stocks from over-exploitation fall into two main groups: *technical* e.g mesh size restrictions, and *direct*, e.g. restricting fishing effort. Technical measures are a necessary and effective means of helping to protect stocks, but not on their own. In countries where technical methods alone have been relied on, stocks have invariably fallen and fisheries have become uneconomic. Minimum mesh sizes, for example, cannot usually be practically set high enough to prevent the depletion of spawning stock by increased fishing effort as the stock declines. It is also necessary to restrict the *quantity* of fish caught by implementing direct methods such as limits on catches (Shepherd, 1993).

Technical methods

Technical methods (Sheperd, 1993; Coffey, 1995) aim to ensure that few young fish are caught before they mature. Methods therefore include:

(a) minimum mesh sizes of nets and sizes of hooks;
(b) minimum landing sizes of fish;
(c) closed areas, e.g. nursery areas;
(d) spawning season closures.

Minimum mesh sizes

It was at one time argued that mesh regulations for trawlers must prove ineffective in protecting young fish because it was thought that, during trawling, the net becomes pulled out lengthways, almost completely closing the meshes so that any small fish entering the cod end must be retained. This was disproved in experiments by Davis and Goodchild (Davis, 1934). They enclosed the cod end of a trawl with a long bag of fine-mesh netting capable of trapping any small fish that might escape through the cod end. The fine-mesh bag was encircled by a noose which after a period of trawling was automatically drawn tight, closing the bag. This demonstrated conclusively that large numbers of small fish passed through the cod end unharmed.

Placing a fine-mesh bag around the cod end also provides a means of studying the extent to which different sizes of fish are retained by different meshes of trawl. By counting the number of each size of fish in the cod end and in the bag, these data can be used to construct a graph showing the percentage of each size of fish that escape through the mesh. This is termed a percentage release curve or selection ogive, and usually has the form shown in Figure 9.36. We see that, for each mesh size, all fish above a certain size are retained in the net. There is also a lower limit of size beneath which virtually all the fish escape. Between these two limits, the proportion of fish which are caught or escape varies with their length, there being a particular size at which 50 per cent of the fish escape. This 50 per cent release length is a convenient index for the comparison of the selective action of different meshes.

There is, of course, no size of mesh that is equally suitable for all species of fish. If the mesh is selected primarily to protect young cod, it will let an undue proportion of haddock escape. A mesh suitable for plaice will let through most of the sole. In the case of large fish such as cod, total reliance on mesh size

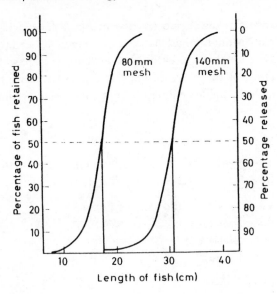

Figure 9.36   Selection ogive for plaice in nets of 80 and 140 mm.
(From Wlmpenny 1953, by courtesy of The Buckland Foundation.)

restrictions for conservation of the stock would necessitate a mesh size of about 200 mm which is much larger than that currently in use (currently 100 mm). Estimates of how the spawning stock biomass of North Sea cod would increase or decrease with changes in mesh size used and in relation to fishing intensity have been calculated (Shepherd 1993). Fisheries biologists widely consider 20 per cent of the unfished level to be the minimum desirable level for spawning stock biomass. Another lower level, the minimum tolerable level, is also calculated. Below this level, biologists consider the stock does not have a better-than-even chance of maintaining its size through recruitment. At current fishing levels, with the current mesh size of 100 mm, the stock would be kept below not only the minimum desirable level but also below the minimum tolerable level. To alleviate this by changing mesh size alone, without restrictions on catch and effort, would require an increase in mesh to at least 140 mm and preferably 180 mm. This is not practical. In the Georges Bank fishery for cod, haddock and yellow flounder, all large species, reliance on a minimum mesh size of 140 mm for the past 10 years has led to a drastic decline (Sheperd, 1993).

Direct methods
Direct methods aim to limit the quantity of the catch and to make sure the total death rate is low enough to ensure stock maintenance even if young fish are caught. Methods include:

(*a*)   limits on catches, e.g. total allowable catch per annum (TAC);

(*b*)   limits on the length of the fishing season or hours of fishing permitted;
(*c*)   limiting the size of the fishing fleet;
(*d*)   temporary bans on the fishing of particular over-exploited species in particular areas.

Thus, although the suitability of different methods of control varies with the nature of the fishery, it is generally necessary to set a limit to the total catch of a particular species, as other protective measures may not be completely effective in preventing overfishing because the fishing effort can often be sufficiently increased to circumvent them.

9.5.2   European Community Common Fisheries Policy (CFP)

In the north-east Atlantic, where most British fishing takes place, the problems of devising satisfactory fishery regulations are especially complex because of the number of nations involved and the range of species sought. The European Economic Community (EEC) was first formed at the signing of the Treaty of Rome in 1957 with six member countries: West Germany, France, Italy, Belgium, Luxemburg and the Netherlands. Since then the UK, Ireland, Denmark, Spain and Portugal have joined and various fisheries regulations have been implemented.

An operational Common Fisheries Policy (CFP) is now in place within the EC. Various regulations have been introduced since 1970 and the legislation is quite complex. The policy covers five aspects of fishing: access, conservation, market management, production and marketing structures, and the organization of international relations. Details are given in Coffey, 1995. The basics of the agreement are outlined below:

(*a*)   National fisheries jurisdictions were extended to 200 miles (1976 Hague Agreement) along the North Sea and Atlantic coasts (but not the Mediterranean, Baltic, Skagerrak or Kattegat). The Exclusive Fisheries Zone (EFZ) thus extends from the baseline out to 200 nautical miles. The baseline is generally low water mark or sometimes a straight line across bays. The EFZ came into force on 1 January 1977. This greatly increases the area over which coastal states can supervise fishing operations. For example, the UK now claims fishing rights over a considerable part of the north-east Atlantic. Non-EC ships can only fish these waters with UK consent and in accordance with UK regulations.

(*b*)   Access to EC waters under the CFP is one of equal access for all member states to each other's fishing zones. However, within the 0–6-mile coastal zone, access is restricted to vessels registered in the coastal state. Between 6 and 12 miles, certain EC states are allowed to fish in particular areas for particular species based on historic rights of usage. This arrangement holds until the year 2002 when the present system comes up for renewal.

(c)   The CFP includes various standards for states fishing in community waters, aimed at conserving stocks. These include minimum mesh sizes, minimum landing sizes, permitted levels of by-catch of edible species with industrial species, and areas where fishing is prohibited or limited at certain times of the year to protect spawning and nursery areas.

(d)   The EC has also introduced a quota system of total allowable catches (TAC) for major commercial species stocks in particular areas. These quotas are specified at the beginning of each year by the Council of Fisheries Ministers. The scientific information on which the TACs are based is provided by various organizations, principally the Advisory Committee for Fisheries Management of the International Council for the Exploration of the Sea (ICES), and the Northwest Atlantic Fisheries Organization (NAFO). The areas used are ICES blocks (Figure 9.37).

9.5.3   Difficulties with fishery policies

The CFP faces a great many difficulties and is by no means a fully effective system. Disputes between member states are common and at times can even lead to violence. Politicians and scientists are often in disagreement particularly concerning TACs and such aspects as discards and bycatch. Nevertheless, despite these difficulties a degree of accord has been reached on aspects such as minimum mesh and landing sizes, closed areas and seasons for certain fisheries, and regulations on size and use of particular types of fishing gear, notably beam trawls, in certain areas. It is the responsibility of each coastal state to supervise and enforce the regulations within its own fishery zone, but the degree of enforcement certainly varies between states.

The problems of framing beneficial fishery policies are made more difficult by the far-reaching effects of fishery regulations, often involving many people not directly employed in fishing. If fishing is reduced by cutting down the number of boats, fishermen are put out of work. Unemployment among fishermen has severe social consequences for local communities where fishing is a traditional way of life. More jobs are also lost in associated occupations, such as in shipbuilding, especially in small yards, in fishing gear manufacture and in enterprises involved in handling, processing and marketing fish. There are probably at least five shore jobs dependent on each fisherman's employment.

However, if fewer boats put to sea, those fortunate enough to remain fishing are likely to make larger catches; and if the total quantity of fish landed is less, the price of fish may rise and profits for some may be greater. If national quotas are allocated, it may well be more profitable to take the permitted catch with a few boats and very large nets than with many smaller boats using less effective gear.

There are consequently many interests to be kept in mind in formulating fishery

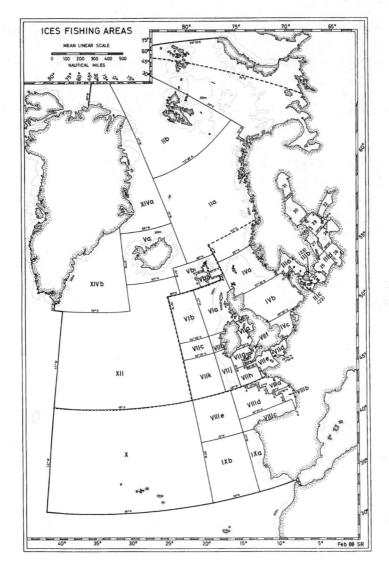

Figure 9.37 The European Community's waters showing ICES blocks.
(Reproduced by permission of the International Council for the Exploration of the Sea.)

policy. The public needs a good supply of fish at reasonable prices. The industry needs to be assured of fair profit. There are some communities for whom fishing is a heritage which should be protected even though it may be done in ways which are less profitable than those which modern technology can offer. None of these interests can in the long term be safeguarded without rational conservation of the fish stocks on which they depend.

9.6   The seas in relation to human food supplies

9.6.1   World population and food supplies

World population in 1992 stood at 5420 million, and is currently increasing by about one thousand million (one billion) every 10 years. UN world population projections made in 1992 estimate 6228 million in 2000 and 8472 million in 2025. The rate of increase is less than had previously been predicted, due to increased efforts at population control using family planning. Even so, these estimates indicate a greatly increasing demand for food. The highest population growth rates are in Asia, Africa and South America where scarcity of food is most prevalent.

While it is extremely difficult to assess the world's food supply and requirements, it is estimated that at present 10–15 per cent of the world's population are underfed, having insufficient food to provide even enough calories for a normally active life. In addition to these hungry millions, there are many more, possibly as many as 30 per cent of the total, whose calorie intake is adequate, but who suffer from malnutrition because their diet lacks various substances essential for health.

The food situation differs greatly in different areas. In Western Europe, North America, Australia and New Zealand the consumption of food is, in general, fully adequate for daily needs, and health problems arise more from over-indulgence or improperly balanced diets than from any insufficiency of quantity. In Africa, South America and the Near East, however, food intake is precariously balanced with requirements, and often inadequate, while serious food shortages occur in parts of Asia.

In all communities, inadequate diet is closely connected with poverty. In such conditions infant mortality rates are usually high, partly the result of malnutrition. High infant mortality encourages high birth rates in compensation, and in some communities family planning is unlikely to reduce birth rates appreciably until there is a reasonable expectation that children will survive. Consequently, effective measures of population control require the raising of standards of nutrition, hygiene and health in the poorest populations. The problems of feeding the increasing world population, stabilizing its size and achieving a satisfactory balance between food supply and demand are therefore closely interrelated. It is not solely a matter of expanding food production but also of ensuring a more equitable distribution of wealth and a proper apportionment of food to bring all diets to adequate levels.

To what extent may we reasonably hope for greater supplies of human food from marine sources? Seventy-one per cent of the earth's surface is covered by seawater. Although production rates in the sea are generally less than on land, overall the organic production of the oceans cannot be very much less than that of the land surface. Yet only about 1 per cent of the total supply of human food comes directly from the sea, and only about 10 per cent of human consumption of animal protein derives from the sea either directly or indirectly via fishmeal fed to livestock. Fish is a high-protein food of excellent nutritive quality, in some respects better than meat, and any increase in fish supplies would be a valuable supplement to the world's sources of protein. Many fish are also rich in edible oils, and fish livers are an important source of vitamins A and D. Whales, molluscs and crustaceans are other marine groups which are useful foods, and might in some cases be more fully exploited, especially the cephalopods. It is also possible that food might be obtained from the sea by unconventional methods, such as direct harvesting of the marine plankton. We will briefly consider some of the ways in which the seas might make a greater contribution to the world's pressing need for more food.

9.6.2 Prospects of increased fish landings

During the past 30 years there has been so great an expansion of world fisheries, and world landings of sea fish have risen so steeply, that virtually all easily accessible fish stocks must now be regarded as fully or over-exploited. This is backed up by the fact that between 1970 and 1980 world landings hardly rose at all, and since then landings have risen only slowly despite greatly increased fishing efforts and now appear to have fallen since peaking in 1989 (Figure 9.25).

The need to relieve pressure on local fish stocks has led to searches further afield. Several nations now have long-distance fleets operating far from home waters. British fishing vessels have travelled to Australian and Antarctic waters in hopes of finding new supplies, though with disappointing results. The pelagic long line fishery by Japanese and Korean ships, mainly for tuna, now extends over all the warmer parts of the major oceans. Fisheries for hake and a variety of pelagic species have been rapidly developed since the mid-1960s, by fleets from Europe, the USSR, Japan and Korea in the productive areas of upwelling along the west coast of Africa. There are already grounds for fearing that some of these relatively newly exploited stocks are being overfished. If there are any remaining underfished stocks, they must be mainly in remote areas, possibly off the southern part of South America or around the East Indies. There may also be some unfamiliar demersal species living well down the continental slope that might be worth fishing if conservative habits of diet are overcome. Alternative methods of presentation, such as 'fish fingers', have certainly helped to extend the market to a wider variety of fish. But prospects of finding major new sources of fish, at present unexploited, do not seem very promising. On the contrary, in the immediate future there may possibly be some contraction of world fishing as coastal states seek to protect their resources by extending their territorial fishing limits, thereby excluding the fishing boats of other nations except by special agreements. As fishing within the 200-mile fishery zones becomes more subject to restrictions, some decline in world landings is to be expected before the benefits of conservation measures become apparent.

In some underdeveloped parts of the world, the productivity of local fisheries is severely limited by their primitive methods of fishing, preserving and marketing. Small craft can only make short voyages, with the result that the close inshore grounds are intensively overfished. Much better landings might be obtained if the fishermen could extend their operations over wider areas. Furthermore, the consumption of sea fish is often restricted to coastal regions because little is done to preserve the catch for distribution inland. The expansion of these primitive fisheries requires large capital investments for the mechanization of fishing, for the training of fishermen in new methods, and for the provision of proper facilities for preservation and distribution.

The difficulty of finding the expertise and large sums of money needed for equipping, crewing and maintaining modern fishing vessels has led some coastal states to negotiate with foreign companies to allow them to fish within their fishing zones. These concessions do not necessarily bring much benefit to coastal states because, although there may be some increase in local fish landings, the greater part of the catch is liable to be removed for marketing in more affluent areas where fish that the local population would use for human consumption are instead converted to fishmeal.

The classic history of the Peruvian fishery for anchoveta (*Engraulis ringens*) has been a remarkable example of the rapid development of a primitive fishery to become the world's major source of fish, unfortunately followed by a dramatic decline. During the period 1948 to 1968 this fishery increased over a hundred-fold to 10 million tonnes per year (more than one-sixth the entire world catch), providing Peru with a larger catch than any other nation. In 1970 the fishery took 12.5 million tonnes, but by 1973 landings had fallen to little more than 2 million tonnes. This failure exemplifies some of the problems of distinguishing the effects of overfishing from those of concurrent environmental changes. The collapse of the fishery occurred at a time when changes of the circulatory pattern of the area reduced the extent of upwelling of deep water along the continental slope, upon which the area's prolific production of marine life depends. There had previously been signs of impending threat to adult stocks because fewer young fish were entering the shoals, probably the result of over-exploitation. Although there has subsequently been some recovery of the fishery, the 1987 catch was only 2.1 million tonnes and stocks remain low to date.

Transplanting young fish

Various suggestions have been made of ways in which the productivity of the more intensively fished areas might be maintained or increased. For example, yields might be increased by transplanting young fish to areas of more abundant food.

Since the early years of the twentieth century Professor Garstang repeatedly advocated the transplantation of young plaice, and demonstrated a three- to fourfold increase in growth rate of young North Sea plaice moved from their crowded coastal nurseries to the Dogger Bank, where bivalves suitable as plaice food are plentiful. Calculations indicated that, on a large scale, this could prove a profitable enterprise, but such a project would require close international cooperation to ensure that the transplanted fish were left long enough in the sea to benefit, and that costs and profits were equitably shared.

Stocking the sea

There have also been hopes that it might become possible to increase the stocks of certain species of marine fish by artificial rearing during the early stages of life. The first few weeks are a period of special danger when extremely heavy fish mortality always occurs. The number of fish surviving in each year-brood seldom bears any close relationship to the number of eggs spawned; generally, the more eggs, the greater the number of casualties. Survival is mainly determined by environmental factors, for example, temperature, salinity, food supply, currents and predation (see Section 9.4.4 Recruitment). It might, however, be possible to increase the stocks of fish if eggs and larvae could be raised in very large numbers in protected conditions, and supplied with ample food until the danger period is passed before setting them free in the sea. The unusually heavy catches made on some of the north-east Atlantic fishing grounds immediately following the two World Wars during which very little fishing took place, indicate that these areas are able to support many more fish than they ordinarily do when fishing is proceeding at peacetime rates.

In experiments during the 1950s, J.E. Shelbourne (1964) and his colleagues of the Lowestoft Fisheries Laboratory (UK) developed successful methods of rearing large numbers of young plaice by stocking the open circulation seawater tanks at the Marine Biological Station at Port Erin, Isle of Man, with plaice eggs spawned in captivity. Methods were devised for bulk preparation of suitable planktonic food for the developing larvae, mainly nauplii of *Artemia salina*. A measure of bacterial control was achieved by treating the water with antibiotics and ultraviolet light. In these conditions many thousands of young plaice have been reared to the completion of metamorphosis, and survival rates of over 30 per cent of the original egg stock have been achieved. However it became apparent in due course that the number of artificially reared fry that could conceivably be released into the sea each year is infinitesimal compared with the number produced naturally, and that no significant contribution to north-east Atlantic plaice stocks was feasible by this means. Release into the sea of hatchery fry in this area therefore ceased and the experiments were reorientated towards the raising of hatchery fish to market sizes in captivity, i.e. fish farming (see below).

Despite the discouraging conclusion of work on the release of plaice larvae into coastal waters around the UK, restocking the seas may be a more realistic proposition with other species in particular areas. In the Sea of Japan artificially cultivated young of red bream (*Sparus major*) and prawns (*Penaeus japonicus*) are regularly released with the objective of improving fishery landings. In the Caspian Sea, sturgeon fisheries have suffered badly from over-exploitation and the damming of vital spawning rivers. Many have been successfully revived by the yearly release of hatchery-reared fry. In Prince William Sound in Alaska, hundreds of millions of pink salmon fry (*Oncorhynchus gorbushca*) are released from hatcheries every year. This species has a relatively simple two-year life cycle and such releases have been quite successful in boosting harvests. In the North Atlantic there are some prospects that, if netting at sea could be adequately controlled and costs of rearing equitably shared, release of hatchery-reared smolts of Atlantic salmon (*Salmo salar*) to feed naturally at sea might prove more profitable than ongrowing them in fish farms on expensive food which might be better used in other ways.

Although we must conclude that there is no immediate prospect of obtaining much larger quantities of food from more intensive exploitation of the natural fish stocks of the oceans, we may reasonably hope that measures to conserve these resources by scientific management of fisheries may eventually lead to some further increase in world landings. Beyond this, there are suggestions that the productivity of fisheries might be enhanced by modification of the ecosystem. For instance, we have previously referred to the competition for food that exists between bottom-feeding fish and the large numbers of predatory invertebrates, and to Thorson's calculations that 'only 1–2% of the fish food is actually eaten by fish, the rest is taken by invertebrates' (see Section 6.4.5). An area could presumably support a greater number of food fish if competitors could be eliminated, or at least considerably reduced. Hardy (1959) in his classic text, suggested that it might eventually become possible to weed the sea-bed of 'unwanted' creatures so that a larger proportion of the food becomes available for fish. He writes:

> If Thorson's calculations are correct, and if man could eliminate just a quarter of the pests and so allow the fish to have some 20%, instead of 2%, of the potential food supply, then he could make a given area support ten times the quantity of fish. How are such pests as star fish to be eliminated? I believe that just as we harrow and roll the land in addition to reaping our crops, we shall in time systematically drag some combing or other devices over the sea floor at intervals to weed out the creatures that take food from the more valuable fish; and the pests themselves may well be ground into meal to feed poultry ashore. It will require much more ecological research to determine just which of the animals we can do with in smaller numbers.

In the present climate, with increased knowledge of the complexity of ecosystems and the interrelationships between species, this approach seems unlikely to gain acceptance even if it became feasible.

9.6.3 Aquaculture

Until relatively recently, man has acted only as a hunter-gatherer in exploiting the oceans. Now, with fish stocks falling and fisheries failing, there is an upsurge in interest in farming the oceans. Freshwater aquaculture already plays a very important part and some estimates suggest that one in five fish that end up on the dinner plate is a farmed fish. Increases in total aquaculture production (freshwater and marine) have nearly doubled in the past 10 years with most of this due to freshwater culture.

On a local scale, freshwater fish culture has been practised from early times, in many parts of the world, particularly in the warmer areas. The Chinese were rearing carp 4000 years ago and in developed countries, before the advent of modern refrigeration techniques, the fish pond provided a ready source of fresh protein. Many, usually fast-growing, vegetarian species are cultivated in shallow ponds. The growth of suitable pond weed for their food is encouraged by enriching the water with sewage or organic refuse. Where sunshine keeps the water temperature high, biological processes proceed very rapidly and remarkably high rates of food production can be obtained from efficiently managed fish ponds. Capital and labour costs may be low, and land unsuitable for ordinary agriculture can often be profitably farmed in this way although there are often problems with water supply.

Cultivating marine fish – *mariculture* – presents greater problems, but these problems are now being addressed. As well as finfish, a wide variety of other groups is cultivated worldwide, of which the most important are molluscs, crustaceans and seaweeds (see Table 9.2). The techniques and species used are described in detail in Barnabe (1990) and specific aspects relevant to the UK, in a number of laboratory leaflets produced by MAFF (see Appendix 4). Currently around 6 per cent of the total worldwide marine harvest comes from cultured sources. In the developed world, efforts are mainly concentrated on species that command a premium market price such as salmon. There has been a dramatic increase in commercial mariculture in Europe and America in the past 10 to 20 years. Cultivation of species such as salmon and sturgeon may take the pressure off wild stocks but does little at present to increase the supply of world fish protein. Most of these fish are carnivorous and are fed on fishmeal prepared from wild-caught fish. The high price they fetch also means they are destined mainly for those who can already afford other sources of protein. As better feeds are developed, possibly from vegetable sources, this situation may change.

At the other end of the scale, particularly in the East Asian seas region, traditional small-scale, low-technology techniques are used to rear a wide range of fish, crustaceans and molluscs. This provides cheap protein and employment for local people. However, systems such as the Indonesian *tambak* ponds take up a lot of space and still rely on the collection of wild-caught fry and, to a large extent, on natural tidal flushing to provide the feed. Intensive pond production

Table 9.2 The most commonly cultivated marine species.

Species	Area	Installation
Atlantic salmon (*Salmo*)	Northern Europe	Anchored floating cages
Pacific salmon (*Oncorhynchus*)	North America, Japan	
Flatfish (various)	Northern Europe, Japan, Korea	Onshore tanks, cages, semi-open coastal waters
Milkfish (*Chanos*)	Asia-Pacific mainly SE Asia	Brackish coastal ponds
Mullet (*Mugil*)		
Yellowtail (*Seriola*)	Japan, Korea, SE Asia	Anchored floating cages
Groupers, bass, bream, snappers, etc.	Mediterranean, Japan	Anchored floating cages; semi-open coastal waters
Oysters (*Ostrea, Crassostrea*)	Worldwide	Natural rocky seabed; hanging ropes
Mussels (*Mytilus*)	Worldwide	Natural rocky seabed; posts (buchot); hanging ropes
Clams (*Tapes, Mercenaria, Anadara, Mactra, Meretrix, Haliotis* (abalone))	Worldwide	Seabed
Shrimps and prawns (*Penaeus, Metapenaeus*)	Asia-Pacific, S. America	Ponds
Seaweeds (various)	Asia-Pacific	Suspended culture

systems using artificial feed and pumped water are being increasingly used. These can create pollution problems and may require clearance of valuable natural habitats such as mangroves.

Mariculture requirements and methods

A primary requirement for animal farming is to keep one's stock within a protected area, where they are safe from predators and can grow under controlled conditions without severe competition for food from unwanted species. Oysters and mussels, once they have settled, remain virtually stationary, and the beds where they are grown can to some extent be protected from enemies and competitors. But fish roam about, often over considerable distances, and fish-pens cannot easily be constructed in the open sea.

A second requirement for efficient husbandry is the selection of stock by controlled breeding for particular desired qualities such as rapid growth, efficient food conversion, taste, appearance, etc. For some marine species this is not yet possible because they do not complete their life cycle in captivity although they may be suitable in other respects for culture, e.g. eels (*Anguilla anguilla*). In others, such as the freshwater *Tilapia*, there has been considerable success in selective breeding.

It might seem a simple matter to make enclosures for cultivating sea fish by erecting dams or screens across coastal inlets. However, as soon as small bodies of seawater are isolated the natural processes of circulation and renewal of water are interfered with. Problems then arise from fluctuating and abnormal hydrographic conditions (temperature, salinity, deficient oxygenation, etc.) which are likely to be unfavourable for many marine species. Even those which may be suited to shallow water when young may require deeper water as they grow larger. Fast-growing marine fish suited throughout life to conditions in enclosed lagoons are mostly insufficiently palatable to fetch good prices in markets where there is much choice.

Enclosure of fish in bays and inlets may also present difficulties of adequately protecting the stock. Nutrients added to enrich the water so as to promote the natural production of fish food may be lost by leakage to the sea, or may stimulate the growth of unwanted or even pathogenic organisms. There may be difficulties of weeding to keep the area free from competitors which take the food needed for the fish, and of safeguarding the potential harvest from predators. There may also be legal problems associated with enclosure of sea areas, involving establishment of ownership, rights of navigation, fishing, etc.

Therefore in most cases, finfish farming is carried out in shallow onshore ponds, in floating cages moored in sheltered bays, inlets or sea lochs and in onshore tanks.

Pond culture

Salt and brackish-water pond culture is the prevalent system in many Asian countries including Malaysia, Indonesia and the Philippines. Ponds are stocked with species of grey mullet (*Mugil*) or with milk fish (*Chanos chanos*). Mullet are also cultivated in various lagoons around the Mediterranean. These are hardy species which can tolerate a wide range of salinity and temperature, and have the further advantage of being vegetarian feeders, mainly on detrital and microscopic plant material. The food chain is therefore short and corrrespondingly efficient, and the fish make rapid growth. In some of the brackish fish-ponds in Indonesia, enriched with sewage, production estimates for *C. chanos* have been as high as 5 tonnes per hectare per year.

This system of marine fish husbandry is fairly primitive. Almost always the fry are obtained from the wild stock at sea and raised in captivity to market size. In some cases the fry can be persuaded to swim upstream from the sea into rearing ponds, but usually they must be collected at sea by fine nets, sometimes necessitating distant searches. The stocking of ponds with milkfish is liable to be hampered by shortage of fry due to uncertain location of spawning grounds. Problems also arise from the destruction of natural habitat to make room for the fish ponds. In the Philippines, mangrove forests now cover only about 140 000 hectares compared with 400 000 hectares in the 1920s. At least 60 per cent of this loss is due to conversion to milkfish farms. This can affect offshore fisheries since many offshore species spawn in mangroves.

Cage culture

Fishes that are successfully grown in cage cultures include Atlantic salmon (*Salmo*) in Europe and America; Pacific salmon (*Oncorhynchus*) in America; the yellowtail (*Seriola*) in Japan; sea bass (*Lates*) and groupers (*Epinephelus*) in Malaysia and Indonesia.

One of the most dramatic examples of recent growth in fish farming has been the exponential increase in the production of salmon in the North Atlantic region. Much of this expansion has been in the siting of floating fish farms in Scottish sea lochs (Lincoln and Howell, 1994; La Tene Maps, 1995). Fifteen years ago, such farms were a novelty. Now there are practically no sea lochs left which do not have at least one fish farm. Annual Scottish production of farmed salmon stands at around 50 000 tonnes whereas ten years ago it was only around 10 000 tonnes (SOAEFD annual surveys; see Appendix 4). However, the greatest production in this region comes from Norwegian farms.

Eggs are hatched from selected stock and the young fish raised in fresh water to the smolt stage, when they would normally go downstream to the sea. The captive fish can then be transferred to seawater in floating cages where they can be grown on to market size. They are fed on pelleted food. In sea lochs off the west coast of Scotland these fish reach 1.5–2 kg after one year in seawater and 4–5 kg after two years.

These farms require a large initial capital outlay and in Scotland many of the new smaller outfits have not survived. One of the problems encountered is poor siting of farms in areas of low water movement. This leads to pollution of the sea-bed with consequent deoxygenation problems and disease. Under crowded conditions, fish lice can be prevalent and this is now being tackled biologically as well as chemically. Experiments are in hand to see if various species of wrasse, known to act as 'cleaners', can control the lice.

Onshore tanks

In the UK, experiments in the rearing of juvenile flatfish in tanks were initiated in the 1950s by the Lowestoft Fisheries Laboratory and the Marine Biological Station in the Isle of Man. This early work was later used as a basis to provide young plaice and Dover sole for further rearing. Fish of marketable size were produced in less than half the time taken in the sea, by keeping them in tanks supplied with warm water from the cooling system of the power station at Hunterston, Ayrshire. Other power stations also supported small farms but they were all uneconomic and none is in production today. However, the use of warm-water effluents is still being pursued in France and the Graveline nuclear power station near Calais supports one of the largest European culture units for bass and turbot.

Attention turned next to turbot (*Scophthalmus maximus*), another high-priced fish though less valuable than sole. The early larvae of turbot require very small food, and this was provided by developing the culture of rotifers (*Brachionus*). When the fry have grown larger they will take *Artemia*. Turbot have proved to be very suitable fish for tank cultivation in warm water as long as the temperature is controlled around an optimum level. They can live at high densities, have a good food conversion efficiency and make relatively rapid growth to about 2 kg in two years in optimal conditions. The life cycle can be completed in captivity, with possibilities of improvement by selective breeding.

Other species which have been successfully grown in tanks are brill and lemon sole. It is also possible in tanks in this country to rear the prawn *Palaemon serratus* and various bivalves, notably oysters (*Crassostrea*) and American clams (*Mercenaria mercenaria*).

Fish factories
Recent 1990s technology has seen the development of prototype 'fish factories' which aim to mass-produce both seawater and freshwater fish. These 'biosystems' are closed units which function on a very efficient re-cycling of seawater and so can be situated well away from the coast on normal industrial estates. The system is being developed by the Danish Institute of Aquaculture Technology and is already producing freshwater eels. Commercial production of marine species such as cod is in development and may well become a reality in future years.

Shellfish
Shellfish such as oysters and mussels need only a fairly primitive form of husbandry to attain worthwhile yields and have the potential to make a significant contribution to world protein production. Some remarkably high rates of meat production are claimed. For example, annual yields of about 250–300 metric tons of meat per hectare have been quoted for raft culture of mussels around the Spanish coast. Japanese raft culture of oysters achieves 50 tonnes of meat per hectare per year. In the UK both mussels and oysters are widely cultivated (Spencer, 1990).

A variety of culture methods is used around the world. In France, mussels are traditionally grown on upright posts or *buchots* and also as beds on the seafloor. Both oysters and mussels are commonly grown on hanging ropes suspended beneath rafts. In some areas such as Xiamen harbour in China, the rafts are so densely packed that it is difficult for boats to find their way between them. Culture depends largely on larvae (spat) settling naturally on the substrate provided by the cultivator. In some cases, the spat is collected from settlement areas and then transferred to trays, mesh bags or beds for growing on. In the Irish Sea, experimental spat collectors for scallops have been developed at the Marine Station on the Isle of Man. The idea here is not to grow the scallops to maturity but to seed fishing areas with young scallops.

Molluscs are cultivated not only to provide food, but also for other products such as pearls. Recently experiments have started in Australia, Micronesia and Papua New Guinea, in rearing giant clams (*Tridacna*), valued both for their meat and their shells. If successful, this may relieve pressure on threatened wild populations. Other cultivated molluscs include cockles, the abalone (*Haliotis*) and various clams.

Crustaceans and seaweeds

Prawns and shrimps are the most widely cultivated crustaceans and are reared mainly in the warmer parts of the world using pond culture techniques. Prawn culture in Asia currently accounts for three-quarters of world production. The giant tiger prawn (*Penaeus monodon*) which weighs up to 500 g is reared in intensive systems both for the markets and as a valuable export commodity. Most large supermarket chains in the UK have this species on sale. Lobsters and crabs are also reared in captivity but on a much smaller scale. There are problems with aggression amongst adults and with successful rearing from eggs.

Seaweeds are widely cultivated in countries such as Japan and the Philippines. Red algae of the genus *Eucheuma* are grown from cuttings in the Philippines, attached to nylon ropes covering the seabed. *Porphyra* is grown on floating rafts in Japan and Korea and sold dried in the markets as *nori*.

Problems

One of the problems of marine fish farming is to find suitable sources of fish food. During their early stages of growth, fish larvae require very small food particles and some take only live food. This need for bulk production of suitable planktonic foods adds to the other problems of fish farming. Even if live food is not essential, minced fish and similar finely divided foods are very prone to bacterial contamination and consequent detrimental effects. There is some hope that the invention of microcapsules as artificial food particles may prove to be a useful contribution to aquaculture. Precise mixtures of food materials suited to the requirements of particular organisms can now be encapsulated within artificial membranes, producing particles of controlled composition and size, some of which are readily accepted as food by certain small organisms or by filter feeders (Jones *et al.*, 1974; 1979).

Many of the currently farmed finfish species around the world, especially those in Europe and America, are carnivorous, requiring relatively expensive animal food. The farming of carnivorous animals obviously reduces rather than augments the supply of animal protein. In addition, most of the fish farming at present undertaken in developed countries converts fish meat from one form to another. The species used in the feed products are not just those that are unpalatable to us. So-called 'industrial' fishing scoops up huge quantities of young edible species and includes herring, anchovies and, in the UK, sandeels. These are converted to fishmeal which is used for feeding chickens and other livestock as well as cultured salmon and other farmed fish. This is obviously not an efficient way of providing protein to the human population. Variations in the supply of wild fish suitable for conversion to fish food sometimes cause changes in the quality of the food over which the farmer has no control and which may be detrimental to his stock.

An ideal fish food would be made from materials unsuitable as human food or for feeding to other livestock, presented in dry, easily stored and handled forms, readily accepted by the fish, nutritionally complete, easily digested and assimilated efficiently, pathogen-free and cheap. No food combining all these properties has yet been developed, although work is being done on producing protein-rich and nutritionally adequate fish foods from soya and other plant material.

Cultivation of vegetarian species avoids the need for expensive, high-protein food. In Asia, vegetarian species such as the milkfish (*Chanos chanos*) do well in captivity. The only European marine fish which feed largely on vegetable matter and are well suited to culture are grey mullet, but their flesh is not widely popular. Bivalve molluscs do not need to be fed artificially as they filter phytoplankton from the water.

Another major problem facing the continued expansion of mariculture as a means of providing more protein to a hungry population is the increasing coastal pollution throughout the world. Whether the mariculture units used are based onshore or in the natural marine environment, they rely on a good supply of clean seawater. Larval stages of many marine species are sensitive to pollution. In addition, the flesh of captive fish can become 'tainted' from pollution incidents such as oil spills. Shellfish can also accumulate heavy metals and toxic chemicals and become dangerous to eat. Blooms of plankton (red tides) caused by eutrophication of coastal waters can also produce toxins that are accumulated by shellfish.

9.6.4 Increasing the fertility of the seas

If fish farming in enclosed bodies of seawater presents difficulties, why not simply raise the productivity of the open sea by enriching the surface waters over wide areas by the addition of plant nutrients? We spread fertilizers on the land to promote the growth of crops; why not spread them on the sea?

To produce any appreciable increase in concentration of plant nutrients in the open sea would require enormous quantities of fertilizer, and the costs would be tremendous. It would be an extremely wasteful process because so small a proportion of the nutrients absorbed by phytoplankton eventually becomes incorporated in fish flesh. Any additional plant growth obtained as a result of fertilization would contribute very largely to the production of unwanted organisms. Calculations of the increase of yield from sea fisheries that might be obtained by large-scale fertilization of seawater do not stand up to comparison with those known to be obtained from the use of equal quantities of fertilizer applied to the land.

None the less, there are unintentional processes of artificial enrichment of seawater going on in some areas, though not with the deliberate aim of benefiting fisheries. Close to large centres of population, great quantities of sewage are discharged into the sea, which decompose to provide nutrients for marine plant growth. For example, the fertility of the southern North Sea is augmented by the outflow of London's sewage via the Thames. It may be doubted, however, whether this form of sewage disposal is really in our long-term interests. Present-day problems with coastal pollution from sewage are currently in the limelight and are described in Chapter 10. In 1948, Dr L.H.N. Cooper, chemist at the Plymouth Laboratory, wrote:

> As matters now stand, very large amounts of nutrients are being poured into the sea, the great sink, as sewage from coastal towns and by way of the rivers from inland towns and farms fertilized and unfertilized. Phosphorus is a very precious commodity which in not so many years will become very scarce. The scale on which phosphorus even now is being dissipated to the sea is more than the world can afford. In years to come the cry will be for more methods for recovering phosphorus from the sea, not for putting it in.

Other measures for raising the productivity of seawater have been proposed which do not involve the addition of valuable fertilizers. The seas already contain vast reserves of plant nutrients in deep water, and it might be possible to devise means of bringing these to the surface. Any measures which increase the mixing of surface and deep water would be likely to lead to increased production. For example, it has been suggested that upwelling could be brought about artificially by sinking atomic reactors in deep water to generate heat and cause convection currents to carry the nutrient-rich deep water to the surface. The Japanese have already created some artificial reefs and breakwaters to redirect coastal currents and create artificial upwelling on a smale scale.

In low latitudes where warm, high-salinity layers overlie water which is much colder and of lower salinity, the temperature and density differences above and below the thermocline can power the raising of deep water to the surface. Water drawn up a pipe from a depth of 1 km or more becomes warmed by conduction as it rises, and the density is correspondingly reduced, so that the system once started continues to provide its own energy for raising the water as long as the temperature and salinity differences between the two ends of the pipe persist.

In some areas it might be possible to alter the natural circulation in ways that could lead to a better supply of nutrients at the surface, or a higher water temperature. Digging out the Strait of Gibraltar to a greater depth has been advocated as a means of raising the productivity of the Mediterranean by allowing the entry of deeper levels of water from the Atlantic. It has also been predicted that a barrage across the Strait of Dover which allowed only a one-way flow of water from the Channel to the North Sea would reduce the entry of colder water into the northern part of the North Sea, and gradually raise the North Sea temperature, thereby promoting higher productivity.

None of these ideas is ever likely to become a practical proposition. The costs would be prohibitive and the long-term consequences would be difficult to predict. There would also most likely be strong opposition from conservationists. However, there are some methods which may work and which are being experimented with today. One of the most promising is briefly described here.

Pumping iron

More than a fifth of the world's oceans have plenty of nutrients but only tiny amounts of iron. This lack of iron has long been suspected as one of the main limiting factors in phytoplankton production in some open ocean areas. Joint British–US experiments have now shown that this is indeed the case. Areas of the Pacific Ocean several kilometres across were 'fertilized' with iron sulphate. This resulted in a stupendous growth of marine algae (Van Scoy and Coale, 1994; Pearce, 1995). As the algae grew they used up large quantities of carbon dioxide from the ocean and the experiments were partly designed to see if this would result in a reduction in atmospheric carbon dioxide. Some success seems to have been achieved in this direction with obvious implications in the control of global warming (see Section 10.2.1). Whether the increased plankton production can be used directly or indirectly as a food source is another challenge (see Section 9.6.5).

9.6.5 Harvesting plankton

The losses of organic material that occur at each stage of a food chain are thought generally to amount to some 80–90 per cent. On this reckoning, 1 kg of phytoplankton provides about 100 g of herbivorous zooplankton, which in turn yields 10 g of first-rank carnivore, 1 g of second-rank carnivore, and so on.

Most of the food that man takes from the sea comes from food chains involving several links, and therefore the harvest can be only a small fraction of the primary production. Some of the most plentiful pelagic fish are first-rank carnivores, but the majority of the most popular species for human food feed at later stages of the chain. Cod, for example, feed largely on other carnivorous fish or on carnivorous benthic animals. It is, therefore, apparent that far larger quantities of food could be obtained from the sea if it were possible to collect the earlier stages of food chains than can ever be provided by fishing. Instead of catching fish, why not directly harvest the plankton itself and process it to extract the food materials?

The practical difficulty of collection presents a major obstacle to obtaining large quantities of food in this way. Usually, plankton is dispersed in a very large volume of water, and even in the most productive areas enormous quantities of water would have to be filtered to obtain plankton in bulk. If the smaller organisms are to be retained, and particularly if the aim is to collect the phytoplankton, very fine filters would be required and the process of filtration could therefore proceed only very slowly. It seems unlikely that direct harvesting of the plankton from the open sea could be carried on economically, except perhaps in a few areas where there are very dense aggregations of the larger zooplankton such as krill (see below).

In recent years, there have been a number of experiments to investigate the possibilities of mass culture of marine phytoplankton. There seems little doubt that methods can be developed for culturing diatoms in large, shallow, seawater tanks enriched with plant nutrients. In dense fast-growing cultures, availability of carbon dioxide becomes a limiting factor; but if the culture tanks are sited near industrial installations, washed flue gases can be used to supply the carbon dioxide for photosynthesis, and waste heat to maintain the optimum water temperature. In this way, a rapid growth of phytoplankton can be maintained, and it might be possible to develop continuous culture methods similar to those now used in brewing or the preparation of antibiotics. However, although diatoms are rich in protein and oil, there are considerable difficulties in the separation of the plants from salt water and the subsequent extraction of the food materials from the cells. These processes are fairly efficiently performed biologically, and it seems likely that mass cultures of phytoplankton will find their chief usefulness in association with the rearing of young fish or the culture of some of the popular species of bivalve molluscs.

Krill

The Southern Ocean at times contains enormous numbers of a small shrimp-like crustacean known as krill (*Euphausia superba*) (see Figure 2.11). This is the food on which the great baleen whales including the largest of all, the blue whale, depend almost entirely. They grow quickly to a huge size simply by sieving this crustacean from the water.

At 5 cm long, krill can be captured using fine-meshed nets. The harvesting of krill probably represents one of the only remaining ways in which we can significantly increase the harvest of the world's oceans. The stock of krill is variously estimated at 50–100 million metric tons and is probably greater now than it was when whales still abounded in these waters. Krill is now harvested by ships from a number of nations including Russia, West Germany, Poland, Japan and Taiwan. The present total annual harvest is around 300 000 tonnes and the stock could most probably sustain a much larger harvest in the region of many millions of tonnes. Shoals of krill are detected using sonic techniques and there are further prospects of using satellite sensors to track the distribution of shoals.

However, there are problems to be overcome, not the least of which is persuading people to eat them! At present, most krill is converted to protein meal on board factory ships (mostly Russian). Krill does not keep well and can only be efficiently exploited by technically advanced fishing fleets. The problems of suitably processing krill into acceptable products for human consumption have yet to be fully solved but it can certainly be used for animal feeding. Whether the returns will justify the high costs of maintaining ships in distant Antarctic waters remains to be determined.

If krill stocks were to become heavily exploited, there would also be environmental costs. In addition to whales, many other animals also feed on krill, including seals, penguins, other birds, squid, and many species of fish. Excessive human harvest of krill could lead to diminished numbers of these predators and could prevent the hoped-for recovery of the baleen whale populations. To monitor the situation a degree of international accord has been provisionally reached on measures to set annual quotas and to assess the effects of krill harvesting in order to establish safe limits.

As whales can collect krill so efficiently, various suggestions have been made for devising 'artificial whales'. These might be constructed as atomic submarines with gaping bows opening to revolving filter drums, and provided with means for the continuous removal, processing and storing of the filtered zooplankton. At present, such ideas remain in the realms of Jules Verne.

Seaweeds

Various marine macroalgae are used as human food in many parts of the world, notably in Japan. In the British Isles Red Laver (*Porphyra umbilicalis*), Green Laver (*Ulva lactuca*), Carrageen Moss (*Chondrus crispus* and *Gigartina stellata*) and Pepperdulse (*Osmundea pinnatifida*) are each eaten in certain localities. However, seaweed does not contribute significant quantities to human food supplies and seems unlikely to increase much in consumption. The chief commercial importance of seaweed is as a source of alginic acid which, in addition to innumerable other uses, has many applications in the food industry as an emulsifier (Booth, 1975). A surprising number of processed foods and household products now contain substances derived from seaweed. Examples include ice cream, toothpaste, cakes, milk desserts, milk chocolates, fruit juice, beer and baby foods.

Conclusions

Apart from the obvious technical difficulties, proposals for increasing the yield of human food from the sea by attempting to bring about widescale alterations of the marine environment must be considered with much caution. Our knowledge of most aspects of the working of marine ecosystems is inadequate for us to be able to make predictions with certainty. There are many risks of unforeseen, detrimental consequences from tampering on a large scale with a vast environment we do not well understand, however well intentioned our actions.

It seems probable that marine fish farming and shellfish culture may eventually become more widespread and intensive than at present. However, the economics of these enterprises seem mainly to require the production of high-priced species. Large additions to our food supplies are not yet possible from these sources.

The immediate prospect of obtaining greater quantities of food from the sea lies mainly in the possibility of wider, controlled exploitation of natural stocks. This requires concurrent developments along several lines, including the utilization of a greater variety of species – especially the pelagic stocks. If krill harvesting proves to be economically viable, there would seem to be a large resource of food in this form which is at present virtually untapped except via whaling. Even if not readily suitable for human diets, use of krill for fishmeal might reduce pressures on fish stocks and facilitate measures for conservation. But as already mentioned, any over-exploitation could have disastrous consequences for wildlife.

International cooperation in fishing and fishery science is a prerequisite for major advance, without which optimum yields cannot be estimated and fishing appropriately regulated. The chief hope of making the best use of the food resources of the sea lies in a wider application of rational methods of control of fishing. Uncontrolled, competitive *laissez-faire* hunting inevitably leads eventually to declining yields from diminishing stocks. In fishing, as in most human affairs, progressive improvement depends upon intelligent control of human be-haviour.

References and further reading

Barnabé, G. (ed.) (1990). *Aquaculture*, Vols 1 and 2. Ellis Horwood.

Blaxter, J.H.S. and Holliday, F.G.T. (1963). The behaviour and physiology of herring and other clupeids. *Adv. Mar. Biol.*, **1**, 261.

Booth, E. (1975). Seaweeds in industry. In *Chemical Oceanography*, Vol. 4. J.P. Riley and G. Skirrow, eds. London, Academic.

Coffey, C. (1995). *Introduction to the Common Fisheries Policy: an Environmental Perspective*. IEEP London background briefing No. 2. Institute for European Environmental Policy, London. (Available from IEEP, see Appendix 4.)

Cooper, L.H.N. (1948). Phosphate and fisheries. *J. Mar. Biol. Ass. UK*, **27**, 326.

Cormack, R.M. (1968). The statistics of capture-recapture methods. *Oceanogr. Mar. Biol. Ann. Rev.*, **6**, 455.

Culley, M. (1971). *The Pilchard*. Oxford, Pergamon.

Cushing, D.H. (1966). *The Arctic Cod*. Oxford, Pergamon.

Cushing, D.H. (1968). *Fisheries Biology. A Study in Population Dynamics*. Madison, Milwaukee and London, University of Wisconsin Press.

Cushing, D.H. (1975). *Fishery Resources of the Sea and their Management*. Oxford, OUP.

Cushing, D.H. (1975). *Marine Ecology and Fisheries*. Cambridge, CUP.

Davis, F.M. (1934). Mesh experiments with trawls. *Fish. Investig. Ser. 11*, 14, No. 1.

de Ligny, W. (1969). Serological and biochemical studies on fish populations. *Oceanogr. Mar. Biol. Ann. Rev.*, 7, 411.

Earll, R.C. and Fowler, S.L. (eds) (1994). Tag and Release Schemes and Shark and Ray Management Plans. Proceedings of the 2nd European Shark and Ray Workshop, 15–16 February 1994. Unpublished report. (Available from JNCC, see Appendix 4.)

Graham, M. (1956). *Sea Fisheries*. London, Arnold.

Gulland, J.A. (1977). *Fish Population Dynamics*. London, Wiley.

Gulland, J.A. and Carroz, J.W. (1968). Management of fishery resources. *Adv. Mar. Biol.*, 6, 1.

Hardy, A.C. (1959). *The Open Sea*. Part II. Fish and Fisheries. London, Collins.

Heath, M. and Richardson, K. (1989). Comparative study of early life survival variability of herring *Clupea harengus* in the north-eastern Atlantic. *J. Fish. Biol.*, 35 (supplement A), 49–57.

Hodgson, W.C. (1967). *The Herring and its Fishery*. London, Routledge and Kegan Paul.

Jakobsson, J. (1970). On fish tags and tagging. *Oceanogr. Mar. Biol. Ann. Rev.*, 8, 457–99.

Jones, D.A., Munford, J.G. and Gabbott, P.A. (1974). Microcapsules as artificial food for aquatic filter feeders. *Nature*, London, 247, 233–5.

Jones, D.A. *et al.* (1979). Artificial diets for rearing larvae of *Penaeus japonicus*. *Aquaculture*, 17, 33–43.

Jones, F.R.H. (ed.) (1974). *Sea Fisheries Research*. London, Elek Science.

Jones, R. (1977). Tagging: Theoretical methods and practical difficulties. In *Fish Population Dynamics*. J.A. Gulland, ed., pp. 46–66. London, Wiley.

La Tene Maps (1995). Major Scottish Finfish Farms – Map. 2nd ed. Oct. 1995. La Tene Maps, Dublin.

Lee, A.J. and Ramster, J.W. (eds) (1981). *Atlas of the Seas around the British Isles*. Ministry of Agriculture, Fisheries and Food (MAFF).

Lincoln, D. and Howell, B. (1994). Finfish farming in the UK. MAFF unpublished Handout 32 (available from MAFF, see Appendix 4).

Lockwood, S.J. (1978). Mackerel. A problem in fish stock assessment. Laboratory Leaflet No. 44. Lowestoft, MAFF.

Lockwood, S.J. (1989). *Mackerel: its biology, assessment and the management of a fishery*. Farnham, Surrey, Fishing News Books.

Lockwood, S.J. and Johnson, P.O. (1976). Mackerel research in the southwest. Laboratory Leaflet No. 32. Lowestoft, MAFF.

Macer, C.T. and Easey, M.W. (1988). The North Sea cod and the English fishery. Laboratory Leaflet No. 61. Lowestoft, MAFF.

Nichols, J.H. and Brander, K.M. (1989). Herring larval studies in the west-central North Sea. *Rapp. P.-v. Reun. Cons. int. Explor. Mer.*, 191, 160–8.

Nikolskii, G.V. (1969). *Theory of Fish Population Dynamics*. Edinburgh, Oliver and Boyd.

Parrish, B.B. and Saville, A. (1965). The biology of the North-East Atlantic herring population. *Oceanogr. Mar. Biol. Ann. Rev.*, 3, 323.

Parrish, B.B. and Saville, A. (1967). Changes in the fisheries of North Sea and Atlanto-Scandian herring stocks and their causes. *Oceanogr. Mar. Biol. Ann. Rev.*, 5, 409.

Pearce, F. (1995). Iron soup feeds algal appetite for carbon dioxide. News report in *New Scientist*, **147** (1984), 1 July 1995, 5.

Rae, B.B. (1965). *The Lemon Sole*. London, Fishing News (Books) Ltd.

Rankine, P.W. (1986). Herring spawning grounds around the Scottish coast. *Intl. Coun. Explor. Sea (ICES)*. Pelagic Fish Committee Paper CM 1986: H15.

Russell, E.S. (1942). *The Overfishing Problem*. Cambridge, CUP.

Sainsbury, J.C. (1996). Commercial Fishing Methods. An introduction to vessels and gears. 3rd edition. Fishing News Books.

Shelbourne, J.E. (1964). The artifical propagation of marine fish. *Adv. Mar. Biol.*, **2**, 1.

Shepherd, J.G. (1990). Stability and the objectives of fisheries management: the scientific background. MAFF Laboratory Leaflet No. 64. Lowestoft, Directorate of Fisheries Research.

Shepherd, J.G. (1993). Why fisheries need to be managed and why technical conservation measures on their own are not enough. MAFF Laboratory Leaflet No. 71. Lowestoft, Directorate of Fisheries Research.

Simpson, A.C. (1959). The spawning of the plaice in the North Sea. *Fishery Invest., Lond., Ser. II*, **22**, No. 7.

Simpson, A.C. (1959). The spawning of the plaice in the Irish Sea. *Fishery Invest., Lond., Ser. II*, **22**, No. 8.

Spencer, B.E. (1990). Cultivation of Pacific oysters. MAFF Laboratory Leaflet No. 63. Lowestoft, Directorate of Fisheries Research.

Steven, G.A. (1948). Contributions to the biology of the mackerel, *Scomber scombrus*. Mackerel migrations in the English Channel and Celtic Sea. *J. Mar. Biol. Ass. UK*, **27**, 517.

Steven, G.A. (1949). A study of the fishery in the South-West of England with special reference to spawning, feeding and fishermen's signs. *J. Mar. Biol. Ass. UK*, **28**, 555.

Van Scoy, K. and Coale, K. (1994). Pumping iron in the Pacific. *New Scientist*, **144** (1954), 3 Dec. 1994, 32–5.

Walsh, M. and Martin, J.K.A. (1986). Recent changes in the distribution and migration of the western mackerel stock in relation to hydrographic changes. *Intl. Coun. Expl. Sea (ICES)*. Pelagic Fish Committee Paper CM 1986: H17.

Wimpenny, R.S. (1953). *The Plaice*. Buckland Lecture. London, Arnold.

Wollaston, H.J.B. (1915). Report on spawning grounds of plaice in the North Sea. *Fishery Invest. Lond., Ser. II*, **2**, No. 4.

10 Human impact on the marine environment

Only a hundred years or so ago, it would have seemed inconceivable that anything we did could significantly affect the vast resources of the oceans or result in large-scale alterations and damage to marine ecosystems. We have already seen in Chapter 9 that this is no longer true for fishery resources. Modern technology has resulted in a huge increase in our ability to catch and store edible marine species. Many or most important fishery stocks are now being exploited at levels considered to be at or over the maximum sustainable level. In this chapter, two other major ways in which human activity is affecting our oceans will be discussed. These are the effects of adding materials to the sea (pollution) and indirect effects on the oceans through human-induced changes in the world's atmosphere (e.g. the greenhouse effect).

10.1 Marine pollution

The word 'pollution' is now widely used as a convenience term for virtually any substance released into the environment by human activities which has a deleterious effect on marine organisms and ecosystems or is a nuisance to mankind. Effects of these materials include smothering and poisoning of organisms, interference with physiology and behaviour, and increase or decrease in biological productivity with consequent effects on organisms in other parts of the food web.

In recent years there has been much concern over the extent to which the oceans may be adversely affected by their use as a dumping ground for an ever-increasing quantity and variety of human and industrial wastes. The amount of literature now available on various aspects of marine pollution is vast. Space can be given here for only a brief review of this important aspect of marine ecology, and for more information students should refer to the key references provided at the end of the chapter.

10.1.1   Capacity of the 'ocean dump'

From the beginnings of civilization sewage and domestic rubbish have been disposed of mainly by spreading them over the ground or discharging them into rivers, lakes or the sea. Hitherto, these simple methods of waste disposal have usually been fairly satisfactory because much of this refuse has been rapidly broken down to inoffensive forms by natural processes of decay; indeed the procedure has been generally beneficial by returning to land and water the constituents necessary for the maintenance of fertility. However, growing towns have always tended to pollute themselves with their own wastes, especially by overtaxing the natural capacity of their streams and rivers to disperse, degrade and recycle the excrement and rubbish poured into them. Open sewers in the crowded cities of mediaeval Europe were a serious health hazard causing massive outbreaks of disease. This is still the case in some developing countries. In later years the much improved health of urban populations in developed countries has been attributable in considerable measure to cleaner towns with much safer methods of sewage and waste disposal.

In contrast to land or fresh water, the enormous volume of the sea has a huge capacity for mineralizing sewage or organic refuse. To the extent that the sea can accept wastes without detriment, it is obviously sensible to take full advantage of this natural sink. It has always been assumed that any noxious substances which are not readily inactivated in the sea by natural processes soon become so diluted as to be quite harmless. Many coastal towns still take advantage of this by pouring their untreated sewage directly into the sea. Industries which produce large volumes of waste have often selected sites on the coast because, in addition to the many other advantages of ready access to the sea, they obtained a simple and inexpensive outlet for their effluents. Advantages of waste disposal at sea include the following:

1   The great capacity of the sea to degrade many substances, especially natural organic wastes.
2   Extreme dilution of toxic materials can be effected in course of time.
3   Solid wastes may be permanently lost within sediments.
4   Dumping at sea is often the cheapest method of disposal.
5   It may sometimes be preferable to pollute the sea rather than land. Sewage farms occupy large areas which may not be available where needed. Poisonous wastes placed on rubbish tips, however rigorous the precautions, may present much greater hazards than if discharged at sea.

However, in recent years it has become increasingly obvious that in many cases we are exceeding the capacity of our seas to cope with our various inputs. Recent rapid growth of populations together with industrialization and new technologies has produced wastes in much greater quantities and variety than ever before. Public health problems have arisen from raw sewage discharges into the sea (see

Section 10.1.5), more land has been needed for rubbish tips and industrial spoil, and more effluents are reaching the sea via rivers and estuaries. An additional problem is that some of the new materials produced in large amounts, notably many plastics, accumulate because they are extremely long lasting, being little if at all subject to biological decay.

Despite the great size of the oceans and their thorough intermixing, water circulation is mostly slow and dispersion of materials in the sea is sometimes a very gradual process. Consequently where large amounts of wastes are discharged into shallow water, especially into enclosed areas like the Baltic, the North Sea and Mediterranean, concentrations have now been reached which present a variety of problems. Even if wastes are disposed of by tipping far out at sea, pollution may become obvious if persistent floating substances drift back to the coast, or if solids which sink to the bottom are later carried inshore by shifting of sands along the sea floor.

10.1.2   Chief routes of entry of marine pollutants

Marine pollutants find their way into the sea not only through deliberate routes such as sewage discharges and dumping (legal and illegal) but also by a variety of other, not always obvious, routes outlined below.

Drainage

Many pollutants reach the sea either through direct drainage from coastal towns and industries or indirectly via rivers. Dilute industrial effluents, treated sewage and cooling water are often discharged into rivers and estuaries. Fertilizers, pesticides and animal wastes may drain into rivers from agricultural land. Huge amounts of silt resulting from rainforest clearance are carried down to the sea by tropical rivers. Rainwater runoff from cities and towns carries oil, heavy metals and other material into rivers. A surprising area of sea-bed around domestic sewage outfall pipes is often contaminated with oil (Dipper, pers. obs.).

Dumping

Coastal towns and cities discharge raw or treated sewage into coastal waters. Tipping at sea is used to dispose of sewage sludge, industrial wastes, dredged materials, ocean incineration wastes, oil platform wastes and rubbish (the latter particularly in undeveloped countries). Ships dispose of many of their day-to-day wastes (including oil tanker washings) by dumping, often illegally. The recent (1995) appearance of dangerous phosphorus bombs on Scottish beaches is thought to be the result of poorly controlled military dumping after World War II.

Airborne pollution

Many airborne pollutants are dissolved by rain which may then fall over the sea. Others are carried into the sea as dust particles or by solution of volatile materials

Table 10.1 Approximate quantities of wastes dumped into the North Sea in 1985 (data from QSR, 1987).

Type of material	*Amount dumped annually*
Dredged spoil	5 million tonnes
Liquid and solid industrial waste	1.9 million tonnes
Sewage sludge	5 million tonnes

from the atmosphere. For example, huge amounts of smoke containing PAHs (polycyclic aromatic hydrocarbons) were created by burning oil wells in the 1991 Gulf War which drifted for many miles over land and sea.

Accidents

A great variety of objects and substances finds its way into the sea through shipwrecks and lost cargoes, from plastic ducks and sneakers to potentially lethal (to humans and marine organisms) chemicals. Nearly 6 tonnes of Lindane, a highly toxic pesticide, still remain on the seabed in the English Channel after being washed overboard from a ship during a storm in 1989. Oil spills are another obvious concern.

10.1.3   Persistence of pollutants in the sea

The long-term effects of waste disposal at sea must obviously depend upon the length of time taken for material to be broken down to a harmless form. The extent to which continuous addition of substances may be cumulative in effect will depend partly on their persistence in the water. Pollutants can be tentatively classified as transient, moderately persistent, very persistent and virtually permanent. Examples of the length of time that various litter items may persist are given in Table 10.2. These are maximum times which may be shortened by wave action and abrasion.

Transient pollutants

Sewage, some domestic rubbish and certain industrial organic wastes are rapidly biodegraded to inorganic form by marine bacteria within a few days unless processes are retarded by oxygen deficiency in slow-mixing waters. Mineral acids are quickly neutralized by the large alkali reserve of seawater.

Moderately persistent pollutants

Oil and many organic industrial effluents, including solid wastes such as organic fibres and pulps, are only slowly degraded by natural processes, often taking many months for complete decomposition. The time taken for untreated oil pollution on rocky shores to disappear depends on the exposure of the shore to wave action.

Table 10.2 Approximate timescales (maximums) for litter breakdown.

Litter item	Persistence
Glass bottle	1 million years (undisturbed)
Plastic bottles	Indefinitely
Aluminium cans and ring pulls	80–100 years
Tin cans	50 years
Leather, e.g. shoes	50 years
Nylon material	30–40 years
Plastic film container	20–30 years
Plastic bags	10–20 years
Plasticized paper	5 years
Wool garments	1–5 years
Cigarette butts	1–5 years
Orange peel and banana skins	2 years

In most cases the majority of the oil will have gone within about two years but thick tarry deposits can remain on the upper areas of sheltered shores for many years (IPIECA, 1995).

Very persistent pollutants

Many artificial organic products are highly stable in seawater, undergoing only very gradual natural degradation. These include some detergents, polychlorinated biphenols (PCBs) and some widely used pesticides including dichloro-diphenyl-trichloroethane (DDT), hexachlorobenzene (HCB) and Dieldrin.

Virtually permanent pollutants

Many plastic waste items, such as plastic bottles, will last indefinitely (see Table 10.2). Glass bottles, if undisturbed by waves, may last a million years! Some radioactive isotopes are very long-lived pollutants. Addition to seawater of certain toxic metals may produce a long-lasting increase of their concentration in the sea. Even if the pollution ceases, the concentration of such metals may fall only very gradually, by dilution and by various processes of sedimentation and precipitation. Solid inorganic residues such as mine spoil may become a permanent part of the sea bottom.

10.1.4 Regulations

Regulation of deliberate dumping and pollution in the world's oceans has been slow in coming and although there are now a number of international and many national regulations, enforcement and compliance remain major problems. Only a few examples of important contemporary legislation can be given here.

International regulations

International regulations require agreement and ratification between countries and are rarely applicable on a truly worldwide basis. Nevertheless some progress has been made in recent years through the United Nations International Maritime Organization (IMO) which is responsible for conventions concerning oil pollution, marine pollution and dumping at sea. Oil and litter both have very visual impacts and have aroused public awareness. The deliberate discharge of oily ballast waters and tank washings from ships is illegal under an international regulation known as MARPOL Annex I, which is fully ratified. However, the practice is still widespread. Under Annex V of the same convention, it is illegal to dump any plastic items over the side of ships. However, this annex is not yet fully ratified so that some nations are not bound by the regulations.

European regulations

Sewage pollution of beaches has become a major topic in Europe in recent years. There are now two main EC directives relating to sewage pollution:

(i)   The Bathing Water Directive (1975)

This aims to protect recreational users of bathing waters from health risks associated with sewage pollution. Bathing waters are designated by each EC member state and are tested annually. Whilst a good concept, there have been considerable problems with the standard tests used and confusion over various beach awards such as the EC Blue Flag award and the Seaside Awards. These awards can be given to beaches with widely differing water quality. In the UK, the Marine Conservation Society now produces an annual Good Beach Guide (MCS, 1997), an independent source of information about the state of Britain's beaches.

(ii)   The Municipal Waste Water Directive (1991)

When fully implemented, this will require all coastal sewage discharges serving populations over 10 000 people to receive secondary treatment (see Section 10.1.5) before discharge. However, discharges to 'less sensitive' waters need only receive primary treatment. Discharge to 'sensitive' waters may require tertiary treatment (removal of nitrates and phosphates). The disposal of sewage sludge from treatment plants, by dumping at sea, should be phased out in UK by 1998.

10.1.5   Sewage

Over the past decade or so, the public has become increasingly aware that in many countries including the UK, sewage is not being disposed of effectively. Conspicuous lumps of faecal material and vaginal tampons deposited on the beach are still a common and unwelcome sight. Sewage-related debris (SRD) remains the second most common category of litter recorded in the UK Marine

Conservation Society's Beachwatch survey. This survey uses volunteers to collect and count litter items along selected stretches of coastline throughout the UK, and is repeated at intervals.

Sewage treatment

Sewage enters the sea via short and long sea outfalls, stormwater drains and rivers. The discharge of raw sewage into coastal waters is still widespread in the UK and in many other countries. In Europe, increasing efforts are being made to treat sewage effectively before discharge. This clearly reduces the impact of the sewage both on the marine environment and on coastal amenities. Sewage pollution of coastal and estuarine waters is usually most severe during the summer when temperatures are raised, river outflows reduced and seaside populations often increased.

Untreated domestic sewage consists mainly of waste water and solids from toilets, sinks and drains, which includes detergents (often containing phosphorus), other chemicals and plastics (e.g. from panty liners). Industrial waste is also dumped into sewers and oil and run-off from road systems may enter from storm drains. Raw sewage discharged into the sea may therefore contain large quantities of metals such as arsenic, cadmium, copper, mercury and lead as well as organic matter, petroleum products, fats, solvents and dyes. Thus there is considerable potential for human health risk and for ecological damage when untreated sewage is discharged into the sea. Currently over 80 per cent of Britain's large coastal discharges (serving more than 10 000 people) receive no treatment or are just screened. Hopefully this situation is changing as new legislation comes into force (see Section 10.1.4).

There are three main levels of sewage treatment:

1   Primary treatment. This involves screening and settling to remove larger solids and addition of chlorine or exposure to UV light to kill bacteria and most pathogens. However, there remains the problem of disposing of the sewage sludge.
2   Secondary treatment where the sewage is digested by bacteria in a treatment plant before chlorination. This can reduce the effluent's biological oxygen demand by as much as 99 per cent.
3   Tertiary treatment in which, in addition to the above, nitrate and phosphate are removed.

Effects of raw or undertreated sewage dumping

The addition of sewage to water can greatly increase fertility by release of nutrients, with possible long-term beneficial effects. However, in most cases the effects are detrimental because of the sheer volume of the discharges. Such effects include foul deposits, deoxygenation, eutrophication, reduced salinity, infection and toxic residues.

Foul deposits

Discharge of untreated sewage may result in strandings of recognizable faecal material and other objectionable objects. Unless fully treated, solid deposits around sewer outfalls form a black sludge, sometimes blanketing the substrate. This generally causes some change in bottom fauna; often bivalves decline and worms increase, and this may reduce the value of the area as a feeding ground for fish. Suspended solids reduce light penetration, and increase scouring and silting effects, with possible adverse effects on both plant and animal life. Stinking mud from sewer outfalls, even though sited some distance below low tide level, may sometimes drift along the sea-bed and become deposited on the shore.

Deoxygenation

Addition of organic solids or solutes accelerates bacterial growth, increases oxygen demand and may lead to deficient oxygenation if the water is not well mixed. Deoxygenation encourages the growth of sulphur bacteria and the production of offensive hydrogen sulphide. Much of the Baltic Sea suffers from low oxygen levels as a result of its enclosed nature, lack of flushing and high input of sewage from its many coastal cities.

Eutrophication

High concentrations of plant nutrients may lead to so rapid a growth of phytoplankton and zooplankton that when the bloom dies off and the material sinks, there is an accumulation of organic debris on the sea-bed. Associated bacterial multiplication causes the water to become deoxygenated and foul with severe effects on the bottom fauna. This condition of eutrophication, though more of a problem in fresh waters than the sea, can occur in sea bays, estuaries and enclosed seas if mixing is slow. It is often indicated by the presence of species such as the polychaete worm *Capitella capitata*, which can survive and do well in polluted conditions.

'Red tides', which are blooms of toxic phytoplankton (see Section 2.2.2), are also associated with excess nutrients supplied by sewage. These can poison fish and shellfish and may indirectly cause paralytic shellfish poisoning in humans. In 1990 and 1991, shellfish farms along large stretches of the north-east coast of England had to be closed due to red tides.

Reduced salinity

Reduced salinity occurs around sewer outfalls, which may be adverse to stenohaline species.

Infection

Sewage-borne infections such as typhoid, viral hepatitis, enteric infections and ear, nose and throat infections have been associated with exposure to sewage-polluted seawater. Infection may be transmitted to humans directly through contact with water or spray, or indirectly by consumption of marine foods. In the UK there have been several recent cases where surfers have suffered severe illness and paralysis thought (but not yet conclusively proved) to be due to infection from faecal viruses. The increased bacterial content of sewage-polluted water favours filter feeders and often results in excellent growth of bivalves such as cockles and mussels. Shellfish from such areas may be much prized but require thorough purification before they are eaten. Some viruses survive longer than bacteria in seawater and may remain in shellfish even after they are bacteriologically clean.

Toxic residues
Sewage is increasingly contaminated with heavy metals, insecticides and various persistent organic poisons which may be transmitted via the food chain to humans.

10.1.6   Inert solid wastes and litter
During the British Steel Round the World Yacht Race in 1992–3, the crews took part in a project called 'Ocean Vigil'. Part of this project involved recording floating rubbish. The crews reported oil slicks, plastic containers and bags, fishing nets, wood, oil drums, shoes and a freezer! Rubbish was present even in the remotest parts of the Southern Ocean. Some estimates suggest that the world's ships dump or lose 6 billion kilograms of rubbish into the oceans every year. Yet more enters through our sewerage systems. Much of this eventually gets washed up on the seashore. The majority of this rubbish (possibly more than 80 per cent) is plastic, including plastic pellets, bags, sheeting and bottles, condoms, monofilament fishing nets and drink pack holders. A further major source of debris on beaches comes from tourist and recreational use.

Most plastic degrades only very slowly and even with the controls now being implemented (see Section 10.1.4) it will take many, many years for the situation

Table 10.3 Sources of marine debris collected from selected beaches by 'Beachwatch '96' (see Figure 10.1). From: Pollard and Parr (1996) Beachwatch '96 nationwide beach-clean and survey report by kind permission of the Marine Conservation Society.

Sources	Total	Per cent
Tourist/recreational	64 272	22.1
Shipping	50 654	17.4
Sewage-related debris	39 240	13.5
Fishing	36 467	12.5
Fly-tipping	2 822	1.0
Medical	214	0.1
Non-sourced	97 559	33.5

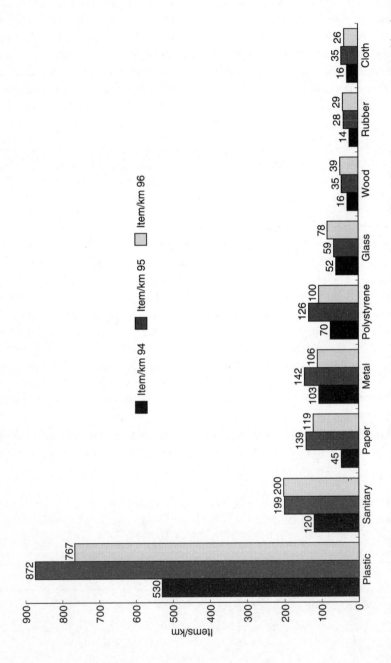

Figure 10.1 *Comparison of marine debris recorded from British beaches in 1994–96 by 'Beachwatch'. This project, organized by the Marine Conservation Society with sponsorship from Reader's Digest, uses volunteers to monitor and clean up litter on Britain's beaches.* (From Pollard and Parr (1996) Beachwatch '96 nationwide beach-clean and survey report by kind permission of the Marine Conservation Society.)

to improve. A beach covered in plastic is very unpleasant for us but it is often fatal to marine life. Animals can be poisoned or starve after eating plastic; or they can become entangled and trapped. Classic examples include air-breathing turtles, seals and cetaceans which are drowned by floating fishing net and plastic sheeting. Turtles, especially the leatherback, eat plastic bags mistaking them for jellyfish. The bags make them feel full so that they stop feeding properly. In addition chemical plasticizers may poison them. Even birds and fish may be poisoned in this way. In the English Channel, dead cod have been found killed by ingesting plastic cups thrown overboard from cross-Channel ferries. Tiny floating plastic pellets are mistaken for fish eggs or other plankton and eaten. Near some industrial centres in New Zealand, up to 100 000 pellets have been found on one square metre of beach.

Artificial reefs

Some types of rubbish can be beneficial to marine life. Divers often report blennies and other small fish living in tin cans and glass jars. Parts of the sea floor in the vicinity of large ports are littered with clinker and ash dating from the time when steamships were powered by coal-burning furnaces. Clinker, tyres and many other objects on the sea-bed can encourage the development of a larger and more diverse population by providing a range of cavities and holes offering protection and concealment. Experiments are being carried out in many parts of the world in deliberately constructing artificial reefs using a variety of materials. The aim is often to encourage a greater variety and abundance of fish life for the benefit of local fishermen or divers (Collins and Jensen, 1996). Shipwrecks are well known for their ability to act as a focus for marine life and new wrecks soon become colonized. Some fish seem to be directly attracted to large objects such as wrecks and will move in well before the wreck can provide anything other than shelter. Groupers have been observed to move into a wreck in the Arabian Gulf only days after it sank and many miles from the nearest rocky area or reef (Dipper, 1991).

The question of dumping de-commissioned oil platforms at sea has become topical recently, as many platforms in the UK are reaching the end of their useful life. Some scientists feel these would make good artificial reefs, whilst others are against the idea. The question of using deep-ocean areas for dumping came to the fore in 1996 with the de-commissioning of the giant Brent Spa platform. Whilst some scientists argued that the deep ocean dumping was an ecologically sound option, others disagreed and public opinion was so strongly against it that disposal on land was finally agreed. The deep-ocean is vast and, in relative terms, sparsely populated by marine organisms. However, water exchange in the depths is extremely slow as are rates of breakdown and decomposition. Deep-sea communities may be easily disrupted since reproductive rates are often very slow. Waste disposal in the deep ocean is discussed by Angel (1996).

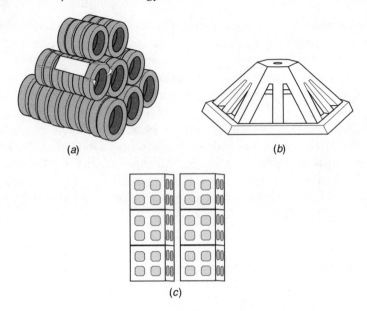

Figure 10.2 Some materials from which artificial reefs may be constructed: (a) scrap tyres securely bound together and weighted; (b) purpose-built Japanese unit made from concrete containing coal fly ash; (c) Taiwanese open structure, coal-ash units designed to attract shoals of fish.

10.1.7 Pesticides

Pesticides enter the sea from agricultural runoff, rivers and by airborne transfer (particularly DDT). In estuaries the quantities of pesticide in the water tend to vary seasonally according to local agricultural practice. Around the British Isles there are often two peaks, one in early summer following springtime dressings on crops and orchards, and a second peak in late autumn following the use of pesticides on autumn-sown wheat.

DDT

When it was first made, DDT was heralded as a wonder chemical capable of killing malarial mosquitoes and many other pests, with consequent saving of human life. It was only much later that the environmental consequences both on land and in the sea were realized. Chlorinated hydrocarbons such as DDT and PCBs have become widely dispersed throughout the oceans, even to the Arctic and Antarctic.

The main problem is that they are only very slowly degraded and are readily available for biomagnification in food chains. Phytoplankton take up the DDT because the large hydrocarbon molecules are not very soluble in water but are very soluble in fats such as the oils in diatom cells. Zooplankton such as copepods eat the phytoplankton with its burden of DDT and do not break down or excrete it. Small fish eating the copepods gain yet more DDT, and so on right up to mammals, birds and the human population. The effects of DDT pollution are varied. Some marine organisms, especially crustacea, are extremely sensitive to both organochlorine and organophosphorus compounds and are killed by very small concentrations. In others, reproductive success is affected. Dramatic reproductive failures in fish-eating seabirds such as pelicans, cormorants, terns and fish eagles and in land raptors such as ospreys in the early 1970s, were finally attributed to a thinning of their egg shells and consequent breakage, caused by DDT. Some animals, including seagulls, have a high resistance to DDT and were not affected.

The use of DDT was banned outright in many countries in 1972 and restrictions placed on its use in others. However, this and other persistent pesticides are still used in some Third World countries.

TBT

A recently identified problem concerns the use of anti-fouling paints containing the organotin, tri-butyl-tin (TBT). The use of paints containing TBT has been banned since 1987 in the UK for boats under 25 metres long and for aquaculture equipment. TBT is still in use for larger boats and can also enter the marine environment when ships' hulls are stripped and re-painted. TBT causes various deformities in molluscs and had such a bad effect on oyster farming that the UK industry practically collapsed. In dogwhelks (*Nucella lapillus*), it causes a condition known as imposex, where females develop male sexual characteristics and breeding is impaired. Dogwhelks live for 5 to 6 years and spend all their lives on the same stretch of shore. They are therefore very good indicators of TBT pollution and were used as such when the case for banning TBT was being investigated.

10.1.8 Toxic (heavy) metals

Many heavy metals such as arsenic, lead, cadmium, mercury and copper are naturally present in seawater at very low concentrations reflecting their low solubility (see Table 4.3). Various organisms need some of these in very small amounts, for normal metabolism. However, increased concentrations resulting from pollution may be harmful both to marine organisms and to humans.

The concentration of heavy metals in the water may be raised locally by discharges from many industrial processes, and in sediments they may become very high. Sewage sludge dumping provides a significant input. Metals may also be released into the water from sediments disturbed by dredging, or by changes in pH or redox potential.

Shellfish such as oysters, mussels and clams bioaccumulate metals but do not seem themselves to be affected. In contrast, most fish and crustaceans excrete any metals they take in with their food. The exceptions are mercury and cadmium. Levels of these metals in top predatory fish such as tuna may exceed levels considered safe for human consumption. High levels have also been found in dead killer whales washed ashore and may have contributed to their deaths.

Mercury is present in the effluents from several industries, for example those involved in the manufacture of chlorine, acetaldehyde, caustic soda and paper. It is contained in many agricultural fungicides and some is released by the burning of fossil fuels. Mercury poisoning was the cause of the most serious human catastrophe yet arising from marine pollution, an outbreak in 1953 at Minamata in Japan of what was at first thought to be a mysterious disease. Later the problem was traced to the consumption of local-caught fish and shellfish contaminated by mercuric wastes dumped into the bay by a nearby chemical factory. Two hundred and thirty people subsequently died and a thousand others suffered permanent cerebral and nerve damage from this source. A second outbreak killing five people and affecting thirty others occurred in Japan in 1965 near the mouth of the river Agano from a similar cause. These tragic events illustrate the folly of assuming that discharges into coastal waters are safely diluted and dispersed.

Metallic mercury is virtually insoluble in water and very little is absorbed into tissues except in those who expose themselves to long and continuous contact, such as dentists. In water, metallic mercury is acted upon by micro-organisms which gradually convert it into a variety of organic compounds, notably to methyl-mercury compounds which are soluble and highly toxic, and readily absorbed and concentrated by organisms.

Arsenic is present in some detergents and was widely used in pesticides and herbicides until at least the 1960s. In seawater it exists mainly as arsenate but a proportion becomes converted to the more highly toxic arsenite. Arsenic compounds are readily concentrated in the tissues of certain marine fish.

Lead is often present in the effluents from mine workings in metalliferous areas, and becomes concentrated in the tissues of some marine species. Other metals which may also be present in marine sediments near river mouths carrying mine-washings include cadmium, chromium, nickel, copper and zinc, all of which have been found in high concentrations in various worms and molluscs from these areas (Clark, 1986). Lead also enters the sea in appreciable quantities from the air due to atmospheric lead pollution from the exhaust fumes of internal combustion engines.

Many marine organisms concentrate heavy metals and it appears that this increases their tolerance to even greater concentrations. Certain metals, such as copper, are essential for normal enzyme activity but may become enzyme inhibitors at high concentrations. Except for the Minamata tragedy there is not much evidence of human detriment from metal contamination of marine foods.

Some fatalities have been reported where chromium and cadmium contamination of shellfish were implicated. There is also the danger that several metals may act additively or synergistically.

10.1.9 Radioactive pollution

Radioactive materials may enter the sea from two main sources: from weapon testing via atmospheric fallout and from atomic power industries. The main contaminants are strontium-90, caesium-137 and plutonium-239. In the UK, a major source of radiation pollution has been via the discharge of cooling water from Sellafield Nuclear Power Station. The level of discharge has been considerably reduced in recent years but caesium-137 (which does not occur naturally) remains in sufficient quantities for it to be used as a tracer for ocean currents around Scotland and into the North Sea.

Some highly dangerous radioactive wastes are also disposed of by dumping into the deep oceans in sealed containers. The containers will eventually corrode and the fate of the residues is uncertain. Although there are few links between the food chains of the deep ocean and the shallow waters used for commercial fishing, it is known that some deep-water currents surface in the Antarctic. There are hopes that incorporation of radioactive wastes into solid glasses from which nothing can leach or leak may prove a safe method of disposal.

The fallout from the explosion of the Chernobyl nuclear reactor in the Soviet Union in 1986 greatly increased the radioactive load of the world's oceans. There is currently great concern that some of the atolls in the South Pacific used for underground testing of atomic devices may collapse and release huge quantities of radioactivity.

Seaweeds can concentrate radioiodine with great rapidity and fish absorb a variety of radioactive substances. In addition radioactive substances can bioaccumulate in marine animals in a similar way to heavy metals. The effects on marine organisms are not fully understood but may include genetic disturbances and increased mortality both in young stages and in adults. Interestingly, many marine invertebrates can withstand radiation doses that would kill people. Some deep-water marine shrimps, exposed to doses of natural radiation sufficient to debilitate people, remain unharmed. A variety of cancers in humans, such as childhood leukaemia, is linked to radiation exposure.

10.1.10 Thermal pollution

Warm effluents discharged into bays and estuaries may raise the water temperature. This reduces the solubility of oxygen, and the oxygenation may be further reduced by increased oxygen consumption by animals and bacteria, and by reduced vertical mixing due to thermal stratification. The effect may be that the underlying layers become deoxygenated and foul. Migratory fish such as salmon may be discouraged from passing through the area. Warm water may favour pests such as shipworm and gribble, accelerating their growth and extending their breeding seasons, and it may also encourage the establishment of foreign pests. The fouling rate on ships' hulls is likely to be raised.

There are also some possible benefits. Growth rates of useful species such as edible bivalves may be improved and the area may be colonized by useful foreign species e.g. *Mercenaria mercenaria* in Southampton water. Use can also be made of the warm water from power stations to rear certain fish such as eels (see Section 9.6.3).

10.1.11   Oil

Of all the many types of pollution to which the oceans are subject, oil is the one to have caught the public's attention the most. This was the first type of large-scale marine pollution to become apparent to the public as beaches became oiled when ships sank during World War II.

Sources of oil pollution

Each decade since the war, there have been dramatic instances of gross oil fouling of long stretches of coastline from tanker accidents – notably the wreck of the *Torrey Canyon* in 1967 fouling beaches in Cornwall and Brittany; the *Amoco Cadiz* in 1978 fouling the north Brittany coast; the *Exxon Valdez* in 1989 affecting beaches in Prince William Sound, Alaska; the deliberate release of oil into the Arabian Gulf during the 1991 Gulf War and most recently, in 1996, the *Sea Empress* which oiled shores in Pembrokeshire including Skomer marine nature reserve. These events have been widely publicized and attention drawn not only to the damage done to amenities but also to the destruction of tens of thousands of seabirds and other marine life, and possible devastation of local fisheries, fish nursery areas and shellfish beds.

However, although such accidents are dramatic, far more oil enters the marine environment from other less publicized sources (Table 10.4). The figures in such estimates must of necessity be very approximate, but it can clearly be seen that pollution from domestic and industrial discharges easily exceeds that from tanker accidents.

Impacts of oil spills

A great deal of research has been done on the effects of oil pollution on the marine environment and there are many books and articles dealing with this subject. Only a few major aspects can be dealt with here but key references are given at the end of the chapter.

Apart from fouling of beaches, which if severe can lead to the destruction of much of the intertidal population, the major threat from oil pollution is to

Table 10.4 Sources of inputs of petroleum hydrocarbons into the world's oceans and estimates of yearly inputs. Data taken from various sources between 1973 and 1981 based on GESAMP, 1993.

Source	Range of estimates (thousands of tonnes)
Urban run-off and discharges	2500–1080
Operational discharges from tankers	1080–600
Tanker accidents	400–300
Non-tanker accidents	750–200
Atmospheric deposition	600–300
Natural seeps	600–200
Coastal refineries	200–60
Other coastal effluents	150–50
Offshore production losses	80–50

seabirds and mammals (see Table 10.5). Oil readily penetrates and mats the plumage of seabirds, making flight impossible and leading to loss of buoyancy and heat insulation. Attempts to preen lead to ingestion of oil and gut irritation. At the present time many hundreds of thousands of seabirds are destroyed annually by oil fouling. The species at greatest risk are those which live mainly on and in the water, e.g. puffins, guillemots, razorbills, shags, cormorants, ducks and divers. Seals and sealions may also suffer as oil fouling of their fur reduces heat insulation. Oil in their eyes causes irritation or blindness.

The impact of oil on marine organisms depends on characteristics of the oil spill such as its toxicity and viscosity, the amount of oil and the time for which it is in contact with the organism. Marine organisms and different life stages of organisms also vary in their sensitivity to oil. For example, on seashores, brown seaweeds are protected by their slimy covering of mucilage such that the oil easily washes off. Barnacles and sea anemones can also survive covered in oil for several days. Grazers such as sea snails and limpets are much more susceptible. Eggs and larval stages of fish, crustaceans and some other groups tend to be more susceptible than adults.

Toxicity

Crude oils and oil products differ widely in their chemical composition and therefore in their toxicity. The direct toxicity of oil to organisms is attributable mainly to the light aromatic components. Because these fractions usually evaporate fairly quickly, oil which reaches the shore soon after spillage is likely to be far more poisonous to the intertidal population than if it had been afloat for a longer time. The greatest toxic effects in the field have been caused by spills of light oil, especially when these have been confined in a small area. In addition to its direct lethal effects, oil may cause death by inducing a state of narcosis in

Table 10.5 Sensitivity of marine animals and plants to oil pollution (modified from IPIECA, 1991).

Group	Comments
Mammals	Whales, dolphins, seals and sealions have rarely been significantly affected. Sea otters are more vulnerable because of their way of life and fur structure.
Birds	Birds using the air–water interface are at risk, particularly auks and divers. Badly oiled birds usually die.
Fish	Eggs and larvae in shallow bays may suffer heavy mortalities under slicks, particularly if dispersants are used. Large kills of adult fish are rare. Adult fish in farm pens may be killed or stressed such that they succumb to disease.
Invertebrates	Invertebrates including molluscs, crustaceans, worms of various kinds, sea urchins and corals, may suffer heavy casualties if coated with fresh crude oil. On the shore, barnacles are more resistant than limpets and snails such as winkles.
Planktonic organisms	Serious effects on plankton have not been observed in the open sea. This is probably because high reproductive rates and immigration from outside the affected area counteract short-term reductions in numbers caused by the oil.
Larger algae	Large algae such as kelps and brown wracks have a mucilaginous coating which often prevents oil sticking to them. Oil sticks better to dry algae such as those high on the shore and these may be broken by waves due to the weight of oil.
Marsh plants	Some species of plant are more susceptible than others. Perennials with robust underground stems and rootstocks tend to be more resistant than annuals and shallow-rooted plants. If, however, perennials such as the grass *Spartina* are killed, the first plants to recolonize the area are likely to be annuals such as glasswort (*Salicornia*). This is because such annuals produce large numbers of tidally dispersed seeds.
Mangroves	The term 'mangrove' applies to several species of tree and bush. They have a variety of forms of aerial 'breathing' root which adapts them for living in fine, poorly oxygenated mud. They are very sensitive to oil, partly because oil films on the breathing roots inhibit the supply of oxygen to the underground root systems.

which animals become dislodged from their substrates. Though some may recover and re-establish themselves, others succumb through being washed into the strandline where they cannot survive.

The heavier fraction which remains after weathering appears not to be particularly poisonous, and, in general, well-weathered crude oil has little effect. However, if oil comes ashore in great quantity, intertidal populations may be killed by smothering with clinging, tarry material.

In the past, many shore animals were killed by the use of detergents sprayed onto the oil to emulsify and disperse it. The present generation of detergents are much less harmful when used properly and in appropriate circumstances.

Tainting of seafoods such as fish, shellfish and crustaceans may be caused by absorption of hydrocarbons into the tissues. This imparts an unpleasant oily taste and there is concern over the dangers to humans from eating polluted seafood since some oil-derived hydrocarbons may be carcinogenic (IPIECA, 1997).

Fate of spilt oil

When oil is first spilt, there is an initial evaporation of the lighter fractions. In the *Amoco Cadiz* incident, the spilt oil was mainly light crude and approximately 30 per cent of it evaporated. Much of the remaining oil formed a stable water-in-oil emulsion known as 'chocolate mousse'. This tends to float and is often the form in which oil ends up on the shore. On some species these emulsions are more adherent and harmful than unemulsified oil.

Oil remaining on the sea or shore after evaporation gradually disappears, mainly as a result of biodegradation by micro-organisms. This is assisted by mechanical breakup of lumps and patches by wave action; for example waves may re-float oil from the shore and break it up into small droplets. This increases the surface area and aeration of the oil and thus assists biodegradation. In the case of the *Braer* spill off Shetland in 1993, the wave action was so severe that most of the oil was dispersed through the water column and very little came ashore. Although most biodegradation is via microbial action, some is ingested by larger animals and at least partially broken down.

Some oil may be partially broken down by chemical degradation catalysed by exposure to sunlight. Remaining products may then be more easily biodegraded. The rate and extent of chemical degradation are affected by light intensity and duration, aeration and oil thickness.

Well-weathered oil at sea may form 'tar balls' which may continue to be washed ashore for many months after the spill, causing a great nuisance to beach users.

Oil pollution of beaches

The fate and effects of oil washed up on beaches depend not only on the variables such as oil type, mentioned above, but also on the energy level of the shore (degree of exposure to wave energy) and on substratum type.

In general, the more exposed the shore, the quicker the biological recovery time of the littoral benthos. This applies both to rocky and sedimentary beaches. Oil does not remain long on wave-battered rocky shores, and where vertical cliffs are present the oil may never reach them due to the action of reflected waves. In contrast, a gradually sloping boulder shore in a sheltered sea loch is likely to trap the oil which will get under the boulders and sink into any sediment beneath.

Once oil gets down into the substratum, it is protected from wave action and will take much longer to disappear. Freely draining sediment shores of coarse sand, gravel or stones allow easy penetration of oil. With time, the oil may become more viscous due to evaporation and weathering and cannot escape. Firm beaches

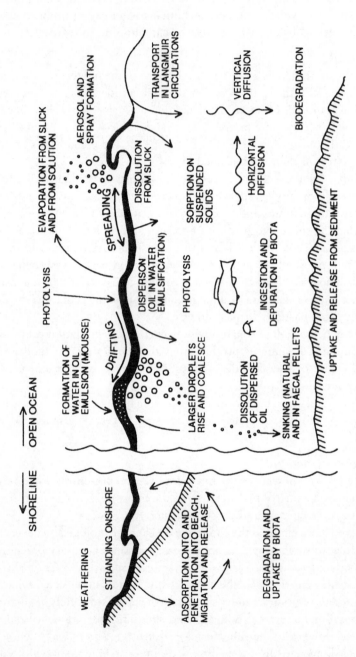

Figure 10.3 Schematic diagram of oil spill processes at sea and shorelines.
In GESAMP 1993 from Mackay (1985). *Petroleum effects in the Arctic environment*, Ed. Englehardt, Elsevier Applied Science.

of waterlogged fine sand or mud resist penetration and most of the oil will be washed away on subsequent tides. However, where there are fine, productive sediments supporting rich populations of burrowing animals and rooted plants, the oil will penetrate deeply through burrows and down root systems. These are the types of conditions found in salt marsh and mangrove systems. Once the oil penetrates such shores, it may remain for many years in an unweathered state. Regeneration of mangrove systems has often been prevented by chronic oiling leaking from such sediments for many years after a spill. Oil that has penetrated shallow sublittoral sediments will similarly be protected to some extent from natural degradation, and may cause persistent fouling on subsequent release from wave-churned beaches long after the original pollution source has ceased.

The general topography of the area and the weather can also modify the effects of spilt oil. If oil drifts onto an irregular coast where there are headlands, inlets, rocks and islands, it will generally form a thinner covering than on a straight shoreline because of the greater distance over which it is spread. Oil on rocky shores tends to be carried upshore towards high water level, whereas on sandy beaches much more of it is spread over a range of shore levels.

Strong onshore winds may carry some of the pollutants inland in windborne spray. This was of particular concern following the *Braer* oil spill in 1993 in Shetland. Here oil was carried inland by gale-force winds, affecting sheep, cattle and humans.

The persistence of oil on a shore can therefore vary from a few months to many years (IPIECA report series).

Recovery of shore populations

The restoration of an intertidal population after destruction by oil pollution generally involves a sequence of stages in which different species are successively dominant. This sequence is now well recognized on British rocky shores.

An oil spill usually kills large numbers of grazers such as limpets. This means that provided no further oil comes ashore, there is usually a rapid settlement and growth of microalgae which turn the rocks green. Fast-growing green macroalgae, especially *Enteromorpha* and *Ulva*, follow. This is often known as the *greening phase*.

In the continued absence of grazers, sporelings of the larger, slower-growing normal intertidal brown algae settle and grow resulting in an abnormally dense cover of seaweeds. This is the *fucoid phase*.

Meanwhile, juvenile limpets and snails have settled out of the plankton in sheltered crevices and start to grow rapidly on the abundance of food. *Patella* spp. often become abnormally numerous and fast-growing. The effect of feeding by this increasing number of herbivores gradually reduces the plant population to normal levels, followed in turn by a decline in number of grazers. Space becomes available for resettlement by barnacles. This is the *Patella phase*.

The shore may look comparatively normal within two to three years. However, the fine balance between grazers and algae may take much longer to stabilize. In the case of the *Torrey Canyon* disaster, the shores around Cornwall took 10 years to return fully to normal mainly because large amounts of toxic detergent were used to clean the beaches.

Beach protection and cleaning

There is no doubt that in the case of oil spills, the old adage 'prevention is better than cure' has never been truer. However, with ever increasing oil exploration and distribution, the opportunities for spills are now greater than ever. Many spills result from human error and from severe weather conditions. However, economic factors also come into play. There is no doubt that in the case of tanker spills, many could be prevented if all tankers were built with double hulls. Several incidences where such ships have run aground and gashed their hulls without spilling any oil bear testament to this.

Protection

Various measures are used for the protection of beaches from oil pollution, when it occurs. Permanent and temporary floating booms and barriers may be used to prevent entry of oil into inlets and estuaries. In Sullom Voe in Shetland, where there is a massive oil terminal, such barriers are kept permanently on station at the various arms of the loch, ready for deployment. Temporary booms were used in the *Exxon Valdez* disaster to prevent the entry of oil to salmon hatcheries. Spilt oil may also be impounded with booms and skimmed up. All such booms are often rendered ineffective by waves or currents.

Clean-up methods

It is now generally accepted that there is no single clean-up method appropriate for all spills. Each spill is different and careful assessment is needed before deciding on a course of action (IPIECA, 1991; IMO/IPIECA, 1996). The action taken must be capable of significantly reducing the recovery time of the shore to below that which natural weathering will achieve. Sometimes cleaning may be undertaken for economic and amenity reasons rather than for the wildlife alone.

Even in recent incidents, clean-up operations have sometimes increased the impacts and extended the recovery time for marine populations. These have usually been where aggressive techniques such as the use of high-pressure hot water (e.g. *Exxon Valdez* spill) and excessive use of dispersants (e.g. *Torrey Canyon*) have been used. These methods often kill off key species that have survived the initial oiling. However, even attempts to use non-aggressive mechanical clean-up methods can be damaging when used inappropriately. When attempting to clean mangrove areas, long-term damage has often been done by the heavy vehicles needed to get into the area.

Mechanical clean-up methods can sometimes be used to remove bulk oil and these generally cause little damage provided they can be used with minimal trampling, and may also prevent further pollution from mobile oil. Suction pumps collect oil from gullies, and rockpools or from trenches dug to collect it. Straw bales and a variety of absorbent materials can be used to mop up small areas of surface oil. A privately developed 'mat' sandwich of straw, which can be unrolled onto the surface of the water, is being used very successfully for small spills, especially in fresh water. Low-pressure flushing with water at ambient temperature washes shores, with little physical damage to marine life but requires booms, skimmers and a large team to collect the oil. Oiled weed and other debris can be collected by hand and oil-saturated sand dug up and dumped elsewhere.

Chemical dispersants have been much used for beach cleaning, especially on British shores. They soon make the beach more pleasant for human enjoyment, but have often proved much more damaging to intertidal organisms than the oil itself. Modern dispersants have low toxicity and may sometimes be appropriate for shore clean-ups. Nowadays they are more usually used on offshore slicks to prevent their coming ashore. They may, however, cause the oil to disperse into the water column, sink and affect the benthos. Even so, dispersants might be used, for example, to prevent oil entering a sensitive mangrove area whilst accepting there might be some detrimental effects to nearby coral reefs (IPIECA, 1992; 1993).

10.1.12 Conclusion

The oceans are so great in volume that overall accumulation of persistent pollutants can only occur very slowly. Organisms have a great capacity to respond to gradual environmental changes by adaptation, acclimatization and evolution. Except where the pollution load is heavy, its effects on marine life are likely mainly to influence the fringes of a species' distribution, where the population is already under environmental stress. In such areas any additional burden may increase mortality.

Despite some scaremongering of impending world catastrophe through oceanic pollution causing the widespread demise of marine life, a reasonable view of the present situation seems to be that these dangers are remote compared with far more imminent threats to human survival arising from our inability to live peaceably together. The oceans do have a capacity to absorb wastes and in some cases may be a much safer place for disposal of certain wastes than storage on land, with the attendant hazards to human populations of contamination of food or freshwater supplies. However, it is important that the effects of waste disposal are looked at in a long-term context and that minimization of waste (see below) should be an ultimate goal.

Although some marine pollutants are remarkably widespread, the immediate dangers of marine pollution tend to be local rather than global, and mostly confined to coastal areas where mixing rates are slower, especially in bays and estuaries. Non-tidal seas such as the Mediterranean and the northern Baltic obviously present special problems as do semi-confined areas such as the North Sea.

Clearly it is necessary to have effective international control of dumping at sea and coastal discharges. Although marine pollution may be mainly a local problem, it does not necessarily stay in one place. All potential sources of pollution must be identified, all coastal areas at risk regularly monitored. Because there are increasing pressures to pollute, continuous and increasing watchfulness and powers for protection are essential. Dangers of pollution by shipwreck and collision must obviously be reduced as far as possible by all feasible safety measures. For example, keeping easily accessible and accurate cargo manifests for dangerous or obnoxious materials, to enable appropriate action to be taken immediately in case of spillage.

Control of pollution often involves conflicts of interests. The requirements of industries to discharge effluents into estuaries may be opposed by those working in local fisheries. The advantages of cheap sewage disposal may be offset by damage to holiday trades. It is therefore sensible that regulations determining the use of inshore waters for the discharge of pollutants should be largely a matter of local responsibility exercised with proper regard to effects elsewhere, rather than of general controls which may be inappropriate in particular localities. Adequate powers of monitoring and enforcement must be provided.

In all matters of waste disposal the question must be asked: to what extent can wastes be put to good use? Waste not, want not. Some materials can be economically recycled. Spoil, rubble and rubbish can sometimes be used to raise land levels for reclamation. Sewage contains valuable fertilizers. Waste heat is an anomaly in an age facing an energy crisis. The term 'waste' is inaptly applied to anything for which some other useful application could be found.

Waste minimization and reduction of pollution at source should be the ultimate goal. Such techniques will only ever be widely adopted if industry can see concrete advantages. In recent years in UK there have been several waste minimization demonstration projects involving local industries. In most cases, the industry concerned found increased business efficiency, lowered costs due to conservation of raw materials and reduced costs and safety risks associated with storage and handling of waste (Earll, 1995).

Future generations will judge this period to have been an Age of Waste. The earth's resources of energy and useful materials are being rapidly consumed and dispersed, often for trivial or destructive purposes, with remarkably little forethought.

10.2 Climate change and global warming

10.2.1 The role of carbon dioxide

Carbon dioxide is stored in the atmosphere, the land and in the oceans. The latter have an immense capacity to contain CO_2, storing by far the largest amount and thus having a major controlling influence on atmospheric CO_2 levels. Physical, biological and chemical factors are all involved in the uptake of CO_2 by the oceans. Dissolved CO_2 in its several forms is conveyed between surface and deeper levels by currents and water-mixing processes. However, phytoplankton plays the leading role in CO_2 transport. Phytoplankton take up CO_2 dissolved in the surface waters and use it for photosynthesis. Herbivores eat the phytoplankton, thus taking up the carbon and transporting it to deeper levels during diurnal migrations. Here it is released during respiration and defaecation or further distributed through the food web. The amount of CO_2 taken up by the oceans is dependent on productivity and is therefore lowest in the tropics and highest in mid- and high latitudes. In the northern hemisphere, uptake of CO_2 by the oceans increases during the spring and summer plankton blooms. In fact, in the tropics there is a net discharge of CO_2 to the atmosphere throughout the year but this is more than made up for in higher latitudes. Anything that lowers the productivity of the open oceans will also lower the oceans' ability to absorb CO_2. For example, increased water stratification resulting from warming of surface waters could prevent upward flow of nutrient-rich water necessary for phytoplankton production.

With the advent of our modern industrial society, the amount of CO_2 directly entering the atmosphere has risen dramatically and now stands at about 350 ppm as compared with 280 ppm in the 1850s. The current rate of increase is about 1.3 ppm per year. Most of this extra CO_2 comes from the burning of fossil fuels and the effects of de-forestation. In the latter, the trees are not only no longer there to photosynthesize and use up CO_2 but the forests are often burnt, thus releasing more CO_2. The oceans are helping to counteract and moderate this build-up of atmospheric CO_2 because as atmospheric levels increase, more CO_2 dissolves at the ocean–atmosphere interface.

The greenhouse effect

This rise in atmospheric CO_2 levels is of increasing concern because of a phenomenon that has come to be known as the 'greenhouse effect' (Gribbin, 1988b). Energy reaches the earth's surface as short-wavelength radiation in sunlight and leaves again as long-wavelength IR (infra-red) radiation. The escape of IR radiation is slowed by the presence in the atmosphere of CO_2, methane and other rare gases whilst incoming radiation is unaffected. Without these gases, the earth would have an average temperature around about freezing point. However, too great a concentration of these gases delays the escape of IR radiation and allows the earth to warm up. The increase in concentration of CO_2 over the past century and a half appears to be having just that effect. The average global temperature has risen by about $0.5°C$ during the present century. However, it is not absolutely certain that this rise has been caused by the greenhouse effect. What climate researchers are certain of is that if we continue to add CO_2 to the atmosphere at the present rate, then global temperatures will rise, perhaps as much as $1.5–4.5°C$ in the next 50 to 100 years. This prospect is now being taken very seriously by governments and target levels for the reduction of CO_2 emissions have been set.

As we have seen above, the absorption of CO_2 by the oceans plays a vital role in mitigating the greenhouse effect. However, if the temperature of the oceans increases due to global warming, this will reduce the solubility of CO_2 and might consequently lead to a net release of CO_2 from ocean to atmosphere thus compounding the problem. Obviously the relationships of these processes with respect to climatic effects are complex and not yet well understood. They are now under intensive study by meteorologists and marine scientists. A recent review by Smith and Hollibaugh (1993) has shown that the role played by coastal ecosystems in the global carbon budget is even more important than had previously been thought.

10.2.2   Ozone and ultraviolet effects

As well as worries about the addition of CO_2 to the atmosphere, scientists are also concerned about the depletion of another gas, ozone. Ozone in the stratosphere absorbs much of the ultraviolet (UV) radiation from the sun. This is important because ultraviolet-B (UV-B) is harmful to most life forms. It is the UV-B rays from the sun that cause sunburn and skin cancer in humans and UV light is used to kill bacteria in sewage treatment plants.

In recent decades, the amount of ozone in the stratosphere has been declining. This was first noticed in the Antarctic in the 1970s and since then large 'holes' in the ozone layer have regularly been recorded there. Losses of about 8 per cent per decade have been estimated at polar latitudes. UV levels naturally rise in spring-time but in recent years these higher levels have been arriving sooner and lasting longer than normal. The destruction of the ozone layer is thought to be mainly due to the release of chlorine into the atmosphere. Chlorine destroys ozone through a series of chain reactions. During the winter, chlorine is concentrated over the Antarctic due to complex and unique climatic conditions. This is why ozone depletion is most marked over the Antarctic (Gribbin, 1988a).

Some chlorine enters the atmosphere from natural sources but the majority results from human activity. A major source is from chemicals called chlorofluorocarbons (CFCs) used in air-conditioning plants, refrigerators, aerosol propellants and various manufacturing processes. Alternatives to CFCs are being increasingly used and sought but the ozone layer at present continues to decrease.

UV-B light can burn and damage plants on land. It can also kill and damage photosynthetic algae in surface waters. In fact it is now thought that UV-B may penetrate as deep as 20 m into the oceans although the amounts fall off rapidly below the first metre or so. Experiments have shown that the UV-B in sunlight inhibits phytoplankton growth. Thus an increase in the amount of UV-B entering the oceans could reduce productivity. This in turn would reduce the oceans' ability to absorb CO_2 (see Section 10.2.1) and so could increase global warming.

Other experiments have indicated that UV light may have a controlling influence on many plants and animals growing in shallow water. The resistance or sensitivity to UV light of different species and different life stages of reef organisms may partially dictate their abundance and distribution. For example, experiments by Wood (1987) suggest that the kelp *Ecklonia radiata* may be prevented from colonizing shallow water by the higher levels of UV radiation occurring there.

10.2.3   Effects on coral reefs

Coral reefs have been in existence for many thousands of years and have survived and evolved through massive climate changes. Climate change (specifically global warming), by itself is therefore unlikely to result in the global elimination of coral reefs and there are few instances where such changes on their own have led to reef destruction. It is likely, however, to result in changes to reefs which will adversely affect local communities dependent on reefs for their livelihood.

Unfortunately global warming is not the only factor currently causing stress to coral reefs. Many reefs are being degraded or destroyed by pollution and sedimentation, over-exploitation of fish and other species and physical destruction. The worry is that these chronic and acute stresses to which coral reefs are not well adapted may act synergistically with climate change and threaten the existence of coral reefs in many parts of the world. Stressed reefs lose their natural ability to recover from disturbances such as cyclones and increased water temperature, both of which could increase in frequency as a result of global climate change.

Specific climate-change threats to reefs identified in a recent report on the implications of climate change on coral reefs (Wilkinson and Buddemeier, 1994) include the following:

(*a*)   Although rising sea levels will not threaten most coral reefs themselves, since they can grow upwards, low-lying islands associated with reefs would be affected. In the Maldives, the government is already seriously considering the implications.

(b)   Frequent episodes of temperature extremes, caused by climate change, will cause coral bleaching (see below), loss of coral cover and a lessening of the ability to withstand other stresses.

(c)   As concentrations of CO_2 in the atmosphere increase, so more CO_2 will dissolve in the surface layers of the sea. The resulting increase in acidity may slow down the rate at which corals and coralline algae can deposit calcium carbonate. In contrast, an increase in the concentration of CO_2 stimulates the growth of algae which might then out-compete the corals. Overfishing and nutrient pollution will exacerbate the latter effect.

(d)   Reefs adjacent to land masses will be affected by increases or changes in rainfall and runoff. Greater sediment loads and an increase in both pollutants and nutrients in coastal water could damage reefs.

(e)   Shifts in current patterns could have serious impacts on reefs.

Coral bleaching

The narrow temperature range within which most corals are at their healthiest is very close to the upper lethal limit. Temperatures of only 1 or 2°C above the usual summer maximum can be lethal and can also trigger an increasingly commonly observed phenomenon known as *coral bleaching*. The beautiful colours of many corals and some of their relatives such as anemones and sea whips mostly come from the symbiotic zooxanthellae living within their tissues. Most or all of these algal cells may be expelled by the coral when it is under stress so that the white skeleton shows through. If the bleaching is severe or prolonged, the coral often dies. Shorter episodes may allow the coral to rebuild its algal population and survive.

Recent large-scale bleaching events on coral reefs seem to be related most consistently to higher-than-normal water temperatures. In 1982–3 there was an unusually severe El Niño southern oscillation event (see Section 1.3.6) which increased water temperature in the eastern Pacific 3 to 4°C above the usual average. Corals became bleached at many places throughout the Indo-Pacific. Up to 90 per cent of corals were killed on some reefs including those in Panama, Costa Rica and the Galapagos. Widespread bleaching occurred again between 1987 and 1988, another strong El Niño year. Some scientists link the apparently increasing severity and occurrence of El Niño to global climate change and suggest that bleaching is one of the first concrete manifestations of global warming. Considerably more detailed and standardized data on seawater temperatures and on the physiological responses of corals to temperature and stress are needed before this conclusion can be justified. However, the Intergovernmental Panel on Climate Change has predicted a 1 or 2°C temperature increase over the next 50 years, and if this proves correct, there is little doubt that the consequences for coral reefs could be disastrous.

10.2.4 Sea level rise

Global warming of 2–4°C such as has been predicted by some scientists would affect sea levels in two main ways: through melting of polar land ice and through expansion of seawater as it warms up. Predictions on the extent of sea level rise vary from about 0.4 m to 1 m and are very difficult to estimate, given the large number of variables involved.

Coastal communities such as saltmarsh and mangrove systems can keep pace with a slowly rising sea level by trapping sediment and growing upwards, and have successfully done so over the ages. They may have to retreat landwards but this is not a problem unless prevented from doing so by sea walls and barriers. Building of extensive walls to keep back the sea as levels rise, could result in the loss of extensive areas of coastal wetlands. Apart from the loss of wildlife, this could also result in the loss of commercially valuable marine fish and other species as nursery areas are lost. Experiments are underway on the east coast of England and in other countries, to remove or lower sea walls and allow and encourage the growth of saltmarsh or mangrove which can act as a natural and adaptable (and cheap) barrier to the sea.

10.2.5 Mitigating effects

Sulphur compounds in the atmosphere tend to promote cooling of the earth because, unlike CO_2, they reflect radiation back into space before it can reach the earth's surface. Over the sea, they can also act as cloud condensation nuclei attracting water vapour, and forming stratus and stratocumulus clouds. These clouds reflect sunlight so that they again reduce the radiation reaching the ground. However, clouds also stop infrared radiation escaping from the earth back into space. Recent work with satellites has indicated that on balance, in a worldwide context, these clouds cool more than they warm the earth.

The major source of the cloud-forming sulphur compounds in the atmosphere is dimethylsulphide (DMS) gas, which is produced by phytoplankton (see Section 4.3.3). Thus phytoplankton may help to cool the climate although it is not known how important this effect is. Volcanic eruptions such as that of Mount Pinatubo in the Philippines in 1991 also release large quantities of sulphur in the form of sulphur dioxide into the atmosphere.

Some scientists have suggested that global warming could be reduced by 'fertilizing' the open ocean with large amounts of iron. Such experiments have been aimed at increasing productivity since lack of iron is one of the limiting factors for phytoplankton growth (see Section 9.6.4). As described in Section 10.2.1, increased productivity can theoretically help to reduce atmospheric CO_2 levels as the phytoplankton utilize CO_2 in their photosynthesis.

The various systems described in Sections 10.2.1 to 10.2.5, including atmospheric CO_2, ozone, and sulphur compounds, interact in complex ways and make it very difficult for scientists and climate specialists to predict the rate and extent of possible climatic changes. There is, however, a general agreement that increased levels of CO_2 in the atmosphere are resulting in a trend towards global warming.

10.3   Management and marine nature reserves

10.3.1   Protected marine areas and legislation in the UK

On land the concept of protecting and conserving areas considered to be of prime importance, in terms of the wildlife and habitats they contain, is well established. The first National Nature Reserve was declared in the UK in 1951 and there are now over 200 with 43 that are over 1000 ha in size (IUCN, 1994). In contrast, as of 1996, there are only two statutory Marine Nature Reserves, Lundy Island in the Bristol Channel and Skomer Island off the Pembrokeshire coast. These were declared in 1986 and 1990 respectively. The area protected amounts to only 3700 ha in total.

Some protection has been afforded to intertidal marine habitats and species through other designations such as Areas of Outstanding Natural Beauty, National Parks and National Scenic Areas where these have included part of the coast. Intertidal areas down to LWM can be included in Sites of Special Scientific Interest (SSSI) but these do not have full statutory protection. Until 1981, there was no provision for declaring Nature Reserves below the LWM. New legislation included in the UK 1981 Wildlife and Countryside Act provided the necessary framework to declare Marine Nature Reserves, but the process is complicated and difficult to implement (Gibson and Warren, 1995). Consequently, the conservation organizations have had to seek other ways of gaining protection for important areas (Gubbay, 1995). Most of these new designations are, like many of the coastal ones, being set up on a voluntary basis. It can be seen from Table 10.5 that there are currently many ideas under consideration which will have to be modified and integrated if they are to be effective. The voluntary approach is not a new idea. Both Lundy and Skomer MNRs started as voluntary Marine Nature Reserves and at least nine others are well established (Gubbay and Welton, 1995).

Currently attention has been once more directed to marine protected areas in the UK because of European Community legislation in the form of the EC Directive on the Protection of Natural and Semi-Natural Habitats of Wild Flora and Fauna (commonly known as the 'Habitats Directive'). This lays down categories of habitats in both marine and terrestrial environments for which each member state must develop proposals for protection. The central aim is to conserve biodiversity across the area of the European Union through a coherent network of Special Areas of Conservation (SACs).

Table 10.6 Types of established and proposed marine protected areas in the UK (modified from Gubbay, 1995).

Title and status	Year (of introduction or proposal)	No. of sites	Details
Voluntary Marine Conservation Areas (*voluntary*)	1970	9	Locally based management groups. Not necessarily at sites of major marine biological importance. Strong educational component.
Marine Nature Reserves (*statutory*)	1981	2	Government conservation agencies. To conserve marine flora and fauna or geological or physiographical features of special interest. Opportunities for study and research.
Marine Consultation Areas (Scotland) (*voluntary*)	1986	29	Scottish Natural Heritage (government agency). A general management arrangement for sites considered to be of particular distinction in respect of the quality and sensitivity of their marine environment. Mainly in response to the rapid growth of marine fish farming in sea lochs.
Marine Protected Areas (*proposed statutory*)	1991	–	Proposal not followed up yet by Marine Protected Areas Working Group (group of conservation organizations and individuals).
Marine Consultation Areas (England & Wales) (*proposed voluntary*)	1992 (withdrawn 1993)	16	Proposed by the DOE and Welsh Office (government depts). Similar idea as for MCAs in Scotland.
Sensitive Marine Areas (*proposed voluntary*)	1993	27	Proposal only at present, by English Nature (government body). Generally similar to MCAs. Possibly the ideas will be combined.
Special Areas for Conservation (*statutory*)	1992 (adopted 1994)	–	EC Habitats and Species Directive (92/43/EEC) Regulation adopted in 1994 by government agencies.

10.3.2 Management schemes

The management of nature reserves, whether terrestrial or marine, is of prime importance if they are to attain their conservation goals. At present, the main management methods for MNRs involve the control of human activities such as fishing and recreation. Active management to maintain or produce certain types of habitat, such as coppicing woodland, is not normally appropriate or possible in MNRs. In most cases it is not necessary or desirable to stop all human activity. In any case, you cannot put a fence around a marine nature reserve and there are a number of 'rights of passage' on the sea. Again an activity such as potting for lobsters and crabs may be considered potentially damaging in one part of a reserve but not in another. One way of dealing with this is the concept of 'zoning schemes' which has been adopted for the management of Lundy Island (Laffoley, 1995). This concept has had considerable success in Australia when applied to the massive Barrier Reef Marine Park.

Zoning schemes identify areas of a reserve or park in terms of their sensitivity to particular human activities. The less sensitive the area, the more activities are likely to be allowed. For such a scheme to work, it is essential that the information concerning which activities are allowed in which areas is disseminated to users of the area in a clear and easily used format. Coloured maps and charts with associated tables are proving quite successful in this context (see Figure 10.4).

Coastal zone management

At the present time, and partly because of the difficulties experienced with marine site protection, more emphasis is being placed on management of the marine environment as a whole. Protected sites are more likely to achieve their aims if

Activity	General Use Zone	Recreational Zone	Refuge Zone	Sanctuary Zone	Archaeological Protection Zone
Recreational					
Diving	Yes	Yes	Yes	Yes	No
Snorkelling[1]	No	Yes	No	No	No
Swimming[1]	No	Yes	No	No	No
Spearfishing	No	No	No	No	No
Commercial					
Trawling	Yes	No	No	No	No
Dredging	Yes	No	No	No	No
Potting	Yes	Yes	Yes[2]	Limited[3]	No
Tangle nets	Yes	No	Limited[4]	No	No
Fixed nets	Yes	No	Limited[4]	No	No
Collecting					
Group educational excursions	Permit	Permit	Permit	Permit	No
Scientific research	Permit	Permit	Permit	Permit	Permit

Figure 10.4   Lundy Marine Nature Reserve zoning scheme. The map distributed to reserve users is in colour which makes interpretation quick and easy.
(By kind permission of English Nature. Map based on Admiralty Chart 1164 with the permission of the controller of Her Majesty's Stationery Office © Crown Copyright.)

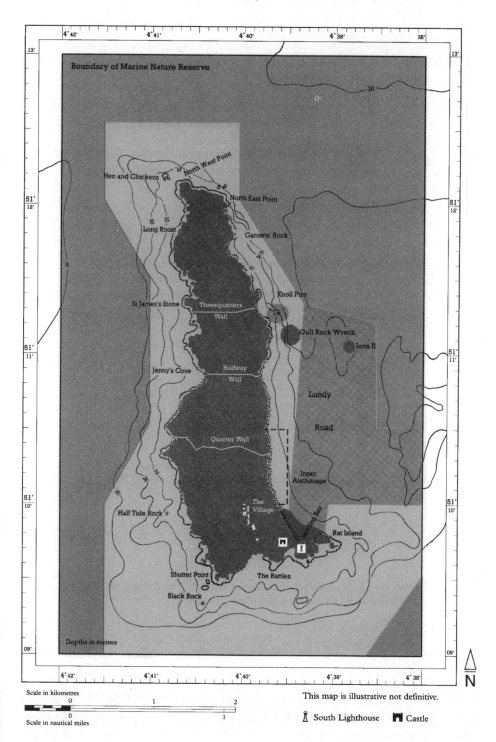

Scale in kilometres

Scale in nautical miles

This map is illustrative not definitive.

South Lighthouse   Castle

N

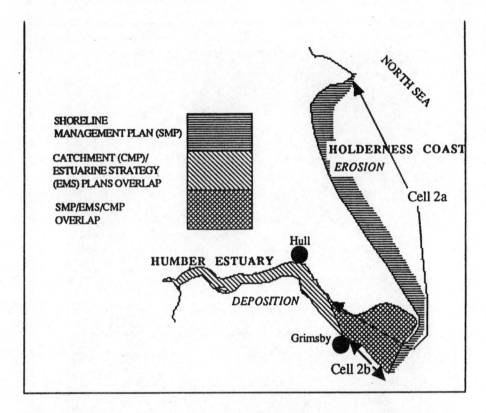

Figure 10.5   Humberside coast and estuary. Despite sediment interchange between coast, estuary and North Sea, management plans are largely independent of each other.
(From Pethick, J. in Earll, R.C. 1995 by kind permission of Dr R.C. Earll.)

they are part of a much wider system of management measures. One such wider system is the concept of coastal zone management (CZM). This aims to look at the coastline and associated areas and devise management policies and action that will integrate human use on a sustainable basis.

The problem with CZM in the UK is that the coastline, including all the major offshore islands, extends for a distance of at least 14 500 km and is extremely diverse in terms of habitats. There is an equally varied human usage of the coast. This makes it difficult to assess the influence of human activity and CZM tends to be fragmented.

Many different local authorities and other agencies are responsible for the management of our coastline and shores. Often a great deal of effort is put in by these authorities and their scientists, to produce, for example, shoreline management plans (SMPs), catchment management plans (CMPs) and estuary management plans (EMPs). However problems arise because each authority tends to look only at their particular stretch of coast and there is a lack of integrated planning. On the Humberside coast and estuary, for example, there is extensive sediment interchange between the Holderness coast, the Humber Estuary and the North Sea, but there are at least three management plans for the area which are largely independent of one another (Pethick in Earll, 1995).

Ever since people first started to build substantial sea defences, there have been classic examples where the building of, for instance, an extensive sea wall to provide safe anchorage along one stretch of coast starves the adjacent stretch of sediment and causes considerable erosion and loss of shore habitats. Examples can still be seen today. There is thus a need in this country, and others with similar problems, to adopt a formal integrated structure for coastal zone management. The key ideas behind CZM and the current situation (1995) in the UK are summarized in MCS (1996).

10.3.3   Species protection

Although habitat management and protection are the mainstays of marine conservation, some rare and vulnerable species are given statutory protection in their own right. This again follows the established pattern for terrestrial conservation. In the UK the 1981 Wildlife and Countryside Act provides full protection for all cetaceans in UK waters. Guidelines have also been produced by the DOE for minimizing acoustic disturbance to small cetaceans. These are intended for use by operators of installations such as oil rigs and ships carrying out seismic surveys (DOE, 1995). Seals are given partial protection (closed seasons) under the Protection of Seals Act 1970.

The EC Habitats Directive (see Section 10.3.1) also provides protection for some marine species including all cetaceans, all marine turtles, otter, walrus and a small number of rare fish and invertebrates. A much fuller list of UK terrestrial, freshwater and marine plants and animals considered to be of conservation concern, along with conservation targets for these species, is given in 'Biodiversity Challenge', a document prepared by a large group of voluntary conservation organizations (Wynne *et al.*, 1995). In a worldwide context, the IUCN produces a list of all threatened animals including marine ones (IUCN, 1994). In many cases, the problem with deciding on protection and management options for species is a lack of information on the status and biology of the species. In the case of vulnerable fishery species such as the basking shark (*Cetorhinus maximus*), many scientists are advocating the adoption of the 'precautionary principle' where exploitation is limited whilst data on the species are being collected.

10.4   Effects of fisheries and aquaculture

10.4.1   Direct effects

The effects of fishing and aquaculture activities and the concepts of overfishing and maximum sustainable yield are discussed in Chapter 9.

10.4.2   Indirect effects

One of the problems with all types of fishing gear is that it always has an incidental or by-catch of non-target species. This may include other commercial and non-commercial fish, seabirds and marine mammals.

Drift or gill nets are of particular concern in this respect. Modern drift nets may extend for several kilometres and are made from monofilament nylon. This makes them virtually invisible to animals using sonar. High seas drift netting was banned in December 1992 by a United Nations resolution, when it was discovered just how huge the incidental catch was. Observers on boats from the Japanese squid fleet counted the animals caught and these figures were extrapolated to give a total for the whole fleet during the whole season, in 1989 and 1990. The totals for 1989 are shown in Table 10.7. When it is considered that Taiwan and Korea run equally large squid fleets, the total effect on the populations of some of the less common by-catch species such as albatross must be very significant.

Drift nets continue to be used, albeit on a much smaller scale, and have been nicknamed 'walls of death'. When drift nets are lost, they may continue to kill many animals by 'ghost fishing'. This is the term used for the vicious cycle whereby animals are caught in the net which eventually sinks under the weight of catch. The dead animals then rot away or are eaten and the net rises up into the water to start fishing all over again. The nylon net itself will not rot for many years.

Tuna fisheries have been a particular problem for dolphins especially in the eastern tropical Pacific. Estimates suggest that at least 6 million dolphins have

Table 10.7 Estimated catch and main incidental catch of the Japanese squid fishing fleet in 1989.

Type of animal	Numbers caught (thousands)
Squid (target species)	70 000
Other fish:	
Pomfrets	32 000
Albacores	1 400
Blue sharks	1 200
Skipjack tuna	200
Birds:	
Albatrosses	14
Shearwaters (various spp.)	200
Mammals:	
Dolphins	24
Northern fur seals	4
Reptiles:	
Turtles	0.5

been killed there in the past 30 years. Dolphins follow schools of yellowfin tuna and fishermen use boats, helicopters and light aircraft to spot the dolphins. The fishermen then herd the tuna using small speedboats and the main fishing boat encircles and catches them using large seine nets (see Section 9.1.3). Unfortunately many dolphins are also caught in the nets and drown. New nets have now been developed which have a fine-mesh area called the Medina panel. As the net tightens, the dolphins can jump over the net rim because the panel makes it visible to them. Sometimes divers and small boats are used to help the dolphins out.

As a result of public pressure, tuna fishing using drift nets has now mostly stopped in the eastern tropical Pacific and some fishing fleets are using so-called 'dolphin friendly' fishing methods. These include longlines and purse seining of tuna schools not associated with dolphins. Longlines are also used for catching other species such as squid and have proved a danger to some seabirds, especially albatrosses which are attracted to the lines and swallow the baited hooks.

Turtles are vulnerable to accidental capture in a variety of nets and trawls. Shrimp trawls in the Gulf of Mexico kill many Kemp's ridley turtles, an endangered species, each year. The Fisheries Service there invented a turtle exclusion device which, when attached to the trawl, allowed any turtles caught to escape through a special trap door. However, the device was fiercely rejected by the shrimp fishermen and is little used.

10.4.3 Non-native species

When a species is accidentally introduced to an area outside its natural geographical range, it can cause considerable environmental problems. The successful establishment of a non-native species usually depends on a combination of circumstances such as a lack of natural predators, favourable physical factors such as temperature and a good food supply. Species usually become established only if they were introduced from similar latitudes. The above conditions and a fast reproductive rate can lead to rapid increases in populations of introduced species. A classic example on land was the introduction of rabbits to Australia. Species can be introduced both from the areas in which they originate or from secondary sources to which they were previously introduced.

In the UK, marine species have been introduced primarily in association with shipping particularly through the discharge of ballast water and fouling of ships' hulls. Some have been deliberately introduced for commercial reasons whilst others have come in unintentionally along with commercial species, e.g. the slipper limpet *Crepidula fornicata*.

A recent government study in the UK has identified 53 non-native marine species of which the majority were red algae, polychaete worms, crustaceans and molluscs (JNCC, 1995). Only a few of these species have so far proved to be a nuisance to sea users or a threat to the environment and it appears that marine communities are more able to accommodate alien species without disruption than are terrestrial ones (see also Canals, below). Well-known examples of species that have become a nuisance include Japanese seaweed (*Sargassum muticum*) which is clogging many inlets in the south of Britain and along North Atlantic coasts, and the slipper limpet (*Crepidula fornicata*) which is now a pest on European oyster beds. The newest reported arrival to become established in British waters is the large brown seaweed *Undaria pinnatifida*, first recorded in the Solent in June 1994. Major effects of such introductions identified by the JNCC study were:

- displacement of native species, e.g. the barnacles *Semibalanus balanoides* and *Chthamalus* spp. by *Elminius modestus* (from New Zealand); the brown seaweed *Halidrys siliquosa* by *Sargassum muticum* (from Japan);
- introduction of new pests, e.g. the slipper limpet *Crepidula fornicata* (from America);
- habitat alteration, e.g. by *Crepidula fornicata* beds;
- degradation of the integrity of the gene pool through hybridization, e.g. of the cord grass *Spartina alterniflora* (from America) with *Spartina maritima* to form *Spartina anglica*;
- associated effects of commercial harvesting, e.g. damage to eel grass (*Zostera*) beds through dredging for the clam *Mercenaria mercenaria* (from America);
- trophic alteration, particularly through dietary competition and predation (most introduced species); and
- improved water quality, e.g. from the filtering action of settlements of the serpulid worm *Ficopomatus enigmaticus*.

Canals

The construction of shipping canals such as the Suez (across Egypt joining the Mediterranean and Red Seas), Panama (across the isthmus of central America joining the Caribbean Sea and the Pacific Ocean), and Kiel (linking the North Sea with the Baltic through Germany), provided unrivalled opportunities for marine organisms greatly to extend their geographical ranges. Physical factors such as freshwater lochs and temperature, and physical barriers, have limited or prevented the spread of species through the Panama and Kiel Canals but around 500 species of Red Sea organisms have spread into the Mediterranean through the Suez Canal. Only about 10 species have moved in the other direction, probably because of prevailing current flows, but the full reasons for this are as yet unknown. Most of the invading Red Sea species have not caused major problems although a few native Mediterranean species have been ousted by their Red Sea counterparts, e.g. the cushion star *Asterina gibbosa* has been replaced in some areas of the Mediterranean by the Red Sea *Asterina wega*. Plankton communities too have been altered by the influx of Red Sea species.

References and further reading

Student texts

Brown, B.E. and Ogden, J.C. (1993). Coral bleaching. *Scientific American*, January 1993, 44–50.

Clark, R.B. (1986). *Marine Pollution*. Oxford, Clarendon Press.

Elder, D. and Pernetta, J. (eds) (1991). *Oceans. A Mitchell Beazley World Conservation Atlas*. Mitchell Beazley in association with IUCN: The World Conservation Union.

Macgarvin, M. (1990). Greenpeace: The Seas of Europe – The North Sea. Collins and Brown. (Note: other seas are covered in this series.)

Marine Conservation Society (undated). Factsheets on Marine Pollution. Available from the Marine Conservation Society (see Appendix 4). (These are useful and concise summaries on a variety of marine pollution topics.)

New Scientist posters. The Greenhouse Effect and The Hole in the Ozone Screen.

Toon, O.B. and Turco, R.P. (1991). Polar stratospheric clouds and ozone depletion. *Scientific American*, **264** (6), June 1991, 68–75.

References

Angel, M.V. (1996). Waste disposal in the deep ocean. In: *Oceanography: An Illustrated Guide*. C.P. Summerhayes and S.A. Thorpe, eds. Southampton Oceanographic Centre, Manson Publishing.

Brown, B.E. (ed.) (1990). Coral bleaching. *Coral Reefs*, special issue, Vol. 8 (4), April 1990.

Clark, R.B. (1986). *Marine Pollution*. Oxford, Clarendon Press.

Collins, K.J. and Jensen, A.C. (1996). Artificial reefs. In: *Oceanography: An Illustrated Guide*. C.P. Summerhayes and S.A. Thorpe, eds. Southampton Oceanographic Centre, Manson Publishing.

Cushing, D.H. and Dickson, R.R. (1976). Biological Response in the Sea to Climatic Changes. *Adv. Mar. Biol.*, **14**, 1–122.

Dipper, F.A. (1991). Colonisation and natural changes in a newly established 'artificial reef' in Gulf waters. In: *Estuaries and Coasts: Spatial and Temporal Intercomparisons*. ECSA 19 symposium. Olsen & Olsen.

DOE (1995). Guidelines for minimising acoustic disturbance to small cetaceans. Department of the Environment, UK. (See Appendix 4.)

Duedall, I.W., Kester, D.R. and Park, P.K. (eds) (1985). *Wastes in the Oceans*. New York, John Wiley and Sons.

Earll, R.C. (ed.) (1994). *Marine Environmental Management: Review of Events in 1993 and Future Trends*. Published by Bob Earll (see Appendix 4).

Earll, R.C. (ed.) (1995). *Marine Environmental Management: Review of Events in 1994 and Future Trends*. Published by Bob Earll (see Appendix 4).

Earll, R.C. (ed.) (1996). *Marine Environmental Management: Review of Events in 1995 and Future Trends*. Published by Bob Earll (see Appendix 4).

GESAMP (1993). Impact of oil and related chemicals and wastes on the marine environment. GESAMP Reports and Studies No. 50. Joint Group of Experts on the Scientific Aspects of Marine Pollution.

Gibson, J. and Warren, L. (1995). Legislative requirements. In: *Marine Protected Areas*. S. Gubbay, ed. Chapman and Hall, London.

Goldberg, E.D. (1976). *The Health of the Oceans*. UNESCO, Paris.

Gribbin, J. (1988a). The Ozone Layer. Inside Science No. 9, in *New Scientist* 5 May 1988.

Gribbin, J. (1988b). The Greenhouse Effect. Inside Science No. 13, in *New Scientist* 22 October 1988.

Gubbay, S. (1988). *A Coastal Directory for Marine Nature Conservation*. Published by the Marine Conservation Society (see Appendix 4).

Gubbay, S. (1995). *Marine protected areas in European waters: The British Isles*. A report for AID Environment from the Marine Conservation Society. (Available from MCS, see Appendix 4.)

Gubbay, S. (ed.) (1995). *Marine Protected Areas. Principles and techniques for management*. London, Chapman and Hall.

Gubbay, S. and Welton, S. (1995). The voluntary approach to conservation of marine areas. In: *Marine Protected Areas. Principles and techniques for management*. S. Gubbay, ed. London, Chapman and Hall.

Holdgate, M. W. (1979). *A Perspective of Environmental Pollution*. Cambridge University Press.

IMO/IPIECA report series. International Maritime Organisation and International Petroleum Industry Environmental Conservation Association, London. Vol. 1. Sensitivity mapping for oil spill response (1996).

IPIECA report series. International Petroleum Industry Environmental Conservation Association, London (an excellent series of booklets):

Vol. 1. Guidelines on biological impacts of oil pollution (1991).

Vol. 2. A guide to contingency planning for oil spills on water (1991).

Vol. 3. Biological impacts of oil pollution: coral reefs (1992).

Vol. 4. Biological impacts of oil pollution: mangroves (1993).

Vol. 5. Dispersants and their role in oil spill response (1993).

Vol. 6. Biological impacts of oil pollution: saltmarshes (1994).

Vol. 7. Biological impacts of oil pollution: rocky shores (1995).

Vol. 8. Biological impacts of oil pollution: fisheries (1997).

IUCN (1994). *1993 United Nations List of National Parks and Protected Areas*. Prepared by WCMC and CNPPA. Gland, Switzerland and Cambridge, UK, IUCN (see Appendix 4).

IUCN (1994). *1994 IUCN Red List of Threatened Animals*. Gland, Switzerland and Cambridge, UK, IUCN.

JNCC (1995). Non-Native Marine Species in British Waters. *Joint Nature Conservation Committee Marine Information Notes*, No. 7, Sept. 1995 (this is a useful series of notes on marine conservation topics available from the JNCC – see Appendix 4).

Johnston, R. (1976). *Marine Pollution*. London, Academic.

Laffoley, D. (1995). Techniques for managing marine protected areas: zoning. In: *Marine Protected Areas. Principles and techniques for management*. S. Gubbay, ed. London, Chapman and Hall.

Marine Conservation Society (1997). *The Reader's Digest Good Beach Guide*. David and Charles.

Marine Conservation Society (1996). CZM (Coastal Zone Management) Information Pack. Available from Marine Conservation Society (see Appendix 4).

Nelson-Smith, A. (1970). The problem of oil pollution at sea. *Adv. Mar. Biol.*, 8, 215.

Nelson-Smith, A. (1972). *Oil Pollution and Marine Ecology*. London, Elek Science.

Pearson, T.H. and Rosenberg, R. (1978). Macrobenthic succession in relation to organic enrichment and pollution of the marine environment. *Oceanogr. Mar. Biol. Ann. Rev.*, **16**, 229–311.

Pernetta, J.C., Leemans, R., Elder, D. and Humphreys, S. (eds) (1994). *Impacts of Climate Change on Ecosystems and Species: Marine and Coastal Ecosystems.* A Marine Conservation and Development Report. Gland, Switzerland, IUCN.

QSR (1987). A report by the scientific and technical working group on the quality status of the North Sea prepared for the Second International Conference on the Protection of the North Sea. London, Department of the Environment.

Schneider, S.H. (1989). The Greenhouse Effect: Science and Policy. *Science,* **243,** 771–81.

Smith, S.V. and Hollibaugh, J.T. (1993). Role of coastal ocean organic metabolism in the oceanic organic carbon balance. *Review of Geophysics,* **31,** 75–89.

Southward, A.J. *et al.* (1975). Recent changes in climate and abundance of marine life. *Nature,* London, **253,** 714–7.

Wilkinson, C.R. and Buddemeier, R.W. (1994). *Global Climate Change and Coral Reefs: Implications for People and Reefs.* Report of the UNEP-IOC-ASPEI-IUCN Global Task Team. Gland, Switzerland, IUCN.

Wood, W.F. (1987). Effect of solar ultraviolet radiation on the kelp *Ecklonia radiata. Marine Biology,* **96,** 143–50.

Wynne, G. *et al.* (1995). *Biodiversity Challenge* (2nd edition). RSPB, Sandy (see Appendix 4).

APPENDIX I
Topics for further study and class discussion or written work

1   Describe in outline the main ocean currents, at the surface and below. How are they set in motion and what factors influence their courses? In what ways is this knowledge of interest to biologists?

2   In general terms describe the overall conditions of life in the marine environment. Giving your reasons, what do you consider to be the major subdivisions of the environment? In what respects do biological conditions in the sea differ from those of freshwater environments?

3   What are the chief processes which bring about vertical water mixing in the seas? Where and when do they occur? Discuss the various effects of vertical water mixing on marine organisms.

4   Discuss the problems involved in measuring the physical and chemical characteristics of the oceans, and indicate why these data are important to biologists.

5   What do you understand by the term 'biomass' and how can this quantity be expressed? Review the methods and difficulties of quantitative sampling of marine plankton, nekton and benthos. Discuss the relevance of biomass estimates in the study of marine ecosystems.

6   Discuss the influence of temperature and salinity upon the distribution of marine species.

7   How do you account for vertical zonation in the sea? Discuss the phenomenon of diurnal vertical migrations, and the advantages and problems associated with this behaviour.

8   By what means do pelagic organisms keep afloat and adjust their depth? Review the mechanisms and problems of buoyancy control.

9   Describe the migrations of a named marine species, explaining how they were

discovered. Discuss the ways in which pelagic animals may be able to navigate.

10 Review the adaptations of the abyssal fauna.

11 If you were the Creator designing a new species for the abyss, with what distinctive attributes would you endow it?

12 Discuss the factors influencing phytoplankton production. Review proposals for increasing the production of phytoplankton in the open sea and discuss their appropriateness.

13 Discuss the reasons why quantities of phytoplankton and zooplankton appear to be in inverse relationship.

14 How do you explain the patchy distribution of marine plankton?

15 Describe the nitrogen and phosphorus cycles of temperate seas and discuss the activities of bacteria in marine cycles.

16 Discuss the value of plankton studies as a means of investigating the movements of water.

17 Review the factors that bring about zonation of shore organisms. Discuss how intertidal zonation is established and maintained.

18 Discuss the effects of crowding on shore populations.

19 Review the causes of periodic changes in shore populations.

20 Give an account of marine wood-boring and rock-boring organisms, and discuss the problems they create for man.

21 All the discoverable specimens of *Gibbula cineraria* were collected within a quadrat of 3 m side at approximately MLWN level on a rocky shore. The maximum shell diameter of each specimen was measured to the nearest 0.5 mm, and the following data obtained:

Max. shell diameter	No. of specimens	Max. shell diameter	No. of specimens
<7.0 mm	0	13.5	2
7.0	1	14.0	5
7.5	3	14.5	9
8.0	5	15.0	11
8.5	2	15.5	6
9.0	2	16.0	4
9.5	9	16.5	7
10.0	5	17.0	4
10.5	1	17.5	4
11.0	0	18.0	2
11.5	0	18.5	0
12.0	1	19.0	2
12.5	4	19.5	1
13.0	0	>19.5	0

Discuss what inferences might be drawn from this information. What problems are posed, and what further investigations would you attempt?

22 Give outline accounts of the biology of any of the following, with special reference to their ecology: fucoids, littorinids, limpets, trochids, barnacles, *Arenicola marina*, *Hediste diversicolor*, *Calanus finmarchicus*, *Euphausia superba*, marine mammals of British coastal waters, whales.

23 Review the ecological conditions and populations of any of the following: rocky shores, intertidal sands, estuaries, the bottom at shallow depths, fast-flowing tidal channels, the English Channel, the North Sea.

24 Give an account of observations you have yourself made on any marine organism or group of organisms.

25 Discuss the role of pelagic larvae in the life cycles of benthic marine animals.

26 Give a general account of selective settlement by marine larvae.

27 Give an account of food webs in the sea.

28 In relation to the populations of the seashore or shallow sea bottom, discuss what meaning you attach to the term 'community'.

29 What do you understand by an 'ecosystem'? Discuss this concept in relation to (*a*) the surface layers of the open sea (*b*) the sea bottom at deep levels (*c*) the neritic province (*d*) the seashore.

30 Give an outline historical account of the development of oceanography and marine biology. What future advances in these fields of study do you foresee?

31 Give an account of camouflage by marine organisms.

32 Outline the chief methods of commercial fishing, explaining how these are related to the habits of the species sought.

33 Give an account of the biology of a named food fish of commercial importance, and indicate how this knowledge can be of value to the fishing industry.

34 Outline a programme of investigations to study the biology of a marine fish.

35 Discuss the problems of overfishing in north-east Atlantic waters, and outline a policy for the regulation of sea fisheries.

36 Give some account of methods of marine fish culture, and the associated problems.

37 Discuss proposals for obtaining more food from the sea.

38 How might past El Niño southern oscillation events have influenced the spread of species within the Pacific Ocean?

APPENDIX 2
Some laboratory exercises

A Salinity measurements by titration and conductimetry.
B pH measurements in seawater and determination of titration alkalinity.
C Estimation of a minor constituent; for example phosphate (Murphy and Riley, 1962).
D Elementary studies on barnacles, for example *Semibalanus balanoides, B. perforatus, B. crenatus, Chthamalus stellatus, Elminius modestus*.

   Diagnostic characters. Measurement of rate of cirral activity over ranges of temperature, salinity and pH. Comparison of the activity ranges of different species in relation to distribution. Observation of the light reflex, and investigation of its sensitivity and fatigue.
E Elementary studies on bivalves, for example *Mytilus, Cardium, Pecten, Tellina, Venerupis, Ensis, Mya*.

   Diagnostic characters. Comparison of structure of siphon, shell, mantle, ctenidium, palps and foot in relation to habitat and mode of life. Use of suspensions or cultures to investigate filtering rates, and pathways of feeding, selection and rejection. Measurement of food particle transport rates over ranges of temperature and salinity.
F Studies on *Ligia*.

   Observation of melanophores. The rate of colour change associated with changes of illumination and background.

   The effects on colour change of covering part or whole of eyes.

   Study of the phototaxis, hydrotaxis and thigmotaxis of *Ligia* and statistical treatment of results.

   Measurement of rate of water loss, and comparison with other shore forms, e.g. *Gammarus, Idotea*.
G Studies on *Corophium* (Barnes *et al.*, 1969).

   Observation of swimming and burrowing behaviour. Substrate selection. Light reactions. Cuticle permeability.

H   Behaviour of *Hydrobia* or *Littorina*. Phototaxes and geotaxes. Substrate selection (Barnes, 1979; Barnes and Greenwood, 1978).
I   Osmotic relationships.

Measurement of weight changes of various animals in relation to changes of salinity, for example *Nereis diversicolor, N. pelagica, Perinereis cultrifera, Arenicola marina, Carcinus maenas.*

Measurement of ionic concentrations in body fluids of *Arenicola* and *Carcinus*, and the changes consequent on changing salinity (Na, K and Ca by flame photometry, CI by titration). Freezing point of blood samples.

J   Examination of named species of planktonic plants and animals for diagnostic features.

Examination of plankton samples from various sources, with special reference to seasonal and geographical differences.

Observations on live plankton – flotation, swimming and filtering activity, phototaxis, etc.

K   Examination of representative collections of benthos from shallow bottoms, shores and estuaries with attention to adaptations, zonation, feeding relationships and community structure.

The use of shell measurement and size/frequency curves for population analysis and determination of mean growth rates.

L   Examination of commercial fish species with reference to recognition features, adaptations, gut contents, parasites, scale and otolith markings. The growth rate curve from herring scale rings.

(In addition to the foregoing exercises, laboratory time is also required following field-work, for sorting collected material, measuring, counting, tabulating and graphically representing results.)

References
Barnes, R.S.K. (1979). Sediment preference of *Hydrobia*. *Est. Coastal Mar. Sci.*, 9, 231–4.

Barnes, R.S.K. and Greenwood, J.G. (1978). Response of *Hydrobia ulvae* to sediments of differing particle size. *J. Exp. Mar. Biol. Ecol.*, 31, 43–54.

Barnes, W.J.P., Burn, J., Meadows, P.S. and McLusky, D.S. (1969). *Corophium volutator* – An intertidal crustacean useful for teaching in schools and universities. *J. Biol. Educ.*, 3, 283.

Murphy, J. and Riley, J.P. (1962). *Analytica Chim. Acta*, 27, 31.

APPENDIX 3
Some field course exercises

The projects summarized below are examples of exercises which may be attempted by students during a course of about a week's duration. These exercises require planning beforehand to avoid wasted time and effort. Transects are usually worked by teams of 4–6 students, and other exercises are allocated to students working in pairs. The value of these projects depends largely upon the provision of adequate class and laboratory time for students to work up their results and present accounts of their work for general discussion and criticism.

An excellent series of class exercises with detailed instructions is given in *Marine Field Course Guide 1, Rocky Shores* (Hawkins and Jones, 1992). These cover aspects of the shore and its biota, zonation patterns, sampling problems and population biology of individual species. Many other useful ideas are explored in a light-hearted manner in the Field Studies Council *Guide for Rocky Shore Investigations* (Archer-Thomson, 1991). Some of the exercises described below are covered in detail in these publications.

Rocky shores
A By plotting the occurrence of organisms on transects, and making reference to tide tables, investigate the relationships of the littoral fringe, eulittoral zone and sublittoral fringe (see Hawkins and Jones, 1992) to the tidal levels of the shore. Compare the extent and levels of these zones on different shores, for example algal-dominated and barnacle-dominated shores.
B Using the notation given below (see page 444), plot the zonation of organisms on shores of various aspects, and attempt to relate to Exposure Scales (Ballantine, 1961; Lewis, 1964).
C Investigate the vital statistics of populations of selected shore molluscs, for example *Gibbula cineraria*, *Monodonta lineata*, *Littorina littorea*, *Nucella lapillus*, *Patella* spp. by plotting size/frequency curves from convenient shell

measurements, for example height or maximum diameter. Compare the populations of different levels and different shores.

D   In appropriate quadrats on various shores and/or levels, investigate the correlations between the following pairs of measurements (a)/(b), and discuss your findings.

  (*a*)   Percentage cover or wet weight of macroflora.
  (*b*)   Numbers or wet weights of *Patella vulgata*, *Littorina obtusata* and *L. littorea*.

  (*a*)   Percentage cover, rough weights or numbers of barnacles.
  (*b*)   Rough weights or numbers of *Nucella lapillus*.

  (*a*)   Ratio of shell height/length in *Patella* occupying surfaces of similar slope and aspect.
  (*b*)   Tidal level.

  (*a*)   Mean size of *Patella* shells.
  (*b*)   Numbers of *Patella* per unit area.

E   Marking experiments. For studying movements and homing tendencies, mark mollusc shells, for example *Patella*, *Nucella*, littorinids and trochids, with waterproof paint. Alternatively, plastic numbers (e.g. 'Dynotape') can be stuck on the shell with epoxy-resin after first drying the shell with alcohol. In the course of these observations, a visit to the shore during a night-time low tide should be made to compare the positions of specimens during day and night.

Sandy and muddy shores

Extensive areas of sand or mudflat offer certain advantages for field course exercises. Digging and sieving are hard work; consequently the environmental damage done by field work on depositing beaches is usually trifling compared with the havoc a class can quickly make of a rocky shore by ill-controlled stone turning and fissure opening. Exercises can be planned to involve teamwork in a number of separate but closely interrelated projects, requiring field collecting and subsequent laboratory investigation.

If all the biological and ecological information inherent in collections obtained from a well-organized dig along a depositing shore transect is to be developed fully, much longer laboratory investigations are required than is usually possible during a field course. Some of this work need not be done immediately if material is preserved and taken back to college. Subsequently as time allows, detailed laboratory exercises can be planned to make the best use of the material, deriving directly from the students' own work in the field.

Shore work
Remember that some mud is treacherously soft, and do not overlook the danger of being cut off on mudflats by a rising tide.

Dig quadrats along line transects, estimating the vertical level of each quadrat. If possible, use coarse sieves to collect macrobenthos; or if this is impractical, collect macrobenthos by hand-sorting. From each quadrat take a sediment sample, about 500 g, and, if obtainable, an interstitial water sample. Keep the complete collections from each station, separately labelled, for later identification of species and quantitative investigations. At each station note the depth of the unblackened layer and any temperature difference between surface and deeper layers.

Laboratory work
A   Identify the macrofauna in each collection, recording numbers and rough weight for each species. Pass part of each sediment sample through 0.5 mm sieves to collect meiofauna. Use menthol or dilute alcohol to detach interstitial species from sand grains, and decant. Plot zonation diagrams for each species, with details for macrofauna of numbers and rough biomass per unit area.

B   Investigate the vital statistics of populations of dominant macrofauna from size/frequency data or from annual rings on shells. Does the age structure of populations vary with shore level? Determine growth rates for size and weight from measurements of age groups. Distinguish rough weights, flesh weights, dry weights and ash-free dry weights. Does the drying technique affect the final figure of dry weight? By combining data on population density, vital statistics and growth rates, attempt to estimate annual production. Construct an energy flow diagram for the shore (Townsend and Phillipson, 1977). Calculate a diversity index for the population.

C   Measure salinities of interstitial water samples. Make particle grade analyses of sediment samples from each station by sedimentation or sieving. Determine organic contents of sediment samples by incineration and/or titration. To what extent are silt content and organic content related, and how do the inorganic parameters correlate with the distribution of each species?

Excursions
The educational value of a marine ecology course is enhanced by any of the following expeditions:

A   Visit to an estuary to observe the distribution of freshwater, estuarine and marine species. Where possible, methods are devised for measuring water depths and collecting water samples from the surface, middle depths and bottom at several stations along the estuary at intervals through the tidal cycle. Salinities, temperatures and oxygen contents are measured. On diagrammatic sections of the estuary the isohalines, isotherms and percentage oxygen saturations are plotted at stages of the tide, and the zonation of organisms recorded.

B   Boatwork at sea for the following purposes:
1   Demonstration of navigation equipment and sonic sounding apparatus.
2   Demonstration of the use of water bottles, CTD instrument packages and rosette samplers, etc.
3   Plankton sampling with various nets.
4   Collection of benthos and fish from various substrates using several types of collecting gear, for example Van-Veen grab, Agassiz trawl, otter trawl. After sorting and identifying benthic material, the sampled communities are related to the various classifications available (see Section 6.3). Fish are examined for external and gill parasites, gut contents, scale and otolith markings and age–length relationship. In the absence of facilities for collecting at sea, water intake screens at coastal power stations are sometimes a useful source of fish.

C   Visit to local fish auction and associated industries, for example fishing vessels, fish curing, processing and freezing installations, net factory and ice factory. Fishery officers and fishing boat skippers are sometimes willing to meet groups of students and tell them about their work.

D   Day tour as far along the coast as practicable to observe biological, geographical and geological features.

E   Visit to a fish hatchery and/or fish farm.

Abundance scale for rocky shore intertidal organisms

'Abundance scales' are semi-quantitative estimates of numbers or cover of conspicuous plants and animals. These scales were first devised by Crisp and Southward (1958) and were used by Ballantine (1961) in deriving his biological exposure scale. Crisp and Southward's six broad categories are still used today and are given below. Since then, other workers have added categories but this does not always result in increased accuracy (Hawkins and Jones, 1992). See also Moyse and Nelson-Smith (1963), and Crothers (1976). Similar scales can be devised for species in sand or mud if their presence is readily apparent from burrow openings, tubes or casts, e.g. *Arenicola marina, Lanice conchilega, Scrobicularia plana, Corophium arenarium*.

Crisp and Southward categories:

A = Abundant
C = Common
F = Frequent
O = Occasional

R = Rare
N = Not found (all cases)

Additional categories:

E = Extremely abundant
S = Superabundant

Algae
E   > 90 per cent cover.
S   60–90 per cent cover.
A   > 30 per cent cover.
C   5–30 per cent cover.
F   < 5 per cent cover but zone still apparent.
O   Scattered individuals. Zone indistinct.
R   Few plants found in 30 min. search.

Lichens, lithothamnia crusts
E   > 80 per cent cover.
S   50–79 per cent cover.
A   > 20 per cent cover at some levels.
C   1–20 per cent cover. Zone well defined.
F   Large scattered patches. Zone ill defined.
O   Widely scattered, small patches.
R   Few small patches found in 30 min. search.

Dogwhelks, topshells, anemones and sea urchins
E   > 100 m^{-2}
S   50–90 m^{-2}
A   > 10 m^{-2}
C   1–10 m^{-2}, sometimes very locally > 10 m^{-2}.
F   < 1 m^{-2}, locally sometimes more.
O   Always < 1 m^{-2}
R   Only 1 or 2 found in 30 min. search.

Large periwinkles, Littorina littorea
E   > 200 m^{-2}
S   100–200 m^{-2}
A   > 50 m^{-2}
C   10–50 m^{-2}
F   1–10 m^{-2}
O   < 1 m^{-2}
R   Only 1 or 2 found in 30 min. search.

Periwinkles, Melaraphe neritoides *and* Littorina saxatilis

E $> 5\,\mathrm{cm}^{-2}$

S $> 3\text{--}5\,\mathrm{cm}^{-2}$

A $> 1\,\mathrm{cm}^{-2}$ at HWN (extending down to mid-littoral)

C $0.1\text{--}1\,\mathrm{cm}^{-2}$ (mainly in littoral fringe)

F $< 0.1\,\mathrm{cm}^{-2}$ (mainly in crevices)

O   A few in most deep crevices

R   Only 1 or 2 found in 30 min. search

Small barnacles

E $> 5\,\mathrm{cm}^{-2}$

S $3\text{--}5\,\mathrm{cm}^{-2}$

A $> 1\,\mathrm{cm}^{-2}$, rocks well covered

C $0.1\text{--}1\,\mathrm{cm}^{-2}$, up to one-third of rock space covered

F $100\text{--}1000\,\mathrm{m}^{-2}$, individuals never $> 10\,\mathrm{cm}$ apart

O $1\text{--}100\,\mathrm{m}^{-2}$, few within 10 cm of each other

R   Only a few found in 30 min. search

*Large barnacles (*B. perforatus*)*

E >300 per $10 \times 10\,\mathrm{cm}$

S $100\text{--}300$ per $10 \times 10\,\mathrm{cm}$

A $10\text{--}100$ per $10 \times 10\,\mathrm{cm}$

C $1\text{--}10$ per $10 \times 10\,\mathrm{cm}$

F $10\text{--}100\,\mathrm{m}^{-2}$

O $1\text{--}9\,\mathrm{m}^{-2}$

R   Only a few found in 30 min. search

Limpets

E $> 200\,\mathrm{m}^{-2}$

S $100\text{--}200\,\mathrm{m}^{-2}$

A $> 50\,\mathrm{m}^{-2}$

C $10\text{--}50\,\mathrm{m}^{-2}$

F $1\text{--}10\,\mathrm{m}^{-2}$

O $< 1\,\mathrm{m}^{-2}$

R   Only a few found in 30 min. search

Mussels and Sabellaria

E $> 80\%$ cover

S $50\text{--}79\%$ cover

A $> 20\%$ cover

C   large patches

F   Many scattered individuals and small patches

O   Scattered individuals, no patches
R   Only a few found in 30 min. search

Tubeworms, Pomatoceros
A $> 500\,\mathrm{m}^{-2}$
C $100–500\,\mathrm{m}^{-2}$
F $10–100\,\mathrm{m}^{-2}$
O $1–9\,\mathrm{m}^{-2}$
R $< 1\,\mathrm{m}^{-2}$

Tubeworms, Spirorbids
A $5+\,\mathrm{m}^{-2}$ on > 50 per cent of suitable surface
C $5+\,\mathrm{m}^{-2}$ on < 50 per cent of suitable surface
F $1–5\,\mathrm{m}^{-2}$
O $< 1\,\mathrm{m}^{-2}$
R   Only a few found in 30 min. search

References
Archer-Thomson, J. (1991). *Guide for rocky shore investigations.* Field Studies Council Occasional Publication 22.

Ballantine, W.J. (1961). A biologically defined exposure scale for the comparative description of rocky shores. *Fld. Stud.,* **1** (3), 1.

Crisp, D.J. and Southward, A. J. (1958). The distribution of intertidal organisms along the coasts of the English Channel. *J. Mar. Biol. Ass. UK,* **37**, 157.

Crothers, J.H. (1976). Common animals and plants along the shores of west Somerset. *Fld. Stud.,* **4**, 369–89.

Hawkins, S.J. and Jones, H.D. (1992). *Marine Field Course Guide 1. Rocky Shores.* Marine Conservation Society. Immel Publishing.

Lewis, J.R. (1964). *The Ecology of Rocky Shores.* London, University Press.

Moyse, J. and Nelson-Smith, A. (1963). Zonation of animals and plants on rocky shores around Dale, Pembrokeshire. *Fld. Stud.,* **1** (5), 1.

Townsend, C. and Phillipson, J. (1977). A field course based on the community energy flow approach. *J. Biol. Educ.,* **11**, 121–32.

A field course book list

The following books are often useful during field courses:

General guides
Barrett, J.H. and Yonge, C.M. (1972). *Pocket Guide to the Sea-Shore* (revised edition). London, Collins.

Campbell, A.C. and Nicholls, J. (1976). *Hamlyn Guide to the Seashore and Shallow Seas of Britain and Europe.* London, Hamlyn.

Cremona, J. (1988). A field atlas of the seashore. Cambridge University Press.

Dipper, F.A. and Powell, A. (1984). *A field guide to the water life of Britain.* Reader's Digest Nature Lover's Library. Reader's Digest Association Ltd.

Eales, N.B. (1967). *The Littoral Fauna of Great Britain*. 4th edition. Cambridge University Press (out of print).
Earll, R.C. (1992). *The Seasearch habitat guide*. An identification guide to the main habitats found in the shallow seas around the British Isles. Marine Biological Consultants Ltd.
Erwin, D. and Picton, B. (1987). *Guide to inshore marine life*. A Marine Conservation Society Guide. London, Immel Publishing.
George, D. and George, J. (1979). Marine Life: *An Illustrated Encyclopaedia of Invertebrates in the Sea*. London, Harrap (out of print).
Hawkins, S.J. and Jones, H.D. (1992). Marine Field Course Guide 1. *Rocky Shores*. Marine Conservation Society. Immel Publishing.
Hayward, P. Nelson-Smith, T. and Shields, C. (1996). *Collins Pocket Guide: Sea Shores of Britain and Northern Europe*. London, Harper Collins.
Hayward, P. and Ryland, J.S. (eds) (1995). *Handbook of the marine fauna of north-west Europe*. Oxford University Press. (Compact version of a two-volume work. Includes keys to most groups.)
Hayward, P.J. (1988). *Animals on seaweed*. Naturalists' Handbooks 9. Richmond Publishing Co. Ltd.
Hayward, P.J. (1994). *Animals of sandy shores*. Naturalists' Handbooks 21. Richmond Publishing Co. Ltd.
Quigley, M. and Crump, R. (1986). *Animals and plants of rocky shores*. Blackwell Habitat Field Guide.

Seaweeds and plankton
Dickinson, C.I. (1963). *British Seaweeds*. London, Eyre and Spottiswoode.
Hiscock, S. (1979). *A field key to the British brown seaweeds*. An AIDGAP key reprinted from Field Studies, 5, 1– 44.
Hiscock, S. (1986). *A field key to the British red seaweeds*. An AIDGAP key. Field Studies Council Occasional Publication No. 13.
Maggs, C.A. and Howson, C.M. (undated). *A photographic guide to some common subtidal seaweeds of the British Isles*. Ross-on-Wye, Marine Conservation Society.
Newell, G.E. and Newell, R.C. (1977). *Marine Plankton, A Practical Guide*. Revised edition. London, Hutchinson.

Sponges
Ackers, G., Moss, D. and Picton, B.E. (1992). *Sponges of the British Isles* ('Sponge V'). A colour guide and working document. Ross-on-Wye, Marine Conservation Society.

Coelenterates
Manuel, R.L. (1980). *The Anthozoa of the British Isles*. Ross-on-Wye, Marine Conservation Society. (Colour photographs and text.)

Polychaetes
Clark, R.B. (1960). *Polychaete fauna of the Clyde Sea Area*. Scottish Marine Biological Association.
Garwood, P.R. (1981). *The marine fauna of the Cullercoates district. No. 9. Polychaeta-Errantia*. Dove Marine Laboratory, University of Newcastle-upon-Tyne.

Garwood, P.R. (1982). *The marine fauna of the Cullercoates district. No. 10. Polychaeta-Sedentaria including Archiannelida.* Dove Marine Laboratory, University of Newcastle-upon-Tyne.

Crustaceans

Allen, J.A. (1967). *The Fauna of the Clyde Sea Area. Crustacea: Euphausiacea and Decapoda, with an Illustrated Key to the British Species.* Millport, Scottish Marine Biological Assoc.

Crothers, J. and Crothers, M. (1988). *A key to the crabs and crab-like animals of British inshore waters.* An AIDGAP key. Reprinted from Field Studies, 5 (1983), 753–806.

Ingle, R.W. (1980). *British crabs.* British Museum (Natural History). Oxford University Press.

King, P.E. (1986). *Sea spiders. A revised key to the adults of littoral pycnogonida in the British Isles.* An AIDGAP key. Reprinted from Field Studies, 6, 493–516.

Molluscs

Beedham, G.E. (1972). *Identification of British Mollusca.* Amersham, Hulton.

Picton, B.E. and Morrow, C.C. (1994). *A field guide to the nudibranchs of the British Isles.* London, Immel Publishing.

Tebble, N. (1966). *British Bivalve Seashells.* London, British Museum (out of print).

Echinoderms and tunicates

Picton, B.E. (1985). Ascidians of the British Isles. A colour guide. Ross-on-Wye, Marine Conservation Society.

Picton, B.E. (1993). *A field guide to the shallow-water echinoderms of the British Isles.* London, Immel Publishing.

Fishes

Dipper, F.A. (1987). *British Sea Fishes.* London, Underwater World Publications.

Lythgoe, J. and Lythgoe, G. (1991). *Fishes of the Sea. The North Atlantic and Mediterranean.* London, Blandford Press.

Muus, B.J. and Dahlstrom, P. (1974, reprinted 1988). *Guide to the Sea Fishes of Britain and Northwestern Europe.* London, Collins.

Vas, P. (1991). *A field guide to the sharks of British coastal waters.* An AIDGAP key. Reprinted from *Field Studies*, 7, 651–86.

Wheeler, A. (1978). *Key to the fishes of northern Europe.* London, Frederick Warne (out of print).

Wheeler, A. (1994). *Field key to the shore fishes of the British Isles.* FSC Publication 225. Reprinted from *Field Studies*, Vol. 8, No. 3.

Synopses of the British Fauna (new series). Linnean Society of London and Estuarine and Brackish-water Sciences Association. Eds D.M. Kermack and R.S.K. Barnes

No. 1. Millar, R.H. (1970). *British Ascidians.*

No. 2. Graham. A. (1988). *British Prosobranchs* (2nd edn)

No. 3. Naylor, E. (1972). *British Marine Isopods.*

No. 5. King, P.E. (1974). *British Sea Spiders.*

No. 7. Jones, N.S. (1976). *British Cumaceans.*

No. 8. Thompson, T.E. (1988). *Molluscs: Benthic Opisthobranchs (Mollusca:Gastropoda)* (2nd edn).

No. 9. Morgan, C.I. and King, P.E. (1976). *British Tardigrades.*

No. 10. Ryland, J.S. and Hayward, P.J. (1977). *British Anascan Bryozoans.*

No. 12. Gibbs, P.E. (1977). *British Sipunculans.*

No. 13. Emig, C.C. (1979). *British and other Phoronids.*

No. 14. Hayward, P.J. and Ryland, J.S. (1979). *British Ascophoran Bryozoans.*

No. 15. Smaldon, G. (1979). *Coastal Shrimps and Prawns.*

No. 16. Murray, J.W. (1979). *British Nearshore Foraminiferids.*

No. 17. Brunton, C.H.C. and Curry, G.B. (1979). *British brachiopods.*

No. 18. Manuel, R.L. (1988). *British anthozoa* (revised).

No. 20. Fraser, J.H. (1982). *British Pelagic Tunicates.*

No. 21. Brinkhurst, R.O. (1982). *British and Other Marine and Estuarine Oligochaetes.*

No. 24. Gibson, R. (1982). *British Nemerteans.*

No. 25. Ingle, R.W. (1983). *Shallow-water Crabs.*

No. 26. Prudhoe, S. (1982). *British Polyclad Turbellarians.*

No. 28. Platt, H.M. and Warwick, R.M. (1983). *Free-living Marine Nematodes: Part I British Enoplids.*

No. 29. Kirkpatrick, P.A. and Pugh, P.R. (1984). *Siphonophores and velellids.*

No. 30. Mauchline, J. (1984). *Euphausiid, Stomatopod and Leptostracan Crustaceans.*

No. 32. George, J.D. and Hartmann-Schroder, G. (1985). *Polychaetes: British Amphinomida, Spintherida and Eunicida.*

No. 33. Hayward, P.J. (1985). *Ctenostome Bryozoans.*

No. 34. Hayward, P.J. and Ryland, J.S. (1985). *Cyclostome Bryozoans.*

No. 37. Jones, A.M. and Baxter, J.M. (1987). *Molluscs: Caudofoveata, Solenogastres, Polyplacophora and Scaphopoda.*

No. 38. Platt, H.M. and Warwick, R.M. (1988). *Free-living Marine Nematodes: Part II British Chromadorids.*

No. 39. Pierrot-Bults, A.C. and Chidgey, K.C. (1988). *Chaetognatha.*

No. 43. Athersuch, J., Horne, D.J. and Whittaker, J.E. (1989). *Marine and Brackish-water Ostracods.*

No. 44. Westheide, W. (1990). *Polychaetes: Interstitial Families.*

No. 45. Pleijel, F. and Dales, B.P. (1991). *Polychaetes: British Phyllodocideans, Typhloscolecoideans and Tomopteroideans.*

No. 48. Angel, M.V. (1995). *Marine Planktonic Ostracods.*

No. 50. Cornelius, P.F.S (1995). *NW European Thecate Hydroids and their Medusae. Parts 1 and 2.*

Statistics

Moroney, M.J. (1965). *Facts from Figures.* Harmondsworth, Penguin.

Parker, R.E. (1979). *Introductory Statistics for Biology.* London, Arnold.

Swinscow, T.D.V. (1977). *Statistics at Square One.* London, British Medical Association.

APPENDIX 4
Addresses of organizations mentioned in the text

Department of the Environment (DOE), European Wildlife Division, Room 902, Tollgate House, Houlton Street, Bristol, BS2 9DJ.

Institute for European Environmental Policy (IEEP), 158 Buckingham Palace Road, London, SW1 9TR.

International Union for Conservation of Nature and Natural Resources (IUCN), 219c Huntingdon Road, Cambridge, CB3 0DL.

Joint Nature Conservation Committee (JNCC), Monkstone House, City Road, Peterborough, PE1 1JY.

Marine Conservation Society (MCS), 9 Gloucester Road, Ross-on-Wye, Herefordshire, HR9 5BU.

Marine Environmental Management and Training, Dr R.C. Earll, Candle Cottage, Kempley, Gloucestershire, GL18 2BU.

Ministry Agriculture Fisheries and Food, Directorate of Fisheries Research, Pakefield Road, Lowestoft, NR33 0HT.

Royal Society for the Protection of Birds (RSPB), The Lodge, Sandy, Bedfordshire, SG19 2DL.

Scottish Office Agriculture, Environment and Fisheries Department (SOAEFD), Marine Laboratory, PO Box 101, Victoria Road, Aberdeen, AB9 8DB.

World Conservation Monitoring Centre (WCMC), New Building, 219c Huntingdon Road, Cambridge, CB3 0DL.

Index